Statistical Thermodynamics

Statistical Thermodynamics

An Information Theory Approach

Christopher Aubin
Fordham University
Bronx, New York
United States

WILEY

Published by John Wiley & Sons, Inc., Hoboken, New Jersey.
Published simultaneously in Canada.

For general information on our other products and services or for technical support, please contact our Customer Care Department within the United States at (800) 762-2974, outside the United States at (317) 572-3993 or fax (317) 572-4002.

Wiley also publishes its books in a variety of electronic formats. Some content that appears in print may not be available in electronic formats. For more information about Wiley products, visit our web site at www.wiley.com.

Library of Congress Cataloging-in-Publication Data is Applied for
Hardback ISBN: 9781394162277

Cover Design: Wiley
Cover Image: © Hector Roqueta Rivero/Getty Images. Image of the handwritten Boltzmann principle provided by Christopher Aubin.

Set in 9.5/12.5pt STIXTwoText by Straive, Chennai, India

SKY10070844_032224

For my father.

Contents

Preface

The first time I took *Statistical Physics* as an undergraduate physics major, I hated the class. This is a pretty strong sentiment, but while I was able to learn the basics of statistical mechanics and thermodynamics, I didn't fully appreciate the beauty of the subject. Then I opted to take the undergraduate course again during my first semester of graduate school (at a different school, taught an entirely different way), and my feelings about the subject changed drastically. I attributed this largely to the organization of the course and how the material was developed.

Since then, I have found that this is a very strange subject in physics. As I discuss in the introduction, this is one of the four "core" fields of physics, but unlike the other three (classical mechanics, electricity and magnetism, and quantum mechanics), there doesn't seem to be anywhere near the same sort of consensus on how to approach the subject. As I have developed this course over the past 13 years, I have worked hard to try to motivate students to find the beauty in statistical mechanics and thermodynamics. There is a plethora of fantastic textbooks out there, all of which approach the subject from different starting points, covering topics in different orders, and of course having different priorities. Many of them have been excellent resources as I developed my class throughout the past 13 years and this book, although much of my inspiration comes from the textbook by Frederick Reif, *Fundamentals of Statistical and Thermal Physics*. Even still, my approach has a slightly different starting point, that of information theory, that has rarely been seen at the undergraduate textbook level. Perhaps this is because it is a more theoretical approach, lacking the direct and immediate applications that a thermodynamic starting point has, but I do believe that this is precisely where the beauty of the field becomes clear.

The approach that I take here (which I first learned about from Claude Bernard, who taught the undergraduate class I took while in graduate school), I find gives a wonderful first-principles approach to the field. While it begins very abstractly, I find it allows the physical applications of thermodynamics to fall out naturally while also setting us up to apply our methods to some very strange ideas. And, this abstract approach allows the broad nature of the subject, that it can be applied to literally any area of physics and engineering, to be made patently clear to the student. Sure the lack of direct practical applications (for example, I do not spend nearly enough time on the topic of heat engines, for reasons I explain in Chapter 7) might cause some to find this less useful. However as is often the case, having an abstract approach provides us with much more power when solving problems.

The basic approach I take is to start from probability in Chapter 2 and information theory in Chapter 3, as the title suggests. From this point, I will move to a discussion of the underlying statistical nature of systems with large numbers of particles in Chapter 4. It is only at this point where I define the concept of temperature in Chapter 5. This is difficult at first, as temperature is such an important concept that everyone has an everyday understanding of, but by waiting for a bit to define it allows a deeper understanding of its meaning, beyond how hot or cold it is outside. This lays the groundwork that, with few assumptions, allows us to formulate all of thermodynamics, which I work through in detail in Chapters 6 and 7.

After the basics of thermodynamics have been laid out, I start fresh with the canonical ensemble in Chapters 8 and 9, thus moving back to our statistical description again. From there, I will work through applications to phase transitions and chemical reactions in Chapter 10 and then quantum statistics in Chapters 11 and its applications in Chapter 12. I will end with a huge detour to discuss the fascinating area of black hole thermodynamics in Chapter 13, not because it is essential for an undergraduate education, but rather because it is a great way to demonstrate how wide-reaching the subject can be.

This textbook is designed to be a one-semester course in undergraduate statistical mechanics. The later topics in the text I have alternated in the times I have taught the course, often due to the interest of the students. Throughout the book are of course examples, but more importantly many of the simpler derivations I have left out. In place of those I have included Exercises that the student should do, to assist with the development of the subject. Additionally, sprinkled throughout the text are references to Jupyter notebooks (using Python) that can be accessed at the companion site to better understand some of the smaller details that we will cover. These are meant on one hand to be a quick way to see some of the results I will derive and also as a way for students to learn a little bit about coding if that is of interest. And of course at the end of each chapter are more detail problems, as any good book has, for the student to master the material and see where this exciting field can lead us.

Christopher Aubin

Acknowledgments

It doesn't matter how many authors are listed on a book, as there are many people who help but are behind the scenes. First and foremost I must thank Claude Bernard for introducing me to this approach to the subject. He was instrumental in two ways, the first of which was showing me this approach to statistical mechanics. Additionally, he and Michael Ogilvie were my PhD advisors, so without them I wouldn't be where I am today. And of course I must thank my students; well over 100 of them have been the unknowing guinea pigs as I developed this course and book.

In terms of this work, I am grateful for Maarten Golterman and Martin Ligare for their useful comments and discussions on the manuscript. I would like to thank Stephen Holler for many discussions during the last 10 years that helped me along the way while preparing this (even if he didn't always realize it). Finally, Vassilios Fessatidis has been a great help in discussions throughout the years and specifically helped me with some Greek to understand the origins of many terms in this subject.

But a book of this sort is not only about the physics, as I must acknowledge the work of my family. Roger, Diane, and Carrie Aubin (my father, mother, and sister) have been my strongest supports throughout my entire life. While my father didn't live to see this book published, he would always love to hear me talk about physics (even if he really didn't care about the subject). And last, but certainly not least, I am so grateful for the love and support of my husband, Corbett, who has put up with me and my physics rambling for more time than anyone not in the field should have to.

Christopher Aubin

About the Companion Website

This book is accompanied by a companion website.

www.wiley.com/go/Aubin/StatisticalThermodynamics

This website includes:

- Computer code in the form of Jupyter notebooks

1 Introduction

Before jumping right into everything, I want to spend a brief moment discussing statistical mechanics and thermodynamics. Why do we need these subjects and what are they? How do they differ and how do they connect with each other? Of course, you can jump to Chapter 2 and start learning the necessary material, but this will be a useful overview to frame our discussion over the next several hundred pages.

For most people, including undergraduate physics students, statistical mechanics tends to be one of the less well-known subjects in physics. Every student knows about classical mechanics (even if not by that name when they first start studying physics), electricity and magnetism, and even quantum mechanics. Additionally, these subjects (along with more "exciting ideas" such as general relativity) are known by many non-physics students as well. However, statistical mechanics is one of the "core subjects" that all physicists and engineers should understand in depth along with classical mechanics, electricity and magnetism, and quantum mechanics.[1]

Students are of course familiar with thermodynamic concepts (heat, temperature, pressure, etc.), and while many schools offer a course with this title, it is also often called *Statistical Physics*, *Statistical Mechanics and Thermodynamics*, or some variation on these. When registering for this class, many times the only idea students have regarding the course is that it "is the hardest class you'll ever take in college, it makes no sense, and you no longer will understand what temperature is."[2] The salient question is then: What is the relationship between the (slightly) more familiar thermodynamics and this unfamiliar statistical mechanics? The quick answer, to some degree, is that they are two different approaches one can take to understand the same subject. I plan to give an overview in this chapter on how they relate and what the difference is when we consider this subject from either point of view. The goal is to understand how they are connected, why we study one over the other, and most importantly, why we're going to start with statistical mechanics before moving on to the study of thermodynamics. After I discuss the two, as well as the approach we will take in this book, we will finally get started on the actual subject.

1 One of the reasons I say this is that a standard physics PhD qualifying exam usually includes these four subjects. More importantly, I say this because the results of statistical mechanics find their way throughout all of physics and engineering, as we shall see.

2 This is a combination of several comments made by my students in the past, and I have edited out some more colorful words used.

Statistical Thermodynamics: An Information Theory Approach, First Edition. Christopher Aubin.

Companion website: www.wiley.com/go/Aubin/StatisticalThermodynamics

1.1 What is Thermodynamics?

Considering the constituent words that make it up, **thermodynamics** is merely the study of the motion of heat. Coming from the Greek, *thermodynamiki*, or θερμοδυναμική, is a combination of *thermótita* (θερμότητα) which means "heat" and *dynamiki* (δυναμική) which means "dynamics." In the early nineteenth century, when this field began to develop, the focus was on heat engines: using heat as a form of energy that can be converted into work (the reverse is easy to do, doing work to create heat). That the study of heat engines, and thus thermodynamics, exploded during this time is not surprising given that this overlaps with the Industrial Revolution. A steam locomotive is an early example of a widely used heat engine: You use the steam produced by burning coal to turn gears (thereby turning the wheels of the train and propelling it forward). In the process, and as the field progressed, four laws of thermodynamics were postulated and these allowed the introduction and development of various concepts such as temperature, entropy, enthalpy, and so forth.[3]

Many of these concepts are familiar from an everyday perspective (like temperature), or from hearing about it colloquially (like entropy). Still others, like enthalpy, are not well known by those who never took a chemistry class. The question remains though: What do these quantities describe physically? The meanings of these ideas can often be muddled (or worse, misunderstood) when first studying thermodynamics. Additionally, we will see that while starting from a thermodynamic perspective is nice from a conceptual (and historical) point of view, it can be very limiting. Finally, it is only natural to ask where thermodynamics fits in with regard to other areas of physics.

As physics developed throughout the centuries, it was necessary to begin to divide it into different subfields we can categorize *very roughly* by the relevant length scales and speeds in a given problem. A depiction of this is shown in Figure 1.1. If the relevant length scale of our system is denoted as L, and the relevant speed is denoted as v, then we can divide physics up as follows. Starting with classical mechanics ("the original physics"), L and v take on "everyday values": values that are common for buildings, cars, people, etc. I'll admit that the term "everyday values" is not a great way

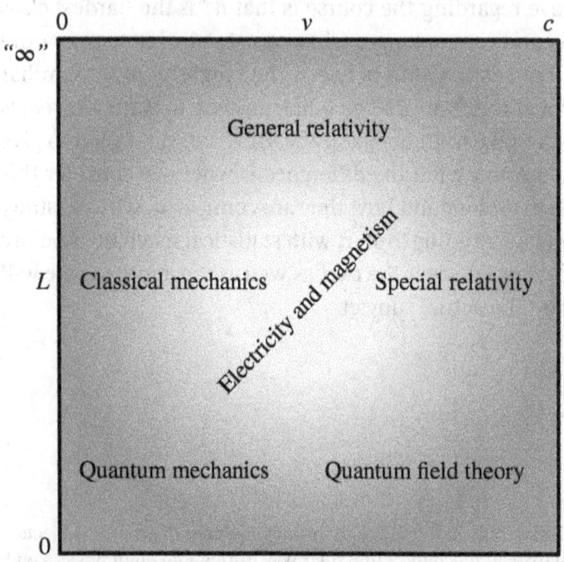

Figure 1.1 Various subfields of physics in relation to each other, when considering a range of length scales L and speeds v of the system under consideration.

3 I will define all of these terms precisely later.

to describe these quantities, as classical physics is valid for a fairly wide range of sizes and speeds. The motion of dust ($L \sim 4 \times 10^{-4}$ m) as well as that of planetary orbits (Neptune is an average distance of about $L \sim 4.5 \times 10^{12}$ m from the sun), both can be described classically. Additionally, planets orbit the sun at high speeds (from our perspective) and they can be described classically (Mercury travels around the sun at almost $v \sim 50\,000$ m/s = 180 000 km/h). The precise length or speed scale is not important here; it just matters that there is a range such that the classical description is valid, and at some point it breaks down.

Around the turn of the twentieth century, experiments were performed at much smaller length scales, and classical mechanics began to fail. Enter quantum mechanics, which becomes relevant as one nears molecular scales, so $L \sim 1\ \mu\text{m} = 10^{-6}$ m and smaller.[4] Around the same time, the theory of special relativity was being developed, which is important to consider when our system approaches high speeds, specifically close to the speed of light c. In each of these cases, we consider one of these quantities to change: In ordinary (non-relativistic) quantum mechanics, speeds are not too high, while in special relativity, the length scales are those of classical mechanics, all shown in Figure 1.1.

Electricity and magnetism, the third of our four "core" subjects, doesn't quite fit into any of these categories, so I added it to the middle of our plot, somewhat spilling over into various other subfields. It belongs a bit more in the special relativity section (as it was important in the development of this theory and in fact is already relativistic by nature), but often can be thought of as a classical topic. One thing to note for clarity though, while it overlaps various categories, it is specifically *not* quantum mechanical.

A couple of other subjects are thrown into the plot (just for some level of completeness) that aren't always studied in an undergraduate curriculum. For large systems, general relativity (the fundamental theory of gravity) takes over for classical mechanics (in the figure, I have infinity in quotes to imply this is for very large systems without needing to specify an actual scale).[5] For very small and fast systems, we combine quantum mechanics and special relativity to formulate quantum field theory (needed to study particle physics, including a quantum description of the electromagnetic field as well as the strong and weak nuclear interactions). While the use of a single length (or speed) scale is overly simplistic, systems with multiple scales that are wildly different are quite difficult to study in physics; this is one of the reasons why it is difficult to combine general relativity and quantum field theory (hence we have no quantum theory of gravity!).

All of the fields above are solvable directly in terms of laws that allow us to determine the equations of motion. For example,

Classical mechanics: We can use Newton's laws of motion to determine the trajectories of particles as long as we know *all* of the forces acting on them given a set of initial conditions.

Electricity and magnetism: With Maxwell's equations (and appropriate initial and boundary conditions), we can determine the electromagnetic field due to any charge and current distribution. Add in the Lorentz force, and we can describe (with classical mechanics) how charged particles move in electromagnetic fields.

Quantum mechanics: We can solve the Schrödinger equation, along with appropriate initial and boundary conditions, to determine the wavefunction which can be used to calculate expectation values of physical observables.

---

4 We will be more concrete with the classical vs. quantum regime when the time comes.

5 Even though most of the universe can be understood with Newtonian gravity, for precise measurements, general relativity must enter the picture.

While more complicated, the same can be said for the other advanced topics, as long as we have the relevant equations (and initial and/or boundary conditions). Keep in mind when we say we can solve these equations, this is all *in principle*: An exact solution is rarely possible in practice, while often approximate solutions are.

But what about thermodynamics, where do we put this in our figure? In Figure 1.1, I have implicitly assumed that the number of objects in our system is small, by which I mean one, two, or *maybe* three. A thermodynamic system, such as the air in the room you're sitting in while reading this or the cup of coffee you are drinking to stay awake while doing so, involves a ridiculously large number N of particles. Many systems we will focus on will have something on the order of a **mole** of molecules, around 6.02×10^{23} (the mole is a base unit in the International System of Units (SI), which, along with other units, I discuss in Appendix A).

Now imagine we had a classical system with this many molecules interacting. Knowing all of the forces between the molecules is difficult, but let us suppose we did, so that we could write down a system of equations for the trajectories of all the molecules. Solving this system of equations would be completely intractable, as even the three-body problem cannot be solved in most cases analytically!

But let's make an absurd assumption to describe a naïve way to think about how intractable this is. We'll assume that we actually *could* solve a problem with such a large number ($\sim 10^{24}$) of molecules. For a three-dimensional system we would have 3×10^{24} equations, and from those would need to determine each particle's trajectory $\mathbf{r}_i(t)$ for $i = 1, \ldots, 10^{24}$. I'll even assume that we are so good at solving these problems that we only need one second to figure out (and write down) each of these 3-D trajectories. Even with these unrealistic assumptions, it would take us

$$10^{24} \text{ sec} \sim 3 \times 10^{16} \text{ yr}$$

to actually solve for all of these trajectories. Not only is this much longer than our average lifespan, the age of the universe is roughly 14×10^9 yr, so *even if we could* solve such a problem, we wouldn't have enough time to do so! This is a ridiculous example, as it is not really a proper measure of how or why to solve physics problems; however, it does illustrate an important limitation that can arise.

More importantly, even if we could determine all of these trajectories, it wouldn't lead to anything useful for practical purposes. For example, consider the air molecules (really just the oxygen molecules) in your room: What about those molecules matters to you? You aren't interested in the individual trajectories of every single molecule, nor do you care *which* of the many O_2 molecules are near you. You really only care that there *are* oxygen molecules near you so that when you inhale, you can capture them. As such, we are more interested in what we call the **bulk properties**, which are large-scale properties of the entire system of oxygen molecules in the room (such as the pressure, density, and temperature, all of which we will discuss in detail when the time comes).

To add thermodynamics to Figure 1.1, we would need to add a third axis, shown in Figure 1.2. The additional dimension is the number of molecules, N, ranging from one to "∞," where I am using the same notation for infinity as I did for L. We will generally just suggest $N \gg 1$, and while we will look at the large-N limit, taking N to infinity will often be troublesome mathematically during intermediate steps.[6] I won't give too many examples of how big N needs to be yet, but we'll see how it comes into play as we go through the subject. That large N corresponds to the subject of thermodynamics actually becomes clearer when we look at it from the point of view of statistical mechanics, which we'll do in Section 1.2.

---

6 Taking the **thermodynamic limit**, when $N \to \infty$, is often ultimately done for many cases, for realistic systems, we will want to keep N finite but very large.

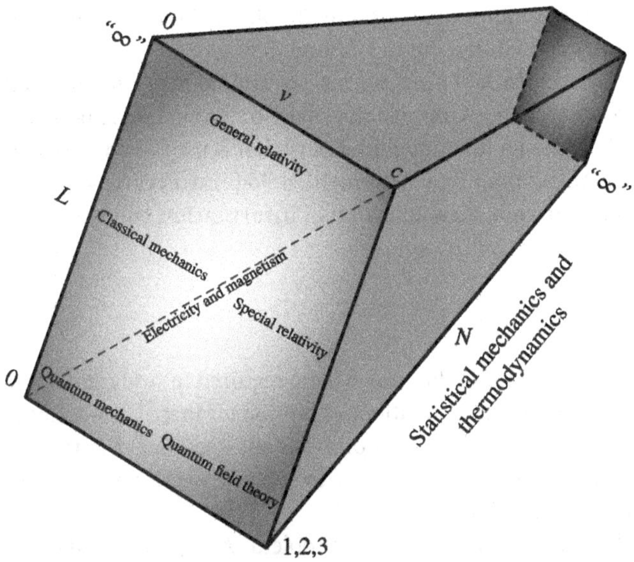

Figure 1.2 Adding a third dimension, N (the number of particles in a system), to Figure 1.1.

One last note before moving on: When we have only one or two particles, of course the problems are (often) relatively easy to solve. When we have a very large number ($\sim 10^{24}$), the problems are also relatively easy to solve, but only because of the questions we will ask.[7] That middle ground, the **few-body problem**—when we have 10 or 20, or maybe a few 100—is actually way more difficult!

1.2 What Is Statistical Mechanics?

How does all of this relate to statistical mechanics? Thermodynamics as we will think of it is an empirical theory of the large systems mentioned above. This arose from centuries of experimental observations of bulk properties such as the number of molecules N, the pressure p, the volume V, the temperature T,[8] the internal energy E,[9] and plenty of other quantities. The four laws of thermodynamics allow us to study these quantities: specifically how they are related, how they change when we make changes to the system, etc. In a sense, this is akin to the study of optics, where so much can be understood without knowing that light results from an electromagnetic wave solution to Maxwell's equations. Geometric optics is a phenomenological theory about how light behaves with a particular set of assumptions, even though it is not the fundamental theory and it has limitations. It is useful but you cannot delve too deeply into the subject without more knowledge of the underlying theory.

Similarly, statistical mechanics is the fundamental theory that gives rise to thermodynamics, and in some sense we can consider it similar to electricity and magnetism. From it, we can derive all of the phenomenological results in thermodynamics (including the four laws), but with fewer, and more general, assumptions. Not only that, but a deeper understanding of the system can be

7 Many students tend to disagree with this statement while taking this class. However, the fact that we even have an undergraduate version of this class shows that these problems are "easy" in some sense!
8 Which you know about from an everyday perspective, but for now forget you know anything about it.
9 Many textbooks use U for this quantity to distinguish it from mechanical energy, but I tend to prefer this notation. If we used U, we would have to contend with a possible confusion with potential energy anyway.

surmised from statistical mechanics, and more importantly, there are physical observations that can only be understood if we start from the more fundamental theory.

This analogy with optics is far from perfect. In that case, we have a well-defined fundamental theory that can be used to derive the results in optics, but it's not really the same here. Statistical mechanics is more of a *methodology* rather than a fundamental theory that is used so that we can make predictions based on a very small set of assumptions. This methodology involves (essentially) one major assumption and, as we shall approach it, an application of **information theory**.

1.3 Our Approach

Given the name, it should be clear that understanding statistics will be required to study statistical mechanics. As we are not able to solve the complicated equations of motion for these large systems (and as mentioned, we don't really want to), instead we use a probabilistic approach. We replace those quantities that we ordinarily care about (such as the trajectories in classical mechanics) with, for example, the bulk properties referred to in Section 1.1.

Some aspects of our approach are the same as in any physics problem. As always we start by precisely defining our system. From that, given that we will take a probabilistic approach, we need to define an **ensemble** of systems: a set of all states that are accessible to the system. This is where our method will diverge from an ordinary physics problem.

When we discuss states, we will now need to distinguish between the possible **microstates** of a system (for example, all of the trajectories $\mathbf{r}_i(t)$ of the 10^{24} particles in our example above) and the possible **macrostates** of our system, which are dependent on what we are interested in measuring. The allowed macrostates will be specified by the ensemble we will use (discussed in Chapters 4, 8, and 11), but initially the macrostates I'll discuss will have a given constant energy E. It's easy to see that generally many different microstates correspond to a single macrostate. As a quick example of this, consider a classical one-dimensional simple harmonic oscillator with mass m. This is a conservative system, so the total energy E is constant. There are an infinite number of microstates (one for each position $x(t)$ and corresponding momentum $p(t) = mdx/dt$), but there is only a single macrostate defined by the total energy of the system.[10]

To calculate physical observables (the primary goal in any physics problem) here, we have to be able to assign a probability that our system will be in a given state, which we can do by considering everything that we know about the system. What we will need to do is figure out how to *quantify* this information, or more importantly, the **missing information** of our system. We will actually find that *not* knowing things isn't necessarily a bad thing![11]

Figure 1.2 makes it apparent that because statistical mechanics is relevant for $N \gg 1$, we can apply these methods to any problem in classical mechanics, electricity, and magnetism, quantum mechanics, etc. That is, it can be applied to any problem in physics, so long as the number of particles in the system is very large; as such, we will need a cursory knowledge of these other subfields.[12] As such, this is really a "capstone course" in physics, as we get to review everything else you studied in other courses. This is one of the reasons that this can be so difficult, as the starting point for our problems can come from anywhere, but is also one of the reasons this field is so rich and interesting!

10 We will return to this example many times, more specifically in Exercise 4.1.
11 Ignorance is bliss, after all.
12 You won't need to know how to solve these problems for this course, just the results of such solutions, specifically the total energy of a state of the system.

As mentioned, we cannot get right into statistical mechanics without covering probability and information theory first. We will cover probability in Chapter 2 at a cursory level; primarily just considering it from an everyday understanding (without much rigor or formalism). In Chapter 3, we will go over the basics of information theory, which will set the foundation for one of the most important concepts in thermodynamics: **entropy**. While it will not be used much beyond setting up the initial stages of our course, it will give us a wonderful foundation upon which to understand what thermodynamics (and more importantly statistical mechanics) can do for us at a deeper level.

2

Introduction to Probability Theory

As discussed in Chapter 1, before we can truly study statistical mechanics, we first will have a brief overview of probabilities. I will not get into the rigor that is needed for a fully satisfactory treatment (there are entire courses for that). Rather I will focus on the basics of the subject that we will need here. Much of what I will discuss in this chapter will be familiar in an everyday sense; what makes it difficult is trying to quantify what we already know.

After completing this chapter, you should be able to

- understand the meaning of probabilities, and calculate them for simple random systems,
- understand the role of mean (average) values in statistical systems, and
- have an idea of the difference between discrete and continuous probability distributions.

2.1 Understanding Probability

Even though it is so prevalent in our daily lives, probability is often misunderstood by many people. In some way, this misunderstanding arises *because* this mathematical concept is used so regularly in a colloquial sense. You may look at the weather forecast and see that it states there is a 65% chance of rain. If it doesn't rain, you might think, "As usual, the weather forecast was wrong!" And if it says there is a 10% chance of rain and it *does* rain, you might think the same thing. It is very easy to have such a misconception of what these percentages (probabilities) are telling us.

Whenever we study probabilistic systems—those for which we don't know what will happen definitively—then there is a specific meaning when we say there is a certain chance of something occurring. If something has a 65% chance to occur, then what we mean by this statement is as follows: If we have a very large[1] number of *identically prepared* systems, then *on average* we would expect this to happen in 65% of those systems. In the weather example, we would need a very large number of identically prepared Earths (which is an odd idea to consider) and in roughly 65% of those Earths, it would rain. This very large number of identical systems this defines an **ensemble**.[2] I have used the terms *on average* and *roughly* here, because we have to be careful as we discuss these concepts, and I'll be more concrete in Section 2.2. What we need to keep in mind is that whenever I refer to a system, I really mean an ensemble of systems.[3]

1 For now I am remaining vague when referring to "large" or "very large" systems. Don't worry, I will clarify this when it is needed.
2 I will define ensembles more concretely in Chapters 3 and 4.
3 It just gets cumbersome to continually write or say "ensemble of" over and over.

Statistical Thermodynamics: An Information Theory Approach, First Edition. Christopher Aubin.

Companion website: www.wiley.com/go/Aubin/StatisticalThermodynamics

2.2 Randomness, Fairness, and Probability

A common application of probability in everyday life is to games of chance, which I broadly define as any activity that involves (among other things) flipping a coin, drawing from a deck of cards, rolling dice, or other sorts of games you might find at a casino. As we discuss probabilities, I will use specific (non-physics) examples to understand the various ideas that will arise.

> What is most important to remember is that you already understand so much about probability without realizing it. You know that if you flip a coin, there is a 50-50 chance of getting "heads," a one out of six chance that you roll a four on an ordinary six-sided die, and with a standard deck of playing cards, there's a 1 in 52 chance you'll draw the ace of spades. However, in order to apply these ideas more generally, and to ensure we don't misunderstand these statements, we need to formalize these ideas more.

We begin by considering a system which can have some number of outcomes or **events**, and while perhaps *in principle* we could determine which of these events could occur, we will say there is no way to know exactly what will happen. In the case of the coin toss, in principle we can account for the initial force which causes the coin to begin to spin as it moves up and down under the force of gravity, and thus could determine which side will be face up upon landing. However, this can be difficult (you can consider the physics of coins, both under the assumption that the coin is very thin so there are only two outcomes in Ref. [1] as well as a thick coin, where there's a significant chance it will land on its side in Ref. [2]). For all intents and purposes we will say that we cannot determine the outcome at all.

Once we have our system and the possible outcomes, then we wish to count how many of these outcomes there are. From this we will be able to determine the probabilities of each outcome.

Example 2.1 Let's consider the coin, the die, and a deck of cards, all examples we will return to several times.

- In the case of the coin, there are two possible outcomes, (usually referred to as heads or tails), for each event, or coin toss.
- If the event under consideration is rolling a six-sided die, then we have six possible outcomes (usually one through six), and each outcome is the side which is facing up after a roll.
- An ordinary deck of playing cards has 52 cards (we are ignoring jokers), where we have 13 different faces (2–10, jack, queen, king, ace) and four suits (spades, hearts, clubs, diamonds) of each. The event in this case is drawing a card, and we have 52 outcomes (if you care only about which *specific* card you draw).

As alluded to in the third example, the outcomes we consider may change depending on what we are interested in—perhaps we only care about the value (say you are hoping for an ace specifically) of the card, not the suit. This must be taken into account as we calculate probabilities.

How then do we determine those probabilities? Unless we have some knowledge about the system (perhaps you know the die is weighted, or you were able to palm cards while shuffling the deck of cards), we will assume that the outcomes are all completely **random**. That is, we cannot assume any one outcome is more or less likely than any other. For our examples above, we will assume (unless proven otherwise) that the coins or dice are **fair** and the deck of cards is thoroughly shuffled that we can assume every outcome is equally likely. A given coin is different on either side and the "pips" on a die are such that each side of the coin or die are *technically* not equally weighted; however, this difference is rarely going to be significant when considering the probability of a particular outcome. The key point is that unless we have some prior knowledge about our system, there won't be any reason to assume any individual outcome is more likely.[4]

Let's now be a bit more precise about the **probability** of an individual outcome, which I will label as r. In this definition I will be a little vague about the variables and terms, as they will become more clear with the exercises and examples below. Suppose a given outcome r can occur n_r times out of a total of N possible events. The individual probability P_r that outcome r will happen is defined by

$$P_r \equiv \frac{\text{number of times outcome } r \text{ can occur}}{\text{total number of possible events}} = \frac{n_r}{N}, \qquad N = \sum_r n_r. \tag{2.1}$$

Keep in mind that r is just an arbitrary label; don't fixate on it too much, and note that the sum over r is over *all* possible outcomes. For example, we could write for the coins (assuming a completely random outcome as discussed above):

$$P_{\text{heads}} = P_h = P_1 = \frac{1}{2}, \qquad P_{\text{tails}} = P_t = P_2 = \frac{1}{2}. \tag{2.2}$$

We could use $r = 1$ or 2, $r = h$ or t, or even just write out heads and tails for the labels. Similarly, for a six-sided fair die:

$$P_1 = P_2 = P_3 = P_4 = P_5 = P_6 = \frac{1}{6}, \tag{2.3}$$

where r refers to the roll of the die. And finally, for any of the possible card draws, we would get $P_r = 1/52$; equal probabilities for any card draw. It's important to note that it *must* be true that summing over all probabilities should give you one:

$$\sum_r P_r = 1, \tag{2.4}$$

and given that $N = \sum_r n_r$, you can easily show this is true from Eq. (2.1). This is because if we have correctly accounted for all possible outcomes, there is a 100% chance that *something* will occur.

However, I must point out that Eq. (2.1) is only true in terms of what we *expect* to occur. That is, just because when you roll a die you expect that 1/6 of the outcomes will result in a four, this doesn't mean that you will *always* get exactly that number of rolls. You might find that only 92 out of 600 rolls of a die result in a four, even though from above you would expect 100 of them. I clarify this when I discuss what I call *experimental probabilities* below in Eq. (2.5), but for now we will only consider this ideal probability.[5]

Exercise 2.1 What if you rolled two six-sided dice, what are the possible outcomes? Is rolling a 1 on one and a 4 on the other more or less likely than rolling a 2 on one and a 6 on the other?

Usually we only care about the sum of the pips on the two dice, so in that case, what is the probability of rolling each allowed *sum* (so a 5 or an 8 in the aforementioned rolls)?

The previous exercise shows how you are familiar with probabilities, even if you don't think so. In Exercise 2.1, you can work out the probabilities two different ways. The simplest (brute force)

---

4 Later we will see situations where each outcome will have a different probability, but specifically *because* we have some prior knowledge.

5 You can consider an extreme case of rolling a die just five times. There is no way to obtain the probability as defined in Eq. (2.1), because there are six possible outcomes and you have only rolled the die five times.

approach is to enumerate all of the possibilities and then apply Eq. (2.1). When rolling two dice, we can get the following:

- a 2 if both dice resulted in a 1,
- a 3 if one die is a 1 and the other is a 2,
- a 4 if one result is 1 and the other is 3 or if both resulted in a 2,
- etc.

In these cases, we see there is one way to obtain a 2, two ways to obtain a 3, three ways to obtain a 4, and so forth. Counting up all of the possible outcomes we find a total of 36 outcomes, and then using Eq. (2.1), we have that the probability, for example, to roll a total of 3 is $P_3 = n_3/N = 2/36 = 1/18$.

Alternatively, we could realize that the outcomes of each die are **statistically independent** and then combine the individual probabilities of each outcome. Two outcomes are statistically independent if a given result does not depend on any prior results. That is, the result of the first die does not rely on the result of the second die.[6] If we want to know the probability of rolling a 3, we would ask, "What is the probability that we would roll a 1 on the first die **and** a 2 on the second die, **or** a 2 on the first die **and** a 1 on the second die?" To determine the probability of two things occurring (so we want this *and* that to happen), we **multiply** the individual probabilities, and to determine the probability of one *or* another event to happen, then we **add** the individual probabilities. In the case of rolling a 3 on two dice,

$$P_3 = \left(\frac{1}{6}\right)\left(\frac{1}{6}\right) + \left(\frac{1}{6}\right)\left(\frac{1}{6}\right) = \frac{1}{18}.$$

Being able to use both approaches is useful; depending on the situation we'll use either. The first approach is often simpler initially, but for more complicated systems can become rather tedious. Keep in mind that this level of rigor will generally be more than sufficient for our purposes; we will not usually need to get more systematic than this.

Again, much of this is familiar, but I do want to make clear that we are only talking about the *probability* of an outcome. If you flip 10 coins you are not guaranteed that exactly five coins will land heads up. There is a chance (albeit small) that all 10 coins will land tails up. Only if you flip a very large number of coins, then *on average* you will see that 50% of the time you will get heads.

Experimentally, we can determine the "probability" by counting the number of times m_r each outcome r actually occurred and dividing by the total number of events M, to get

$$P_r^{(\text{exp})} = \frac{\text{number of times outcome } r \text{ actually occurs}}{\text{total number of events}} = \frac{m_r}{M}. \tag{2.5}$$

I like to call this the **experimental probability**[7] to connect it to our discussion here, but this is not the same as P_r from Eq. (2.1)! We can relate the two by

$$P_r = \lim_{M\to\infty} P_r^{(\text{exp})}. \tag{2.6}$$

That is, for the case of the coin flip, if you flip a coin an *infinite* number of times will you get precisely 1/2 of your results to be heads.

Exercise 2.2 Recall our standard deck of cards, with 13 cards of each suit (spades, hearts, clubs, and diamonds) and those 13 cards include face values of 2 through 9, a jack, a queen, a king, and an ace. What is the probability of drawing from the deck:

6 This is true if we roll the two dice at the same time or if we roll one die two times in a row.
7 Although this is just the fraction of times outcome r occurs.

- a 2 of any suit?
- any face card (a jack, queen, king, or ace)? or
- five cards all of the same suit regardless of their face value (a "flush" in poker)?

Wikipedia has a nice entry [3] on the probabilities of various poker hands, which were calculated long ago to determine the ranking of winning hands.

Exercise 2.3 There are many other dice used in tabletop games with different numbers of sides. The common dice we're most familiar with are the six-sided dice discussed above, or "d6," but other common dice have four sides (d4), eight sides (d8), 10 sides (d10), 12 sides (d12), and 20 sides (d20). Calculate the outcomes and probabilities when you roll two of each (2 d4's, 2 d8's, etc.).

Computer Exercise 2.1 You can visit the companion site to verify Eq. (2.6). Go to the Testing Probabilities section and view the Jupyter notebook which allows you to test systems with variable numbers of outcomes (two for a coin, six for a die, etc.) and compare the experimental probabilities with the theoretical probabilities for larger numbers of experiments.

Our examples above all considered the individual probabilities to be equal. This is simple example of a **probability distribution**, which is simply the set of probabilities for all possible outcomes of an event. For the cases above, the probability distribution is **uniform**, so that P_r = constant for all r. However this will not always be true for more general systems. Perhaps a die is weighted or you have a thick coin so that there is a reasonable chance that the coin may land on its side (but not with the same probability that it will land heads or tails). In these cases, the probability distribution of the system may be more complicated (and more difficult to determine), as P_r will be different for different outcomes r. However, as long as we know the probability distribution for a given system then we can set about doing some calculations, which I'll turn to in Section 2.2.

Combinatorics 2.1

As we saw when considering the probabilities of an outcome when rolling two fair six-sided dice, we needed to count the number of times n_r each outcome r would occur to apply Eq. (2.1). We will have to do this many times as we study statistical mechanics, so it's essential to have a basic understanding of **combinatorics**, which is really just a fancy way to count. As with our discussion of probability, I won't include a formal treatment of combinatorics, just some of the basics that will be useful for our work here. And to do so, I will move away from the examples above for our initial discussion.

Consider the following: We have N marbles, and we want to line them up, one after another. The question we wish to answer is: How many distinct ways can we arrange these marbles in this line? There are (usually) two answers to this question, and which answer we are interested in depends on whether or not any of the marbles are identical.

We start by choosing any one of the N marbles and placing it in the first position. Once that choice has been made, there are $N-1$ remaining marbles that we can choose from to place in the second position. As we continue, there are $N-2$ for the third position, $N-3$ for the fourth, etc. Ultimately, we have

$$N(N-1)(N-2)\cdots 2\cdot 1 = N! \tag{2.7}$$

(Continued)

(Continued)

ways to arrange the marbles. As a quick example, if we had $N = 4$, and labeled them 1, 2, 3, and 4, all of the possibilities are

(1)(2)(3)(4) | (2)(1)(3)(4) | (2)(3)(1)(4) | (2)(3)(4)(1)
(1)(2)(4)(3) | (2)(1)(4)(3) | (2)(4)(1)(3) | (2)(4)(3)(1)
(1)(3)(2)(4) | (3)(1)(2)(4) | (3)(2)(1)(4) | (3)(2)(4)(1)
(1)(3)(4)(2) | (3)(1)(4)(2) | (3)(4)(1)(2) | (3)(4)(2)(1)
(1)(4)(2)(3) | (4)(1)(2)(3) | (4)(2)(1)(3) | (4)(2)(3)(1)
(1)(4)(3)(2) | (4)(1)(3)(2) | (4)(3)(1)(2) | (4)(3)(2)(1)

(The lines are drawn to separate the columns to not confuse the different arrangements.) Counting these, we see there are $4! = 24$ ways to order these four marbles.

That would be the end of the story if we were always arranging a set of distinct objects, but what if some objects are identical (this is going to be the case with most of the physical situations we will apply this to later)? Suppose one of the marbles is red (which we will label with an "r"), and the other three are blue (labeled with a "b"), and there was no other way to distinguish them. Then we have only four possible arrangements:

(r)(b)(b)(b) | (b)(r)(b)(b) | (b)(b)(r)(b) | (b)(b)(b)(r)

This is not different than before: If we consider marbles 2, 3, and 4 to be the blue marbles, then each of the columns above corresponds to a distinct arrangement.

For small systems such as this, writing out the arrangements is simple and quick, but as N increases, this will get harder to do. As we are generally going to be interested in the number of arrangements only, how can we calculate this for this second situation more generally? Clearly when there are some identical marbles, there are many duplicates in the $N!$ outcomes, which will lead to the reduction in the number of possibilities. If n of the N marbles are identical, then for a particular arrangement, we can rearrange the identical marbles $n!$ ways using the same argument we used to get to Eq. (2.7). The "rbbb" arrangement, for example, corresponds to $3! = 6$ of the 24 original options (the first column of the original 24 arrangements of the marbles). The same can be said for the other outcomes: brbb, bbrb, and bbbr. Mathematically, we have

$$\frac{N!}{n!} \tag{2.8}$$

distinct ways to arrange the N marbles if n are identical. This corrects for overcounting the identical states coming from the $N!$ arrangements. For the $N = 4$ case with $n = 3$ identical marbles, we have $4!/3! = 24/6 = 4$, agreeing with what we saw explicitly.

What makes this formalism powerful is to be able to generalize this to the case where there are multiple sets of identical objects. Suppose in our group of N marbles, we have $n_1, n_2, \dots, n_m$ marbles each of m different colors, with

$$\sum_{r=1}^{m} n_r = N$$

total marbles. As before, there are $N!$ ways to arrange the marbles, and if we wish to omit identical arrangements, we divide by $n_r!$ for each r, so that we have

$$\frac{N!}{n_1!n_2!\cdots n_m!} = \frac{N!}{\prod_{r=1}^{m} n_r!} \tag{2.9}$$

distinct ways to arrange our marbles. This equation is useful even when there are zero or one marbles of a given color since $0! = 1! = 1$.

Exercise 2.4 Consider $N = 5$ marbles, with two red marbles, two blue marbles, and one green marble. Show explicitly (by listing them all) that there are 30 arrangements as dictated by Eq. (2.9).

We will return to this idea of counting the possible outcomes throughout this book. There is plenty more to consider when studying combinatorics; however, this elementary understanding will be sufficient for most of our purposes.

2.3 Mean Values

As discussed earlier, the probability of an outcome is only useful when you consider an ensemble of identical experiments. Suppose we have a system with N possible outcomes, and we know the (not necessarily uniform) probability distribution P_r for $r = 1, \dots, N$. These probabilities correspond to what would result if we had a large number M of identically prepared systems such that each may have one of the N possible outcomes. Considering the example of rolling a die, we would say that one experiment corresponds to rolling N dice, and we would do this M times. In the limit that $M \to \infty$, the probability P_r would correspond to the fraction of these experiments would have outcome r (as in Eq. (2.6)).

Consider some quantity y which relates to the system (such as the roll of a six-sided die, the sum of the roll of two dice, etc.), where I'll denote the possible results for y in outcome r by y_r. For example, if y corresponds to the roll of a die we have $y_r = r$, with $r = 1, \dots, 6$. In doing so, I may want to know what we expect to obtain, *on average*, out of those N possibilities. In other words, we want to calculate the **mean**, or **average**, value of y.

I'll denote mean values with a bar over the variable: $\bar{y}$, and to calculate it, we need to calculate the average of y_r weighted by n_r. Given that the probability of a given outcome r is given by Eq. (2.1), we can write the mean value as

$$\bar{y} \equiv \frac{n_1y_1 + n_2y_2 + \cdots + n_Ny_N}{N} = \sum_{r=1}^{N} y_rP_r. \tag{2.10}$$

Thus, if we know the probability distribution, the mean value of any quantity is (in principle) easy to calculate.

Exercise 2.5 Go back to our fair six-sided die, where $P_r = 1/6$ for $r = 1, 2, 3, 4, 5,$ and 6. What is the average value of the roll of one die? How about the average value of the roll of two dice?

To compare this with an experimental setup with a specific example, we imagine rolling N dice, and count the number of times each value is rolled. In this case, I would denote the experimental mean value as $\overline{y}^{(exp)}$, and this would be given by the first equality in Eq. (2.10). As $N \to \infty$, then $\overline{y}^{(exp)} \to \overline{y}$, the theoretical mean value.[8] This is an important point because of the aforementioned misunderstanding of probabilities—if you rolled a die 100 times, it is likely you would not obtain the average value from Exercise 2.5. Only in the $N \to \infty$ limit are they expected to be precisely equal.

Some important points about mean values:

- The mean value of a constant c (a quantity that does not depend on the outcome r) is that constant, $\overline{c} = c$.
- Calculating the mean value is a linear operation, so the mean value of a sum is the sum of the mean values: $\overline{y_1 + y_2} = \overline{y}_1 + \overline{y}_2$.
- The mean value of a constant times a variable is just that constant times the mean of that variable: $\overline{cy} = c\overline{y}$.

Exercise 2.6 Use the fact that the mean value of a constant is just that constant and $P_r = n_r/N$ to show that the average value of n_r is

$$\overline{n}_r = NP_r. \tag{2.11}$$

It's important to realize that the mean value by itself doesn't tell us much. If I told you that the average exam score in a class of 10 students was 50, that doesn't say anything about the distribution of the grades. All 10 students could have gotten a 50, half of them could have gotten a 25 and the other half a 75, and so forth. Thus, in addition to knowing the mean value of a quantity, we are also interested in how individual outcomes fluctuate around the mean. These fluctuations are defined as

$$\Delta y \equiv y - \overline{y}. \tag{2.12}$$

This isn't very useful in and of itself, because $\overline{\Delta y} = 0$. (Note that the mean value of a mean value is just that mean value, $\overline{(\overline{y})} = \overline{y}$.)

Exercise 2.7 Show that $\overline{\Delta y} = 0$ by using the rules above, regardless of the probabilities P_r for each outcome r.

This makes sense because, on average, y should deviate equally above and below the mean value. Thus, Δy itself is not useful to quantify the spread of y_r about the mean; we want a positive quantity that can give us information about these fluctuations. To this end we define the **variance** of y about the mean,

$$\overline{(\Delta y)^2} \equiv \sum_r (y_r - \overline{y})^2 P_r \geq 0. \tag{2.13}$$

Calculating the variance in this form is a bit of a pain, but we can simplify this expression a bit to obtain

$$\overline{(\Delta y)^2} = \overline{y^2} - (\overline{y})^2. \tag{2.14}$$

Therefore we just have to evaluate $\overline{y^2}$ and $\overline{y}$ to extract the variance.

8 If we wanted to be more clear we would call this an *expectation value* as opposed to a mean value. The difference is that the expectation value is what we would expect to obtain after an infinite number of trials, while the mean value is what is actually obtained after some number of trials. This distinction, while important, won't matter for us.

Exercise 2.8 Using the rules for mean values above, derive Eq. (2.14).

The square root of the variance is called the **standard deviation**,

$$\sigma_y \equiv \sqrt{\overline{(\Delta y)^2}}, \tag{2.15}$$

which gives us a way to measure the spread of values of y about the mean (the square root is used to ensure that σ_y has the same units as y if this is a physical quantity). This is the most common method to measure the spread of values around the mean, and we can see that it is never negative. Additionally, the larger the value of σ_y, the more our individual results spread around the mean.

Exercise 2.9 Returning to our fair six-sided die, what is the standard deviation when you roll one die? What is the standard deviation when you roll two dice?

Computer Exercise 2.2 You can visit the companion site in the Mean Values section to study a simple case of equal probabilities for any number of outcomes. Consider a simple system with five possible outcomes, and run an increasing number of "experiments," say 10, 100, 10^3, 10^4, and 10^5. How quickly does the actual mean value approach the expected mean value? This notebook also includes a calculation of the standard deviation, so consider how quickly the measured standard deviation compares with the theoretical value.

If you're thinking carefully, you might ask: Why do we choose to use the standard deviation in Eq. (2.15) (from the variance, Eq. (2.13)) to describe the spread of values about the mean? An equally good measure comes from taking the absolute value of the deviations from the mean, or

$$\overline{|\Delta y|} \equiv \sum_r |y_r - \bar{y}|\, P_r, \tag{2.16}$$

because this too is never negative. For one, it is less convenient mathematically because we cannot simplify this as we did for the variance in Eq. (2.14) (which makes the standard deviation easier to calculate). The variance is also directly relevant for the Gaussian distribution, which is a special probability distribution we will discuss in Section 2.5.2. For some simple cases you can calculate Eq. (2.16) and see that it generally is on the same order as the standard deviation.

For many probability distributions, calculating mean values can become tedious. Luckily when we begin applying this to physical problems, there will usually be tricks we can employ to simplify the calculations.

As one last comment, the mean value and variance are the simplest common **moments of a probability distribution**. The kth moment is defined as

$$\overline{y^k} \equiv \frac{1}{N}\sum_{r=1}^{N} y_r^k P_r. \tag{2.17}$$

You have already seen three moments:

- The zeroth moment, $k = 0$, corresponds to our normalization condition for probabilities.
- The first moment, $k = 1$, is the mean value.
- The second moment, $k = 2$, is used to determine the variance.

Higher moments are used in more advanced statistical problems, but will likely not be needed for our purposes. While they are not going to be needed very much in our elementary treatment of statistical mechanics, they find their ways into various areas of physics in general (most notably the multipole expansions of the potentials in electromagnetic theory) that it is useful to mention them briefly here. It's nice to see how mathematics and physics can overlap in such ways.

2.4 Continuous Probability Distributions

In all of the examples discussed above, I focused on *discrete* probability distributions—that is, those systems which have a finite number of outcomes each with a discrete value (such as one through six, for the die) and an associated probability. However, in many cases we will come across examples where we will have a *continuous* distribution, so the outcomes are labeled by a variable which can take on any real value (possibly within a finite range). The basic ideas of Section 2.3 still hold, but there are some subtle differences I should discuss before moving on.

To see the problem by merely applying the results from above, imagine a system with N outcomes that are all equally likely, so $P_r = 1/N$ for all r. As a mildly contrived example, perhaps you divide up a box into N compartments and toss a marble into one of them as in Figure 2.1. Let's assume we don't try to calculate the trajectory of the toss to know precisely where it will land; then we can say the box it lands in is random (and we will assume it is proportional to its area). As we divide the box into more (and smaller) compartments, as in Figure 2.2, the probability for landing in each box decreases, and if we continue to make the boxes smaller, N will go to ∞ while $P_r \to 0$. Given that the marble has a finite size, let's assume we have just painted lines in the box to create these

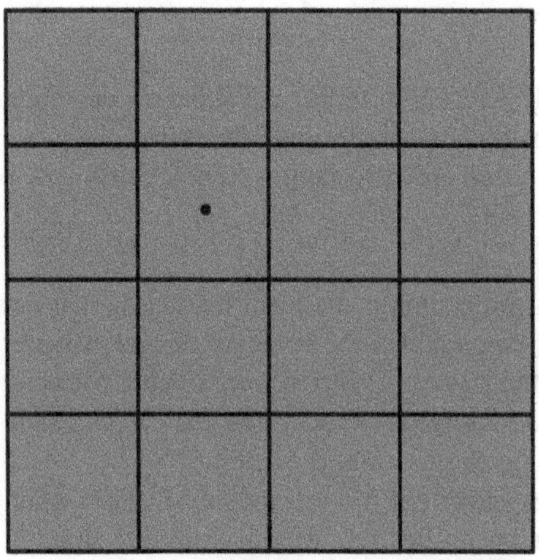

Figure 2.1 A box divided into $N = 16$ boxes with a marble tossed into one of them.

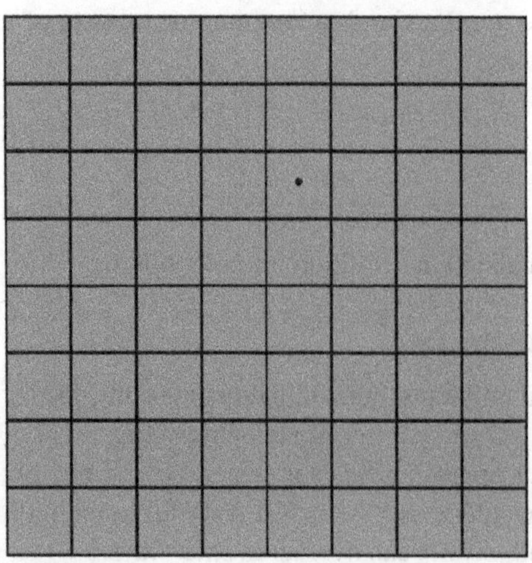

Figure 2.2 A box divided into $N = 64$ boxes with a marble tossed into one of them.

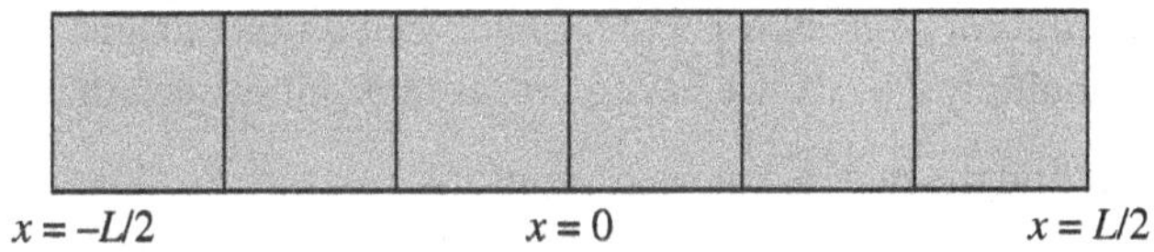

Figure 2.3 A box divided into $N = 6$ boxes along the x-axis with a marble to be tossed into one of them.

compartments that the marble lands. In this limit where these lines get closer together, so $N \to \infty$, this is equivalent to knowing the *exact* position where the marble lands, and since $P_r \to 0$, this is impossible. At best you can say where it lands within some range of values.

We can use this limiting procedure to modify our results from Section 2.3 for a continuous probability distribution. This modification will be something we will use often later, so it is worth the time go through this in some detail now. To make this more clear, I'll work through a more specific example and generalize our results afterwards.

Let's start with a simpler setup as shown in Figure 2.3, where we consider a marble to be constrained at a position in a finite range $-L/2 \leq x \leq L/2$ along a single dimension. Initially we subdivide each axis into six regions, each with a width of $L/6$, and we know the probability the marble will be in one of those regions, $P_r^{(6)}$, where the superscript (6) indicates the number of regions we've subdivided the x-axis. We will subdivide the axis more and more so we eventually have N boxes, and for each N, we can calculate a new set of probabilities $P_r^{(N)}$. The average location of the marble after a large number of measurements would be given by Eq. (2.10),

$$\overline{x} = \sum_{r=1}^{N} x_r P_r^{(N)}. \tag{2.18}$$

We have to be careful as we take the limit $N \to \infty$, as we know that the probabilities themselves will vanish. We can define the position x_r in terms of r, N, and L by

$$x_r = \frac{2r - N - 1}{2N} L, \quad r = 1, \ldots, N. \tag{2.19}$$

This is chosen so that x_r will be the position at the midpoint of the rth box (although the precise value won't matter in the limit $N \to \infty$ so long as x_r is chosen to be *somewhere* inside the rth box).

Exercise 2.10 Verify Eq. (2.19) for the $N = 6$ case explicitly, and convince yourself that this is true for general systems.

The distance between the possible location of the marble in neighboring boxes is

$$\Delta x_r = \frac{L}{N} \Delta r,$$

with $\Delta r = 1$ the difference between two neighboring integers. With this, we can multiply our expression for $\overline{x}$ by 1 in the form of Δr to get

$$\overline{x} = \sum_{r=1}^{N} x_r P_r^{(N)} \Delta r = \sum_{r=1}^{N} x_r P_r^{(N)} \frac{N}{L} \Delta x_r. \tag{2.20}$$

This can be considered a Riemann sum, so we can take the limits $N \to \infty$ and $\Delta x_r \to 0$ to convert this into an integral. We have already argued that $P_r \to 0$ as $N \to \infty$, but if the product of this with $N/(2L)$ remains finite we can define $\mathcal{P}(x_r) \equiv N P_r^{(N)}/L$ to get

$$\overline{x} = \int_{-L/2}^{L/2} x \mathcal{P}(x) dx. \tag{2.21}$$

In this case, notice that $\mathcal{P}(x)$ actually has units of *probability per unit length*, so it is best thought of as a **probability density**, and when we multiply it by dx, we have a proper probability, such that

$$\mathcal{P}(x)dx = \text{the probability that we'll find the marble between } x \text{ and } x + dx.$$

I should really say this is the probability that we'll find the marble in region of width dx centered on x (so between $x - dx/2$ and $x + dx/2$). However for practical considerations, dx will usually be considered to be very small compared with x, so the distinction will not matter much, and this definition will be similar to our later conventions.

With this in mind, a more appropriate question to ask instead of "What is the probability that our marble is at a position x?" is "What is the probability that our marble is between $x = a$ and $x = b$?" This can be found by integrating[9]

$$P(a \le x \le b) = \int_a^b \mathcal{P}(x)dx. \tag{2.22}$$

Our expression in Eq. (2.21) can easily be generalized to any quantity that is a function of x,

$$\overline{f(x)} = \int f(x)\mathcal{P}(x)dx, \tag{2.23}$$

where I am omitting the limits of integration; the integration will implicitly be over all allowed values for the particular system. Additionally, x could be any variable, it does not need to be a position. Additionally, the probability must still be normalized,

$$\int \mathcal{P}(x)dx = 1. \tag{2.24}$$

In Section 2.5 we will go through some examples of continuous distributions to work with these equations.

2.5 Common Probability Distributions

It's important to consider some common probability distributions, so we'll discuss them in this section. The derivations will be left for you in the problems.

2.5.1 Binomial Distribution

The **binomial distribution** is used to describe the probability of an outcome when there are only two possibilities. This is often discussed in the context of the **random walk problem**,[10] and can readily be applied to many different situations. Any number of (discrete) outcomes can be described using the binomial distribution, even if it doesn't seem this way initially. For example, rolling a six-sided die could be described this way, if we consider one outcome to be rolling a "1," and the other outcome is any of the other five possibilities (not rolling a 1).

We'll consider the two outcomes of some experiment to be true or false, where t is the probability to obtain true, and $f = 1 - t$ to be the probability to obtain false (we require $f + t = 1$, as usual, since

9 Remember, we are asking if the marble is at this point, OR another point, OR another point, etc., and when you "OR" probabilities you add them. Addition turns into integration when we move from a discrete to a continuous system.

10 An old school example to derive this distribution is to consider someone having a bit too much to drink after leaving a bar, so they randomly move in one direction or another.

one of these must occur). Then we will run N experiments and ask for the probability of obtaining n_t true outcomes. You can show (see Problem 2.10) that this is given by

$$P_N(n_t) = \frac{N!}{n_t!(N-n_t)!} t^{n_t} f^{N-n_t}. \tag{2.25}$$

This is useful for many situations: flipping a coin, randomly moving in one of two directions, etc.

Exercise 2.11 Show that the binomial distribution is properly normalized, that is,

$$\sum_{n_t=0}^{N} P_N(n_t) = 1.$$

Hint: Consider the expansion of $(x+y)^N$, known as the **binomial expansion**.

We can calculate several mean values using the binomial distribution. The mean number of true outcomes is

$$\overline{n}_t = Nt, \tag{2.26}$$

and the variance is given by

$$\overline{(n_t - \overline{n}_t)^2} = Ntf. \tag{2.27}$$

The first of these could have been guessed without calculation given Eq. (2.6). You will derive these and several other mean values in Problems 2.11 and 2.12 at the end of this chapter.

Exercise 2.12 Consider flipping five fair coins, where a t outcome is tails and f is heads. Using Eq. (2.25), what are the probabilities of obtaining zero, one, …, five tails?

2.5.2 Gaussian Distribution

In the large-N limit for general probabilities, the binomial distribution becomes the **Gaussian distribution** or **normal distribution**. This is a continuous distribution, so the discussion of Section 2.4 is relevant. Let us consider what the probability is of obtaining a result of some variable in the range from x to $x + dx$, if the mean value is given by $\overline{x} = \mu$ and the variance is σ^2. This is given by

$$\mathcal{P}(x)dx = \frac{1}{\sqrt{2\pi\sigma^2}} \exp\left[-\frac{(x-\mu)^2}{2\sigma^2}\right] dx. \tag{2.28}$$

In Problem 2.13, you can follow the steps to derive this expression from the binomial distribution.

Exercise 2.13 Show that the Gaussian distribution is properly normalized, that is,

$$\int_{-\infty}^{\infty} \mathcal{P}(x)dx = 1.$$

Also show that the mean value of x is given by μ and the variance by σ^2. (The properties of Gaussian integrals in Appendix C will be helpful here.)

Computer Exercise 2.3 You can visit the Binomial Distribution section of the companion site to see how the binomial distribution becomes the Gaussian distribution for large values of N. Specifically this simulates the random walk problem, where the possible outcomes are taking a step to the right or left in lieu of true and false. Study the outcomes for larger and larger numbers of steps and experiments to see how the binomial distribution becomes the Gaussian distribution.

To show that the binomial distribution becomes the Gaussian distribution in the large-N limit (in Problem 2.13), one needs to use **Stirling's formula**. This is such an important formula that we'll use many times, it will be good to discuss it now. As with the various distributions, I won't derive it, but instead will show it to be valid in several examples.

There are several forms that you might see for Stirling's formula. First we have the approximation for the factorial of large N,

$$N! \sim \sqrt{2\pi N}N^N e^{-N}, \tag{2.29}$$

and, more commonly for our purposes, the approximation for $\ln N!$,

$$\ln N! \sim N \ln N - N + \frac{1}{2}\ln(2\pi N). \tag{2.30}$$

As the logarithm comes up frequently in our calculations (as you'll first see in Chapter 3), we will use the second form of this formula more often. Additionally, N will generally be so large, say on the order of Avogadro's number ($N \sim 10^{24}$),[11] that the second form has terms that can be neglected. So usually we will apply Stirling's formula in its **simplest form**,

$$\ln N! \sim N \ln N - N. \tag{2.31}$$

Not only is the right-hand side of this expression simpler to deal with than the left-hand side, but it allows us to treat N as a continuous variable. This means that taking the derivative of the right-hand side is more feasible than the left-hand side.

Exercise 2.14 Use Eqs. (2.30) and (2.31) to determine, with each formula, when the approximation is within 1% of the exact value.

Exercise 2.15 At what value of N does the term omitted from Eq. (2.30) to obtain Eq. (2.31) become less than 0.1% of the remaining term? That is, when is

$$\frac{\frac{1}{2}\ln(2\pi N)}{N \ln N - N} < 0.001?$$

Considering the values for N you obtain for these two exercises, keep in mind that N will often be on the order of 10^{24} for physical systems of interest, so it is clear that Stirling's formula will be more than precise enough for our purposes.

2.6 Summary

- We can determine the individual probabilities for simple systems by counting the different ways each outcome can occur. When combining probabilities, remember AND = multiply and OR = add (so the probability of this *and* that to occur is the product of the individual probabilities and the probability of this *or* that to occur is the sum).

---

11 Amedeo Avogadro (176-1856). For the precise value, see Eq. (A.1).

- Given a probability distribution, we can calculate average values of quantities to learn about the expected outcomes. The mean value of some quantity y is given by

$$\bar{y} = \sum_r y_r P_r$$

for a discrete probability distribution and

$$\bar{y} = \int y(x)\mathcal{P}(x)dx$$

for a continuous one.

Problems

2.1 As seen in Exercise 2.5, a standard six-sided die has a different number (1 through 6) on each side, with the average roll being a 3.5. *Grime dice* [4], on the other hand, have a different set of numbers on each side, with the same average roll (3.5) as shown in the table. As we have done regularly throughout this chapter, we treat the dice as *fair*, such that any side is equally likely to be face up when rolled.

Die	Side 1	Side 2	Side 3	Side 4	Side 5	Side 6	Mean Roll
Normal	1	2	3	4	5	6	3.5
Red Grime	3	3	3	3	3	6	3.5
Green Grime	2	2	2	5	5	5	3.5
Blue Grime	1	4	4	4	4	4	3.5

(a) Suppose you and a friend each roll one of the three Grime dice, and the higher roller wins. As you are clever, you ask your friend to pick a die first, then you pick one of the remaining dice, to guarantee *on average* you will win. Calculate the probabilities that red beats blue, blue beats green, and green beats red. (The odd result is why these are an example of **intransitive dice**, also known as *Efron dice*, which you can read about in Refs. [5, 6].)

(b) Calculate these probabilities (e.g., that red beats blue) if you roll two identically colored dice.

(c) What about if you roll three identically colored dice?

2.2 For the Grime dice of Problem 2.1, do the following.

(a) Calculate the average roll of two similarly colored Grime dice.

(b) What is the standard deviation when you roll one Grime die? What is the standard deviation when you roll two Grime dice?

2.3 Recall the results from Exercise 2.3, where you evaluated the probabilities for rolling one or two dice with different numbers of sides (d4, d8, d10, d12, and d20). For each of these dice,

(a) Calculate the average result when rolling one die.

(b) Calculate the average result when rolling two of these dice.

(c) What is the standard deviation when you roll one die? When you roll two dice?

2.4 Some niche boardgames might require you to roll two six-sided dice and use the *product* of the two outcomes instead of the sum. What are the possible results in this case, and what is the probability of each outcome?

2.5 Suppose you have a large collection of marbles, having accumulated 10 000 over the years, and you store them in 100 boxes, with 100 marbles in each box. One of the marbles in each box is red, and the other 99 are blue, and you reach in to blindly pick out a single marble from each of your 100 boxes.

(a) What is the chance that you never pick out a red marble?

(b) What if both 100s are replaced by n?

(c) If $n \to \infty$, what is the limit of your result in part (b)?

Hint: You should find a result in terms of a well-known constant.

2.6 On my family road trips, we have a cooler filled with Diet Coke and seltzer cans. Suppose I pull out two cans blindly (and at random), and both were Diet Cokes. Having known how many cans were in the cooler, I knew that the probability that this would occur was 1/2.

(a) What is the fewest possible number of cans in the cooler?

(b) What is the fewest number of cans if there is an even number of seltzer cans?

2.7 Suppose you and your two best friends want to go on vacation, and are deciding between going to the beach or to the mountains. From what you know about your friends, you figure they each independently have a probability p of choosing the beach, while you don't really care, so you flip a coin for your vote. As with any good democratic friendship, the majority will decide where you go. Alternatively, I (going on vacation alone) am making the same choice and will also choose the beach with probability p.

Who will be more likely to go to the beach, your group or me?

2.8 In a game of Yahtzee, you roll five six-sided dice and need to get sets of 1s, 2s, etc., as well as straights (four or five dice in numerical order, such as "2, 3, 4, and 5"), full houses (one pair and one three-of-a-kind of a different number), and so forth. You get three rolls of the dice and are able to save or reroll any of them each time. A "Yahtzee" will grant you the most points and is when you get five of a kind within those three rolls of the dice.

(a) What is the probability that you will get a Yahtzee on the first roll (where the five of a kind can be of any of the numbers on the dice)?

(b) If you want, you can roll any number of dice on your second roll. Suppose you roll a two of a kind (of any number) on your first roll. What then is the probability that you will get a Yahtzee on the second roll, if you reroll the other three dice?

(c) What is the probability you'll get a full house on the second roll after rolling a two of a kind on the first roll, if you reroll the other three?

2.9 A number is chosen at random between zero and one. What is the probability that exactly five of its first 10 decimal places (to the right of the decimal point) consist of digits less than five? Assume that (for example) a number such as 0.5 will be treated as 0.5000000000, and this number has nine digits less than five in it.

2.10 Let us derive the binomial distribution. As in Section 2.5.1, we will consider the two outcomes, true and false, to have individual probabilities t and f, respectively, with $t + f = 1$. Assume we have n_t true outcomes and n_f false outcomes, and $N = n_t + n_f$ total outcomes.

(a) Argue that the probability that we obtain n_t true outcomes and then n_f false outcomes is given by

$$\underbrace{ttt\cdots}_{n_t \text{ factors}} \times \underbrace{fff\cdots}_{n_f \text{ factors}} = t^{n_t} f^{n_f}.$$

(b) If we only care about how many true outcomes we have, then the order in which these appear above doesn't matter. (The standard problem here is the **random walk problem**, where you wish to determine where someone will end up after some number of steps to the left and some to the right, with each step being random. As such, it doesn't matter the order you take these steps in, as you will always end up at the same point if you've taken a fixed number of steps to the left.) Determine how many ways we can obtain n_t true outcomes and n_f false outcomes.

(c) Use the results of the previous two parts to derive Eq. (2.25).

2.11 Calculate the first four moments of the binomial distribution beyond the zeroth moment. That is, calculate

(a) $\overline{n_t}$,

(b) $\overline{n_t^2}$,

(c) $\overline{n_t^3}$, and

(d) $\overline{n_t^4}$.

For this problem, use the following trick:

$$\sum x^k t^x f^y = \left(t\frac{\partial}{\partial t}\right)^k \left(\sum t^x f^y\right). \tag{2.32}$$

On the right-hand side, the notation with the derivative means you should differentiate the result of the summation, then multiply by t, then repeat this pair of operations k times. When applied to the binomial distribution, $t + f = 1$ holds, but do not use this fact until the end of the calculation.

2.12 Consider the difference in true and false outcomes in the binomial distribution, $m = n_t - n_f$. Calculate

(a) $\overline{m}$,

(b) $\overline{m^2}$,

(c) $\overline{m^3}$, and

(d) $\overline{m^4}$.

Hint: Use the results of Problem 2.11 and the fact that $n_f = N - n_t$ to simplify this problem.

2.13 When N becomes large in the binomial distribution, then we can show that it becomes the Gaussian distribution.

(a) First, argue that for large values of N the probability drops off rapidly when n_t moves away from $\overline{n}_t = Nt$. There's no need to do this rigorously—choose some values of N and n_t that are large but still manageable to see that $P_N(n_t)$ quickly becomes negligible away from its mean. Because of this, you can write the (natural log of the) probability as a Taylor expansion about the maximum (when $n_t = n_t^0$),

$$\ln P_N(n_t) \approx \ln P_N(n_t^0) + \frac{d \ln P_N}{dn_t}\left(n_t - n_t^0\right) + \frac{1}{2}\frac{d^2 \ln P_N}{dn_t^2}\left(n_t - n_t^0\right)^2 + \cdots .$$

The logarithm is used as it has a better behaved Taylor series—the probability itself is too rapidly changing for the Taylor expansion to converge quickly enough.

(b) Take the natural logarithm of Eq. (2.25) and assume that N and n_t are large enough that Stirling's formula can be used in its simplest form to obtain

$$\ln P_N(n_t) \sim N \ln N - n_t \ln n_t - (N - n_t)\ln(N - n_t) + n_t \ln t + (N - n_t)\ln f.$$

(c) Now differentiate $\ln P_N$ with respect to n_t to determine what n_t^0 is—that is, where does the maximum of this expression occur (when this derivative vanishes)?

(d) Calculate the second derivative of $\ln P_N$. Show that it is negative (to ensure we are expanding about a maximum) and write it in terms of the variance, $\sigma^2 = Ntf$.

(e) Calculate the third derivative of $\ln P_N$ to show that it vanishes in the limit $N \to \infty$.

(f) Putting everything together, show that upon exponentiating $\ln P_N$, you obtain a Gaussian of the form of Eq. (2.28).

2.14 (a) Evaluate the third and fourth moments of the Gaussian distribution ($\overline{x^3}$ and $\overline{x^4}$).

(b) Just as the variance is a more useful form of the second moment, there are common forms of higher moments, by comparing them to the mean. We have

i. **skewness:** $\overline{(x - \overline{x})^3}$ and

ii. **kurtosis:** $\overline{(x - \overline{x})^4}$.

Evaluate both of these for the Gaussian distribution.[12]

2.15 A cellular tower is to be installed in a small town for the first time. The town has a population of 2000 people, and the tower would need 2000 "traffic channels" for all 2000 people to make calls simultaneously. Since the town cannot afford this, you are asked to determine the best number of traffic channels: the fewest number of channels needed to have a low likelihood of a failed call due to the system being too busy. Suppose that during the busiest hour of the day, each mobile user makes a two-minute call,[13] and we'll assume these calls are made at random. Find the minimum number of traffic channels N which the tower must include so that at most only 1% of the callers will fail to make a connection.

2.16 In a different limit, the binomial distribution can be reduced to a simpler form which is not the Gaussian distribution but is known as the **Poisson distribution.**

Let's consider a situation where the probability t in the binomial distribution of Eq. (2.25) is small ($t \ll 1$) and we have only a few number of occurrences, so that $n_t \ll N$. (Some examples of such rare occurrences are those of radioactive decay, the chance of a meteorite striking the Earth, or being struck by lightning in a given year.)

12 Often these are defined by normalizing them with the standard deviation, such as $\overline{[(x - \overline{x})/\sigma]^3}$.

13 Probably unlikely nowadays as calling is so old fashioned.

(a) First, only assume $t \ll 1$. Show that $P_N(n_t)$ in Eq. (2.25) is only non-negligible if $n_t \ll N$, so we can assume this as well (and thus $f^{N-n_t} = (1-t)^{N-n_t} \approx (1-t)^N$).
(b) Expand $\ln(1-t)$ to leading non-zero order in t for small t to show that $(1-t)^N \approx e^{-Nt}$.
(c) Use Stirling's formula in its simplest form to show that $N!/(N-n_t)! \approx N^{n_t}$.
(d) Put the above results together to show that

$$P_N(n_t) \rightarrow \frac{\tau^{n_t}}{n_t!}e^{-\tau},$$

where $\tau = Nt$ is the mean number of events.

2.17 Recall the town mentioned in Problem 2.15. After it is built, you study the network over the course of two years (730 days) and found that there were 730 cases where a person tried to make a call, but it failed on the first attempt. You want to determine the probability that there were failed calls on a given day.
(a) Justify why using the Poisson distribution is appropriate for this problem.
(b) Calculate the probability that no failed calls happened on a day.
(c) Calculate the probability that there were *at least* three failed calls on one day.

2.18 Suppose you randomly choose two numbers within the unit square (shown) so $0 \leq a \leq 1$ and $0 \leq b \leq 1$.

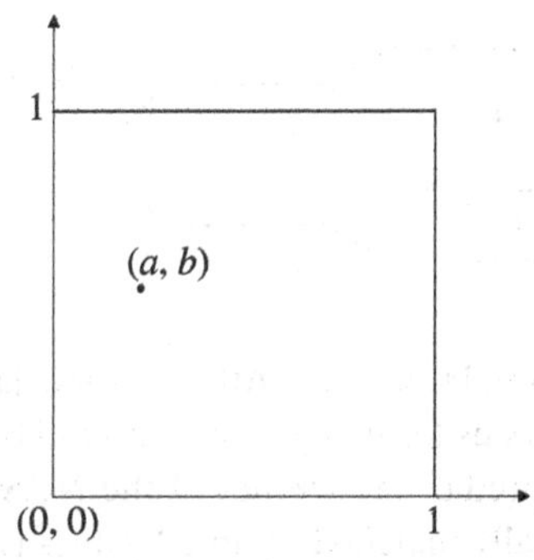

What is the probability that

$$ax^2 + x + b = 0$$

has real solutions?

2.19 Craps is a casino game with perhaps the most even odds of winning. The rules are simple if you are the "shooter": You throw two (fair) six-sided dice and care about the following outcomes:

- If you roll a 7 or 11 you win immediately,
- if you roll a 2, 3, or 12, you lose immediately, and
- any other throw is called your *point*, which allows to you continue rolling the dice. In this case, you win by throwing your point again or lose by throwing a 7.
  (a) What is the probability that you win on the first throw?
  (b) What is the probability that you throw a point?
  Hint: Work out the probability of throwing a 4, then a 5, etc., then remember that you add probabilities for when this OR that can occur.
  (c) What is your chance of winning by throwing a point?

(d) Using your results for parts (a) and (c), what is your overall chance of winning? (The odds that the casino wins are still greater than 50%, but the chance that you win should be *very* close to 50%.)

2.20 At the end of a college sports season, eight teams are paired off in a tournament, to (ostensibly) determine the best college team in a given year. Let's assume each team is so evenly matched at this point that one could flip a coin to determine the winner of a given game. See the tournament ladder below, where the teams competing have not been entered yet. We would like to focus on two schools that have entered the tournament: Statistical University and Thermo College.

(a) What is the chance that these two schools will meet up during a game in the tournament?

(b) How would this change if we had 2^n teams in the tournament?

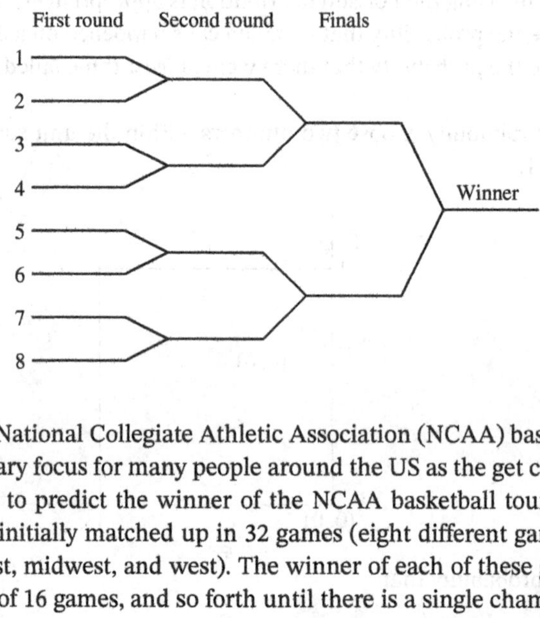

2.21 Every March, the National Collegiate Athletic Association (NCAA) basketball tournament becomes the primary focus for many people around the US as the get caught up in "March Madness" and try to predict the winner of the NCAA basketball tournament. There are 64 teams that are initially matched up in 32 games (eight different games in the different regions: south, east, midwest, and west). The winner of each of these games then goes on to the next round of 16 games, and so forth until there is a single champion. The madness comes into play as everyone attempts to fill out their "bracket," to predict who will win each game (and ultimately the winner of the tournament). The bracket would essentially be as in Problem 2.20 but with $n = 6$ (not three).

(a) What is the probability that you correctly predict the winner of each region?

(b) What is the probability that you correctly predict the winner of the entire tournament?

(c) What are the odds that you would produce a *perfect bracket*, where you actually have predicted the winner of every game?

References

1 L. Mahadevan and E. H. Yong. Probability, physics, and the coin toss. *Physics Today*, 64(7):66–67, 2011.

2 E. H. Yong and L. Mahadevan. Probability, geometry, and dynamics in the toss of a thick coin. *American Journal of Physics*, 79(12):1195–1201, 2011.

3 Poker probability. https://en.wikipedia.org/wiki/Poker_probability. Accessed: 30 January 2023.

4 J. Grime. Non-transitive dice. http://singingbanana.com/dice/article.htm. Accessed: 31 January 2023.

5 R. P. Savage. The paradox of nontransitive dice. *The American Mathematical Monthly*, 101(5):429–436, 1994.

6 C. M. Rump. Strategies for rolling the Efron dice. *Mathematics Magazine*, 74(3):212–216, 2001.

3

Introduction to Information Theory

In Chapter 2 we discussed probabilities, one of the two important ingredients in our development of statistical mechanics. In this chapter we now turn to **information theory** as the second pillar upon which our subject will be built. Not only is this subject useful to begin a study of statistical mechanics, but it has important applications in computer science and quantum computing.

After completing this chapter, you should be able to

- understand how to quantify missing information—what we don't know about a given system—by deriving a numerical expression for it, and
- apply this concept to various systems, both physical and otherwise.

3.1 Missing Information

As stated before, in order to understand systems with a large number of molecules, we have to give something up. Specifically, we give up knowing the exact trajectory of every single molecule in lieu of bulk properties of the system. As such, the question we wish to ask is: How do we *quantify* what we don't know about a system? In other words, how do we derive an expression that relates the amount of **missing information** in a system with a given number of states? This is a perfect example of how we can start from essentially nothing (that is, no empirical evidence) and eventually understand quite a bit about a system (even though that won't really happen until the next chapter).

Let's begin by imagining that we have to choose between n possible outcomes, or what we will now call **states**. For a physical system, we will define a state specifically in Chapter 4, but for now we will avoid too much rigor. Once we define our system, usually the possible states are clear. For example:

- If our system is a single six-sided die, then we have six states, corresponding to the six possible rolls of the die.
- For a coin toss, we have two states (whichever face is pointing up after the toss): heads or tails.
- For a classical object (e.g., a box sliding down an inclined plane) the state would be denoted by its position and momentum.

Statistical Thermodynamics: An Information Theory Approach, First Edition. Christopher Aubin.

Companion website: www.wiley.com/go/Aubin/StatisticalThermodynamics

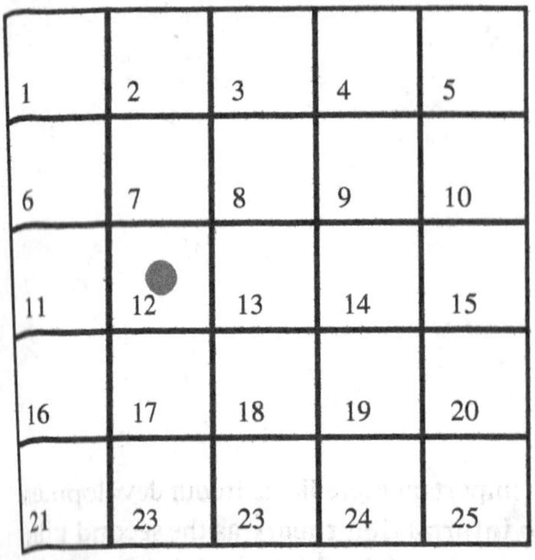

Figure 3.1 A system with $n = 25$ outcomes, with a specific state shown where the ball is in box #12.

For systems with more than one object (ultimately the goal in this course), the number of states increases rapidly and the enumeration of them quickly becomes difficult. We'll see some of these cases and how to deal with them shortly.

While we can keep in mind these examples, for now let's use a concrete example: a container with n boxes and a single ball, such that the ball must be in one of the boxes. In Figure 3.1, we show this system for $n = 25$ (the labels on the boxes are arbitrary), and in the figure we show one specific case—that in which the ball is in box #12. If we enumerated all possible states, we would have a figure with 25 such images, with the ball in each of the boxes exactly one time.

We wish to quantify what we don't know about a system (the missing information), where all we assume is that we don't know which box the ball is in, but we do require that it *must* be in one of them. (Perhaps we turned our back to the container and tossed the ball blindly into one of the boxes, or someone else placed the ball into a box without telling us which one before we arrived.) If there were no missing information, then I would say we know everything about this system, meaning we know exactly which box the ball is in at a given moment. As we first work through this, we will also assume that the ball is equally likely to be in any one of the boxes.[1]

Our goal here is to derive a function $I(n)$ that quantifies the amount of missing information in a system with n possible outcomes, or states. Initially this will be measured in **bits**, which can take on a value of one or zero (on or off), both for simplicity of understanding and for allowing a connection with computer science. As data is stored in bits on a computer, we can instead say that $I(n)$ quantifies the number of bits required to digitally store the complete information of a system. We will make four assumptions that allow us to derive a useful expression for $I(n)$. A couple of these come from straightforward observations, while the other two are less obvious.

Assumption 3.1 First, if we have more possible states (or boxes), then we expect there to be more missing information. That is,

$$I(n) > I(m) \quad \text{if } n > m. \tag{3.1}$$

If you think about how likely you are to guess where the ball is in some extreme cases makes this fairly clear. If you had two boxes, then you would have a 50% chance of guessing which box the ball is in. On the other hand, if there were a million boxes, you would have a 0.0001% chance of guessing correctly.

1 We will loosen this constraint in Section 3.2.

Assumption 3.2 The next assumption is also straightforward. If we only have one state (box), with the insistence that the ball is definitely in one of the boxes, then we have complete knowledge of the system. As such, there is no missing information, or

$$I(1) = 0. \tag{3.2}$$

Assumption 3.3 Now let us imagine we can write $n = n_1 n_2$, that is, n can be written as a product of two other integers. This can be obtained by dividing our n_1 boxes into n_2 compartments, as shown in Figure 3.2 for the $n_1 = 25, n_2 = 36$ case. Equation (3.1) can be written as

$$I(n) = I(n_1 n_2), \tag{3.3}$$

and we would like to relate this to $I(n_1)$ and $I(n_2)$, the missing information contained in a system with n_1 and n_2 states, respectively. In order to fully specify the location of the ball, we could first determine which of the n_1 boxes the ball is in, and then which of the n_2 compartments in that one box the ball is in. We should be able to provide this information piece by piece, without repeating any information—that is, we assume $I(n)$ to be a universal function. As such, we require

$$I(n) = I(n_1 n_2) = I(n_1) + I(n_2). \tag{3.4}$$

$I(n_1)$ is the missing information associated with the box the ball is in (one of the 25 regions segmented by the thick lines in Figure 3.2), while $I(n_2)$ is that associated with the compartment (one of the 36 sections) inside that particular box.

Assumption 3.4 Finally, for any system, n should be an integer (it comes from counting the number of states, after all); however, we can loosen this restriction to help derive an expression for $I(n)$. We can do this by rewriting Eq. (3.4) as

$$I(n_1) = I(n_1 n_2) - I(n_2) \Rightarrow I(n/n_2) = I(n) - I(n_2). \tag{3.5}$$

This implies that it is not unreasonable to allow I to be valid for all positive rational numbers and not just integers.

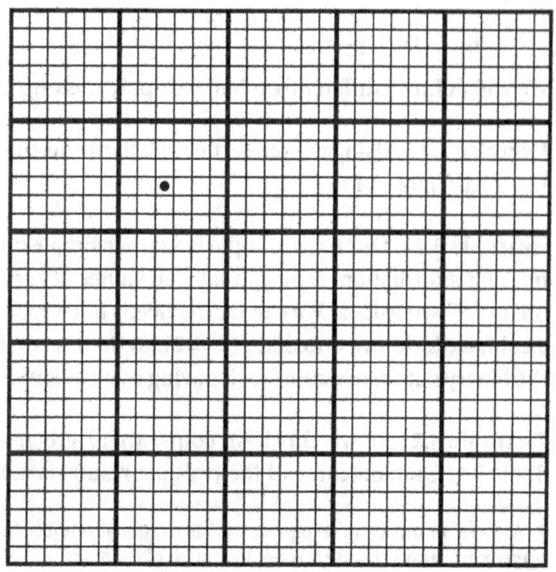

Figure 3.2 A system with $n_1 = 25$ boxes and $n_2 = 36$ compartments per box, with the ball shown in one of them (in this case I won't bother labeling them).

It is clear from Assumption 3.1 that $I(n)$ is a *monotonic* function of n, and that combined with Assumption 3.2, it is never negative. Again, we are only going to consider $I(n)$ to be a function of integers, and while Assumption 3.4 allows us to extend this to any rational number, we'll take this one step further and allow $I(n)$ to be a function of any real (positive) number.[2] This is just so that we can come up with a concrete expression for missing information. With this in mind, we replace

$$I(n) \to I(x) \tag{3.6}$$

to indicate clearly that we are considering it a function of a positive real number x. Specifically, we will set $x = e^{n_1/n_2}$ (so $x \geq 1$, consistent with Assumption 3.2), with positive integers n_1 and n_2, so that $x^{n_2} = e^{n_1}$. Repeated use of Eqs. (3.1) and (3.4) allows us to write

$$I(e^{n_1}) = \underbrace{I(e) + I(e) + \cdots + I(e)}_{n_1 \text{ terms}} = n_1 I(e), \tag{3.7}$$

and

$$I(x^{n_2}) = \underbrace{I(x) + I(x) + \cdots + I(x)}_{n_2 \text{ terms}} = n_2 I(x). \tag{3.8}$$

Setting these two expressions equal, we obtain

$$I(x) = \frac{n_1}{n_2} I(e) = I(e) \ln x. \tag{3.9}$$

$I(e)$ is some constant, and we see the natural logarithm satisfies all of our assumptions above. At this point we now want to revert back to I being a function only of integers, $x \to n$, to get

$$I(n) = k \ln n. \tag{3.10}$$

In this expression, the constant $k \equiv I(e)$ will be specified by choosing the units that we wish to use to measure missing information. As mentioned above, we will first measure I in bits, and to do so, we choose

$$k \equiv \log_2 e, \tag{3.11}$$

so that

$$I_{\text{bits}}(n) = \log_2 n. \tag{3.12}$$

We will define k differently when we apply missing information to statistical mechanics.

Exercise 3.1 Using the properties of logarithms that you learned long ago, derive Eq. (3.12) using Eqs. (3.10) and (3.11).

Example 3.1 Our first example is a rather simple one, but will justify the use of bits for measuring missing information. A bit can take on values of zero or one, much like a fair coin which can take on one of two values: We can define heads to be represented by a one and tails by a zero. If we flip a coin, we have two outcomes ($n = 2$), so there's one bit of missing information. In fact, because of the binary nature of a coin (each having two outcomes), we can think of the coin itself as a single bit.

Example 3.2 If we flip four coins, there are $n = 2^4 = 16$ outcomes, which I enumerate in Table 3.1. Each outcome of the possible results is labeled by a number for reference. These labels

2 Any real number can be represented as a rational number to arbitrary precision using a **Diophantine approximation** [1], so this is not too much of a stretch.

Table 3.1 Combinations of bits corresponding to each coin flip in Example 3.2, assigning each distinct outcome to a box from Figure 3.3.

Box #	Coin 1 2 3 4	Box #	Coin 1 2 3 4	Box #	Coin 1 2 3 4	Box #	Coin 1 2 3 4
1	HHHH	5	THHH	9	TTHH	13	THTT
2	HTHH	6	HTTH	10	THTH	14	TTHT
3	HHTH	7	HTHT	11	THHT	15	TTTH
4	HHHT	8	HHTT	12	HTTT	16	TTTT

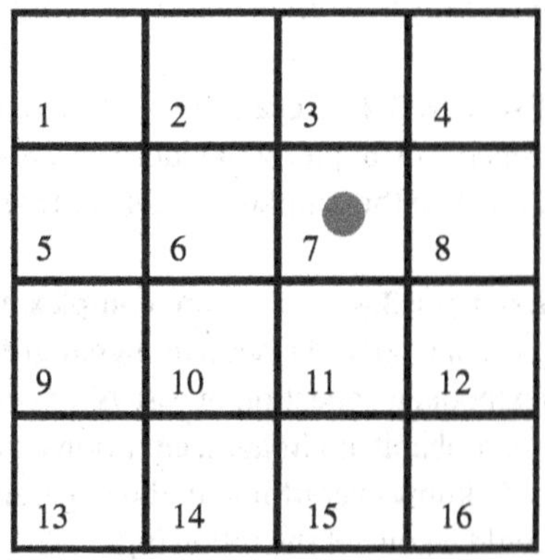

Figure 3.3 Representing the flipping of four coins using our ball-in-a-box example, where the box number could be used to identify each outcome. This particular result would correspond to flipping HTHT in Example 3.2.

allow us to relate this example to our original example used to derive $I(n)$, throwing a ball into (in this case) 16 boxes, which is shown in Figure 3.3.

We can either flip the four coins and see the (random) outcome, or we blindly toss a ball into this set of boxes: The box in which the ball lands corresponds to a particular outcome of flipping four coins. The outcome in the figure shows the ball landed in box 7, so this would correspond to the first and third coins resulting in heads and the second and fourth coins resulting in tails.[3]

This example is a good way to understand information storage in bits. Each coin can correspond to a bit, and we can use these four coins (bits) to completely specify which box our ball is in, of the 16 possible boxes.

Example 3.3 Let's look at a case where we have a number of outcomes that isn't a power of two, so the number of bits of missing information won't result in a whole number. For example, rolling a fair six-sided die, we have

$$I_{\text{bits}}(6) = \log_2 6 = 1 + \log_2 3 \approx 2.6\,\text{bits} \tag{3.13}$$

of missing information. How do we understand a fractional number of bits? If we think of this as the required memory to store this information, then we could assign each result to a specific combination of bits. Given that we can only have an integral number of bits and $2 < I_{\text{bits}}(6) < 3$, we would need three bits, although all combinations of bits are not needed, as shown in Table 3.2.

3 Here we are considering the coins to be distinguishable, which is OK for such a small number of coins (it's easy to keep an eye on each particular coin). Usually we wouldn't consider any of the outcomes 6–11 to be distinct as they each have two heads and two tails.

Table 3.2 Combinations of bits corresponding to each die roll in Example 3.3.

Bit 1	Bit 2	Bit 3	Die roll	Bit 1	Bit 2	Bit 3	Die roll
0	0	0	1	1	0	0	5
0	0	1	2	1	0	1	6
0	1	0	3	1	1	0	—
0	1	1	4	1	1	1	—

We need more than two bits to fully include all outcomes but two combinations of bits are not used.

We see that two bits aren't enough to be able to label all possible outcomes (just two could be used if we only had outcomes of one through four), but also two of the eight combinations are not used.

Exercise 3.2 Recall the other dice used in tabletop games with different numbers of sides mentioned in Chapter 2: 4 sides (d4), 8 sides (d8), 10 sides (d10), 12 sides (d12), and 20 sides (d20). Calculate the number of bits of missing information for rolling just one of each die.

Example 3.4 As a more complex example, let's estimate the amount of information on a typed page of a book. This is somewhat difficult: What type of book are we discussing: a novel, a physics textbook, a collection of essays, or something else? For now let's assume we have a novel, so there are probably no images, equations, or special characters.

To know everything on the page, suppose it suffices to know every character, and each character could be one of the following:

- 26 lowercase letters,
- 26 uppercase letters,
- 10 numerals, and
- perhaps about 10 possible punctuation symbols (including spaces).

This would give us about 72 possibilities for each character in total. On a given page, we might expect 300-500 words per page (let's stick with 400), and an average word is around five characters. That gives us $n = 72^{2000} \approx 5 \times 10^{3714}$ possible combinations of characters, which is quite an intractable number! But the amount of missing information in bits is

$$I_{\text{bits}} = 2000 \log_2 72 \approx 12\,340 \text{ bits}. \tag{3.14}$$

We can see quickly that $I(n)$ as we have defined it must be an upper bound for the amount of missing information on a given page. First, the characters must form words and sentences; random arrangements of characters like "jdai.x;pl" would not be allowed. Additionally, there are grammar and spelling rules, which will depend upon the specific language the novel is written in, so that there are even fewer allowed combinations of these characters. The more we know about the page, then the *less missing information* there is, so it might be better to say

$$I_{\text{bits}} \leq 2000 \log_2 72 \approx 12\,340 \text{ bits}. \tag{3.15}$$

We will see in Section 3.2 that this is true and come up with better estimates for such cases as this. Additionally, as we increase the number of characters, the number of states increases way too rapidly for us to easily deal with, but as the missing information is the logarithm of the number of states, it is much more manageable.

3.2 Missing Information for a General Probability Distribution

In many realistic cases, our n outcomes are not going to be equally likely, so let us extend our understanding of missing information to account for this. Continuing with our "ball-in-a-box" example, we now consider the boxes in our container to be different sizes as shown in Figure 3.4 (without labeling the boxes this time, to avoid clutter). The ball can be in any one of the boxes, and this figure shows six possible outcomes after tossing the ball to the container. Because the boxes are differently sized, we assume the probabilities that the ball will land in each box are not necessarily equal: It is safe to assume that we are more likely to find the ball in the larger boxes than the smaller ones. We assume we have a set of probabilities P_r, such that $0 \leq P_r \leq 1$ and $r = 1, \ldots, n$ ($n = 16$ in Figure 3.4). As usual, we require they are normalized, that is,

$$\sum_{r=1}^{n} P_r = 1, \tag{3.16}$$

which again is just a way of stating that the ball must be in one of the boxes.

Recall from Chapter 2, when we discuss probabilistic systems, we never discuss a single system; we must consider a set of *identically prepared* systems. So for our ball-in-a-box example, a system would correspond to one container with a ball in any one of the allowed boxes (six of which are shown in Figure 3.4). But if we want to consider the probability that the ball is in a given box, then we would need to consider a large number N of such containers, with the ball randomly tossed into one of the boxes (or toss the ball a large number of times in the same box, recording each result), and count how many times n_r the ball lands in the rth box. Then we could use Eq. (2.1) to determine the probability that the ball is in the rth box.[4]

This set of identical systems is known as an **ensemble** of systems, which is a term I will use in two different ways: First we have a **statistical ensemble**, which is the set of all possible outcomes

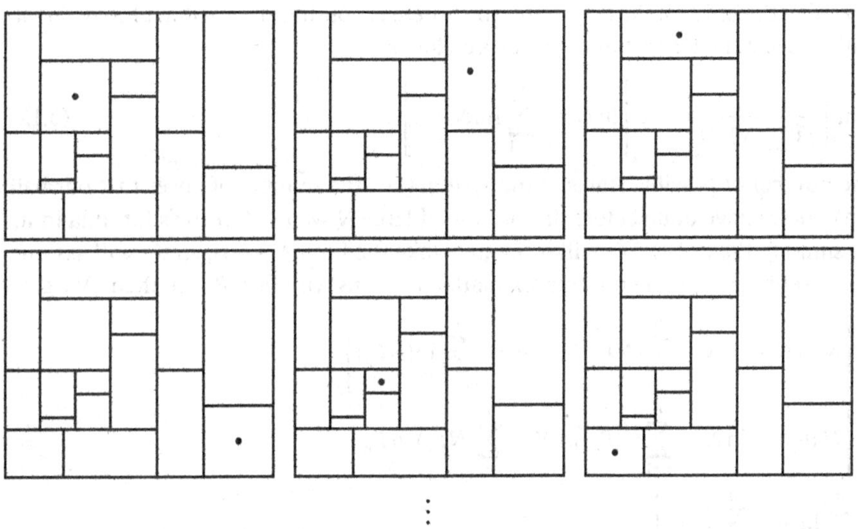

Figure 3.4 Six out of the sixteen possible outcomes in an ensemble of systems with $n = 16$ boxes of different sizes; the ball has a different probability to be in each one.

4 Again, the careful student would realize this is only true when $N \to \infty$, because in this case we are considering our "experimental" probabilities. For our purposes here, this is still useful for conceptually understanding what is going on.

such that each outcome appears once. Such an ensemble is akin to a complete set of states in a vector space, for example: We want to include each possibility exactly one time, with each outcome having a probability P_r. In Figure 3.4, we show only a portion of the statistical ensemble: six of the 16 possible states that our system could be in. If our outcomes are not equally likely, we would need to know P_r from some other means (we couldn't just use Eq. (2.1) to calculate it).

In practice, if we wanted to determine P_r, we could do so experimentally, and use an **experimental ensemble**. Such an ensemble can be thought of as a set of N (again identically prepared) systems, but in this case N is much larger than the number of possible outcomes (in our above example, we would require $N \gg 16$). This way we can properly apply Eqs. (2.5) and (2.6) to obtain the probabilities P_r for the rth state. It can be confusing to consider these two different ensembles (and I will not usually distinguish the two), but for the purposes of deriving an expression for the missing information in a system with unequal probabilities, I will assume we are using the latter definition.

Let's assume we have N identically prepared systems laid out in some order, perhaps corresponding to the order in which we prepared them, with N much larger than the number of possible outcomes. We then count how many ways there are to arrange these identical systems to determine the number of states to put into Eq. (3.10) to find $I(N)$.

There are $N!$ ways to arrange our N systems in such a way and still obtain the same results for the number of systems in each possible state. For large N, we know from the binomial distribution that on average we have NP_r systems in our ensemble where the ball is in the rth box. But since each of the NP_r systems are interchangeable and we don't want to overcount, then we have to divide by $(NP_r)!$ for each r (recall the box on page 9).[5] Thus, the number of distinct outcomes when we have listed all N states is

$$\frac{N!}{(NP_1)!(NP_2)!(NP_3)!\cdots} = \frac{N!}{\prod_{r=1}^{n}(NP_r)!}. \tag{3.17}$$

Once we have specified the order of these systems, then there is nothing else we don't know, so the missing information is just the logarithm of this expression, so

$$I(N) = k\ln\left[\frac{N!}{\prod_{r=1}^{n}(NP_r)!}\right] = k\left[\ln N! - \sum_{r=1}^{n}\ln(NP_r)!\right]. \tag{3.18}$$

Note that n is the number of possible states of the system (i.e., the number of boxes that our ball can be thrown in). As we have done before, in the case of large N we use Stirling's formula in its simplest form to simplify Eq. (3.18). We will ultimately take the limit $N \to \infty$, so we will assume that NP_r is also large enough to apply this approximation to terms with $\ln(NP_r)!$ in them. We get

$$\begin{aligned} I(N) &= k\left[N\ln N - \ln N - \sum_{r=1}^{n}(NP_r)\ln(NP_r) + \sum_{r=1}^{n}\ln(NP_r)\right] \\ &= k\left[N\ln N - \ln N - \sum_{r=1}^{n}(NP_r)\ln N - \sum_{r=1}^{n}(NP_r)\ln P_r \right. \\ &\qquad \left. + \sum_{r=1}^{n}\ln N + \sum_{r=1}^{n}\ln P_r\right], \end{aligned}$$

5 If you're paying attention, you might want to call foul here, as NP_r may not be an integer, and as such $(NP_r)!$ doesn't seem like a valid expression. In practice, N will be large enough such that NP_r can be very well approximated by an integer, as we shall see. (Of course, the even more careful reader may remind me of the Γ function, Eq. (D.7), which allows us to generalize the factorial function to non-integer values.)

and we can use $\sum_{r=1}^{n} P_r = 1$ to cancel the first and third terms to get

$$I(N) = k\left[-\ln N - \sum_{r=1}^{n}(NP_r)\ln P_r + n\ln N + \sum_{r=1}^{n}\ln P_r\right]. \tag{3.19}$$

For large N, the second term dominates the others (as $N \gg n$), and we are left with

$$I(N) = -Nk\sum_{r=1}^{n} P_r \ln P_r. \tag{3.20}$$

The minus sign out front is necessary for I to be positive: We know that P_r is always less than or equal to one, so we have $\ln P_r \leq 0$. This expression makes sense, because as N increases so does the amount of missing information. We can set $k = \log_2 e$ to rewrite these expressions with the base-2 logarithm instead of the natural logarithm to again measure the amount of missing information in bits.

Example 3.5 How much of an error are we making when replacing Eq. (3.19) for Eq. (3.20)? Let's consider a system with a ball in one of four boxes, where the probability of the ball being in each box is given by

$$P_1 = 1/2, \quad P_2 = 1/4, \quad P_3 = 1/8, \quad P_4 = 1/8.$$

How small would N have to be for Eqs. (3.20) and (3.19) to differ by about 1%?

To do this, we want to calculate the fractional difference,

$$\text{frac. diff.} = \left|\frac{-\ln N + 4\ln N + \sum_{r=1}^{4}\ln P_r}{-\ln N - \sum_{r=1}^{4}(NP_r)\ln P_r + 4\ln N + \sum_{r=1}^{4}\ln P_r}\right|$$

$$= \frac{12\ln N - 36\ln 2}{(7N - 36)\ln 2 + 12\ln N}$$

and we want this to be less than 1%. While we could be rigorous about this, we can plot this to see where this function reaches 1%:

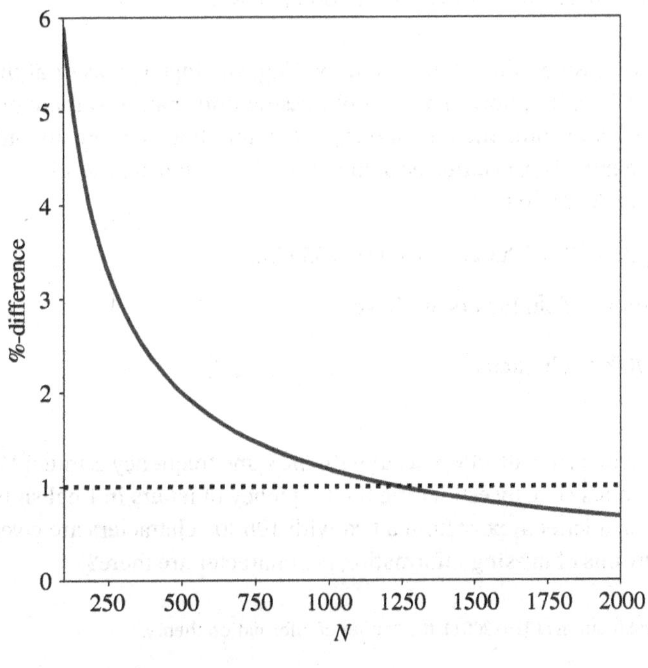

Looking at this plot, when N is greater than around 1200 we reach 1% (the dotted line). A numerical solution shows that this is about 1% when $N = 1\,234$. Most of our systems of interest will have N on the order of Avogadro's number, and at this value of N, the percent difference here is $\approx 10^{-20}\%$. The neglected terms are clearly irrelevant here.

Computer Exercise 3.1 You can go to the companion site, under Missing Information, to see how I made the figure in Example 3.5. This Jupyter notebook allows you to create your own probability distribution and see how the two expressions in Eqs. (3.19) and (3.20) differ as a function of N, and how quickly this difference becomes negligible. For example, consider how large N must be for the percent difference to reach 0.1%, 0.01%, or 0.001%.

For statistical systems, we are interested in the expression above in the $N \to \infty$ limit (the **thermodynamic limit,** which I have mentioned before), but of course I also goes to infinity then. Thus, in this case, we instead consider the missing information *per system*, $I(N)/N$, which will remain finite in the limit $N \to \infty$. To this end, we define the **Shannon entropy,**[6]

$$S \equiv \lim_{N\to\infty} \frac{1}{N} I(N) = -k \sum_{r=1}^{n} P_r \ln P_r. \tag{3.21}$$

We will see this will become our definition for the physical quantity **entropy**, which (you probably already have heard) has important implications in thermodynamics.

Exercise 3.3 Show that when $P_r = 1/n$ for all r (so all states are equally likely), $S = k \ln n$.

Exercise 3.4 What is S when we know the precise state of the system?

As before, if we set $k = \log_2 e$, then the Shannon entropy becomes $S_{\text{bits}} = -\sum_r P_r \log_2 P_r$, and this is the average number of *bits* of missing information per system.

Example 3.6 If you assume that all letters in the English language occur at the same rate (that is, with the same probability), how many bits of missing information is there on a page with 500 characters? (Ignore spaces, punctuation, and capitalization; that is, there are only 26 characters.) Additionally, how many bits *per character* of missing information are there?

For this we use Eq. (3.12), so that

$$I(500) = \log_2(26^{500}) = 500(\log_2 13 + 1) \approx 2350 \text{ bits}.$$

Dividing by the number of characters, we have

$$\frac{I(500)}{500} = 4.70 \text{ bits/character}.$$

Example 3.7 Of course, not all letters occur with the same frequency. Samuel Morse (1791–1872) invented his eponymous code by estimating the frequency of letters in English type. The number of times he found each letter appear from a text with 106 400 characters are given in Table 3.3. In this case, how many bits of missing information per character are there?

---

6 Named after Claude Shannon (1916–2001), the "father of information theory."

Table 3.3 The number of times in a text of 106 400 characters each letter in the English alphabet occurs, as calculated by Samuel Morse.

Letter(s)	Counts per 106 400	Letter(s)	Counts per 106 400	Letter(s)	Counts per 106 400
E	12 000	L	4 000	B	1 600
T	9 000	U	3 400	V	1 200
A, I, N, O, S	8 000	C, M	3 000	K	800
H	6 400	F	2 500	Q	500
R	6 200	W, Y	2 000	J, X	400
D	4 400	G, P	1 700	Z	200

Here it is easiest to use a spreadsheet for our calculation. With the data given, we could list the various probabilities as

$$P_E = 0.112782, \quad P_T = 0.0845865, \ \ldots,$$

and insert into Eq. (3.21) with $k = \log_2 e$ to get

$$S = 4.224 \text{ bits/character}. \tag{3.22}$$

We see this is less than the number of bits per character when we treat all characters as equally likely. This is expected—having more information means that we have *less missing information* about the system!

A practical application of Shannon entropy is in data compression, which is related to the previous example. If you want to transmit data with the fewest number of bits (using only ones and zeros), then you would like to find the most efficient way (that is, the fewest number of bits per character) to do so, which would correspond to S. A naive way to store data would be to use a fixed number of bits for each character (so if we had eight characters, we would need three bits to store every possibility). However, a more efficient approach is to consider which of those eight characters are more likely to appear, and use *fewer* bits to store those characters, while those which are less likely would be stored with more bits (Morse code is a very simple example of this). On average, the number of bits per character will be less than three in this case; the optimal value is given by S (see Problem 3.10). A nice discussion of this can be found in the context of quantum computation in chapter 13 of Ref. [2].

We saw in Example 3.7 that the missing information per system was larger when we treated every outcome as equally likely than if not. As we discussed, this makes sense conceptually because we have more information (thus *less* missing information) about a system if we can assign a probability distribution to the outcomes. Moreover, S as defined in Eq. (3.21) is maximized when all outcomes are equally likely, resulting in $S = k \ln n$, which you can prove generally in Problem 3.8.

3.3 Summary

- We can quantify the **missing information** of a system based solely on the number of possible states and the probability distribution.

- The missing information per system, or **Shannon entropy** for a system with n states where each state has a probability of P_r can be written as

$$S = -k\sum_{r=1}^{n} P_r \ln P_r,$$

or, in bits,

$$S_{\text{bits}} = -\sum_{r=1}^{n} P_r \log_2 P_r.$$

This is maximized when all n states are equally likely, where $P_r = 1/n$ for all r, and thus $S = k \ln n$ or in bits, $S_{\text{bits}} = \log_2 n$.
- The missing information of a system not only has applications to other fields (such as computer science and quantum computing) but also gives us a tractable way to manage the large numbers that will ultimately arise as we begin to treat the large systems needed for thermodynamics.

Problems

3.1 Determine in which case more information is missing. The answer is highly insensitive to approximations.
(a) You do not know the contents of the approximately 55 million books in all branches of the New York Public Library system (in this case knowing all the symbols on all the pages would constitute complete information).
(b) You have one milligram of a paramagnetic substance whose atoms may have their spins oriented "up" or "down," but you lack information about individual spin orientations.

3.2 Suppose you roll two six-sided dice, but only care about the sum of the dice.
(a) How many different states are there? List them all, and label how many ways you can produce each of them.
(b) How many bits of missing information are there? How does this compare to the case where we treat all of the states to be distinct (perhaps the two dice are different colors and you care specifically about what is on each die)?

3.3 A gambler rolls four fair six-sided dice: one white, one red, one blue, and one green.
(a) Describe what a "state" would be and the ensemble that represents the situation. How many bits of information are missing?
(b) If the sum of the faces of the four dice is known to be 6, what ensemble best represents the situation, and how many bits of information are still missing?
(c) Assuming the total roll is 6 as in part (b), what are the probabilities $P_1, \ldots, P_6$ for the various faces 1–6 of the white die? How much missing information is associated with this probability distribution?
(d) Find a numerical value for the ratio of the answer to part (c) to the answer of part (b). You should find this ratio is not equal to 1/4; explain why it makes sense that the missing information associated with 1 of the 4 dice is more (or less) than 1/4 of the total.

3.4 Redo Problem 3.3 with the following setup: you roll four dice: one white regular six-sided die, and each of the three Grime dice (one red, one blue, and one green), with the faces of each die shown in the table in Problem 2.1. When redoing parts (b) and (c), assume the total roll is 14 instead of six.

3.5 Calculate the amount of missing information (in bits) if you rolled 5 d10s. What about if you rolled 5 d20s?

3.6 While Samuel Morse's estimates for the frequency of letters in English words were fairly good, we could redo Example 3.7 using the frequency of letters in English words as given by the Concise Oxford Dictionary [3]:

Letter	Percentage	Letter	Percentage	Letter	Percentage
E	11.1607	C	4.5388	Y	1.7779
A	8.4966	U	3.6308	W	1.2899
R	7.5809	D	3.3844	K	1.1016
I	7.5448	P	3.1671	V	1.0074
O	7.1635	M	3.0129	X	0.2902
T	6.9509	H	3.0034	Z	0.2722
N	6.6544	G	2.4705	J	0.1965
S	5.7351	B	2.0720	Q	0.1962
L	5.4893	F	1.8121		

(a) Use this table to recalculate the number of bits of missing information per character in this case. How similar or different is your result to that in Eq. (3.22)?
(b) Assume now you have a section of a novel written in English. How does having this information change the amount of missing information? In this case, the probabilities are no longer statistically independent (as described on page 7), so working this out quantitatively is non-trivial. As such, give a qualitative description.

3.7 In the 1994 film *Stargate* (and several subsequent TV series), a large circular device (the "stargate") is found. There are 39 symbols (or "glyphs") corresponding to constellations that appear on the stargate, which act similar to numerals on an old rotary phone: You can "dial" them to open a wormhole to another planet. It is discovered that you need seven distinct symbols in order to possibly open a wormhole.
(a) How many possible combinations of 39 distinct glyphs are there, if you need seven? How many bits of missing information are there, assuming they are all equally likely?
(b) It is later discovered that one of the symbols corresponds to Earth, which must be the final glyph dialed. In this case, how many possible combinations of distinct symbols are there? How many bits of missing information?

3.8 Prove that the quantity

$$S = -k \sum_{r=1}^{n} P_r \ln P_r$$

is a maximum when $P_r = 1/n$ for all r. You can do this one of two ways:

(a) Show that $S - k \ln n \leq 0$.
Hint: The inequality $\ln[1/(nP_r)] \leq 1/(nP_r) - 1$ will be helpful. To arrive at an expression in which you can use this inequality, you will have to insert $\sum_{r=1}^{n} P_r = 1$ in an appropriate place.

(b) Use a Lagrange multiplier λ to include the constraint that $\sum_{r=1}^{n} P_r = 1$, by maximizing $-k \sum_{r=1}^{n} P_r \ln P_r - \lambda \left(\sum_{r=1}^{n} P_r - 1\right)$. If you do it this way, you should also take the second derivative to ensure you found a maximum and not a minimum.

3.9 You can find weighted dice online quite easily. Suppose you find a set of three such weighted six-sided dice, so that the probability of rolling a 1 is 1/100, rolling a 6 is 3/4, and rolling a 2, 3, 4, or 5 is 3/50.

(a) Calculate the Shannon entropy in bits when you roll one of these dice.
(b) Calculate the Shannon entropy in bits when you roll two of these dice.
(c) Calculate the Shannon entropy in bits when you roll three of these dice.

3.10 Consider an *unfair* four-sided die (d4), such that[7]

$$P_1 = \frac{1}{7}, \quad P_2 = \frac{3}{7}, \quad P_3 = \frac{2}{7}, \quad P_4 = \frac{1}{7}.$$

(a) Calculate the number of bits of missing information for this system.

(b) Imagine you want to transmit the following sequence of numbers (results from die rolls): 1223423 (which has exactly the probability distribution as given above). How many bits would be needed to transmit this if you assign the bits as follows: 00 = 1, 01 = 2, 10 = 3, and 11 = 4? How many bits per character?

(c) As long as we have an even number of bits, this is a nice simple solution to our data transmission problem. But what if we want to start transmitting separate "words," that is, different combinations of numbers? We can allow each character to be represented by a different number of bits, with the most common characters written with shorter bitsets, such as:

$$001 = 1$$
$$1 = 2$$
$$01 = 3$$
$$000 = 4$$

These are chosen so that a given combination of bits will give a unique set, because some combinations of bits do not correspond to one of our four numbers (such as 100 or 10). Write down 1223423 in binary with these definitions. How many bits are required now? How many bits per character?
Compare your answer for to that for (a). Should they be the same? What could explain any difference?

(d) Compare your results for the number of bits per character for (b) and (c). Which is larger?

3.11 Suppose that a square dart board is divided up into regions as suggested by the picture below. First it is divided into two triangular regions by bisecting the square along the diagonal.

7 This is similar to a problem worked out in chapter 13 of Ref. [2].

Then one of the resulting smaller regions is bisected again, and this is continued a total of N times (the figure shows the case $N = 8$). After the division process is stopped, a dart is thrown at the board. If nothing is known about the dart except that the probability for it to strike a given region is proportional to its area, how many bits of information are missing concerning what region it landed in? What happens in the limit $N \to \infty$?
Hint: The trick $\sum_{r=0}^{N} rx^r = x\frac{d}{dx}\sum_{r=0}^{N} x^r$ can be useful.

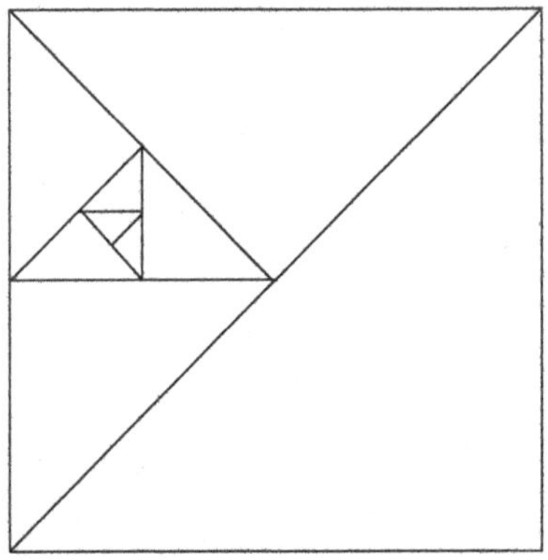

References

1 I. Niven. *Diophantine Approximations*. Dover Books on Mathematics. Dover Publications, 2013.
2 A. Berera and L. D. Debbio. *Quantum Mechanics*. Cambridge University Press, 2021.
3 H. W. Fowler, D. Thompson, and F. G. Fowler. *The Concise Oxford Dictionary of Current English*. Clarendon Press, 1995.

Further Reading

While we are using information theory here to frame our approach to statistical mechanics, the following textbook works through it in a much more rigorous way.

- A. Katz. *Principles of Statistical Mechanics: The Information Theory Approach*. W. H. Freeman, 1967.

4

Statistical Systems and the Microcanonical Ensemble

After the setup in Chapters 2 and 3, we are finally in a position to set the stage in order to study statistical mechanics and thermodynamics. This will first involve discussing how we solve problems in this manner, and more specifically how we can extract physical meaning from the number of states in a system and such abstract ideas as the Shannon entropy.

After completing this chapter, you should be able to

- learn how to identify a system and the states of a system (specifically the distinction between microstates and macrostates),
- understand the setup of the microcanonical ensemble,
- learn how to describe thermal and mechanical interactions of systems, and
- understand quasistatic processes.

4.1 From Probability and Information Theory to Physics

In order to apply our understanding of information theory to physics, we need to figure out how to apply Eq. (3.21) to a physical system. To do this, we first must determine the probability density for the different outcomes of the system. But before we can even consider doing that, we have to understand more about what an "outcome" is in the context of a physical system. As we began to do in Chapter 3, we will refer to these outcomes as the **states** of the system. Additionally, as discussed in Chapter 2, probability is only a meaningful concept when applied to a large number of identically prepared systems, known as an **ensemble**. In this chapter we will discuss each of these in turn and then use our results to begin to calculate physical quantities of interest.

> One thing to keep in mind before we get started: We are only interested in discussing systems in (or very close to and approaching) equilibrium! Nonequilibrium processes are quite interesting, but we have to learn how to walk before we can run. Equilibrium statistical mechanics and thermodynamics is already rich (and hard!) enough as we shall see, so we will stick with this for now.

Finally one last comment before we get into the details: A lot of what we will say will appear to require some form of faith; many assumptions will seemingly come out of the blue. However remember the most important rule here (and in all of science): If we make an assumption, then

Statistical Thermodynamics: An Information Theory Approach, First Edition. Christopher Aubin.

Companion website: www.wiley.com/go/Aubin/StatisticalThermodynamics

any results that come from that assumption **must be supported by observation**! While I won't always show it explicitly, keep in mind that any results I show have been supported experimentally, even if our assumptions seem a little strange initially.

4.2 States in Statistical Systems

The state of a system is clearly an important concept in physics, and is not usually difficult to describe (in principle) in classical or quantum mechanics. That is because in most subfields, we tend to only study systems with one or two (*maybe* three) objects, so the state of a system is easy to figure out. There is a distinction between different *types* of states that will be essential in statistical mechanics but often not as important in other areas of physics. When the number of objects becomes large, we have to be more precise and distinguish between the **microstates** and the **macrostates** of a system.

The microstate of a physical system can be specified in one of two ways, depending on if the system is classical or quantum in nature. For a classical system with N particles, we say that a microstate is fully known if we know the positions and momenta[1] of all N particles in the system. Thus, for each particle in our system we have

$$\{\mathbf{r}_i, \mathbf{p}_i\}, \tag{4.1}$$

with $1 \leq i \leq N$, labeling each particle. Each position and momentum component is known as a **degree of freedom**, and thus in a three-dimensional classical system with N particles, we have $f = 6N$ degrees of freedom. Of course the position and momentum are not independent—if I know $\mathbf{r}_i$ and the mass of the ith particle, then $\mathbf{p}_i = m_i d\mathbf{r}_i/dt$. However, we are not solving the equations of motion to obtain these variables, but rather we consider that we are measuring them so they can be considered as independent.

A couple of comments regarding the classical mechanics problem.

- Solving the equations of motion also requires knowledge of the initial conditions (initial positions and momenta of all particles), so at least initially there is a level of independence among these two quantities.
- In classical mechanics, it is standard to consider **generalized coordinates** $\mathbf{q}_i$ and the **generalized momenta** $\mathbf{p}_i$ conjugate to these coordinates. Such a generalization allows for a simpler approach using Hamiltonian or Lagrangian mechanics, but I will leave that discussion for a classical mechanics course (see chapter 7 of Ref. [1] for example). For our purposes, I will usually not use this generalization as it's not essential to make the points needed here.

For a quantum mechanical system, instead of knowing the position and momentum, we specify the state using a *wavefunction*, Ψ, which is the solution to the Schrödinger equation (with appropriate initial and boundary conditions). For an N-body system, Ψ is a function of the N positions (or momenta, but not both) of all of the individual particles. Determining the solution to the Schrödinger equation can be quite difficult even for one-body systems, and essentially

1 In introductory physics velocities are often considered instead, but conventionally we consider momenta, primarily because that is a relevant quantity in quantum mechanics.

impossible for N-body systems. Often we will look at systems where the particles do not interact strongly, so sometimes just specifying the quantum states of each individual particle in the system will be sufficient.

Example 4.1 Let's return to a non-physics example: rolling some six-sided dice. The microstate of each die is given by whichever value (one through six) is facing up after a random roll. For two dice, there are 36 possible microstates, because for each of the six possible rolls of one die, there are six rolls on the other die; we simply multiply the number of states allowed in each system. In this case it is simple to enumerate all of the possible microstates: We could write them down simply, but let's not waste the space.

However, in many cases, we aren't interested in the individual microstates of each particle (or die in this example) in our system. Usually we are more interested in the **macrostates** of the system, which depend upon what we might be interested in knowing. For example, as we've discussed before, often when you roll two dice you only care about the *sum* of the states of the two dice (as in Problem 3.2). This sum defines a macrostate of this system, so while there are 36 microstates, there are only 11 macrostates: 2, 3, 4, 5, 6, 7, 8, 9, 10, 11, and 12.

Example 4.2 Let's now consider a physical example: a classical one-dimensional simple harmonic oscillator with mass m and natural frequency ω. Here, a microstate r is given by $r = \{x, p\}$, and of course there are plenty of microstates (an infinite number of them in fact). But if we are interested in the total energy of the system in the state r, $E_r = p^2/(2m) + m\omega^2 x^2/2$, then the microstates of interest correspond to values of $\{x, p\}$ that fall along the ellipse defined by this equation. Later we will consider states such that $E < E_r < E + \delta E$, in which case the microstates fill the shaded region shown in Figure 4.1.

Exercise 4.1 Now consider you have two (uncoupled) classical one-dimensional simple harmonic oscillators each with mass m and natural frequency ω. How would you describe the microstate r, and what would the energy E_r of such a state be?

If we consider states such that $E < E_r < E + \delta E$, it is a bit more difficult to sketch this as we did above, as it would require four dimensions. Luckily this is not going to be necessary (even though it would be nice to visualize this!).

In general, it is usually simple to specify the states (microstates or macrostates) of a system, so long as we know the allowed energies, even for multi-particle systems. Of course, calculating

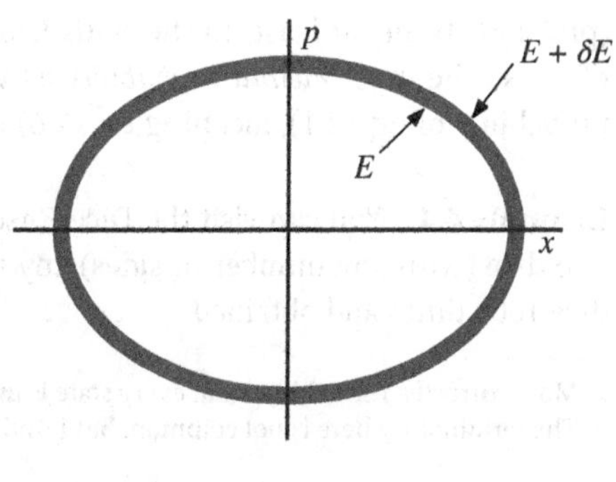

Figure 4.1 A plot of p vs. x for the classical one-dimensional simple harmonic oscillator. The shaded region corresponds to energies in the range from E to $E + \delta E$ as discussed in Example 4.2.

physical observables that depend upon these states becomes tricky very quickly. However, if we can determine the allowed energies, we will see that it is not terribly difficult to determine bulk properties of a system of N particles where N is very large. In the rest of this chapter, we will work through the steps needed to get to this point.

4.3 Ensembles in Statistical Systems

Now that we have an idea of how to define the states of various systems (and more importantly the difference between microstates and macrostates), let's return to our discussion of an **ensemble** of the system: In the context of a statistical mechanics problem, an ensemble is the set of all microstates that are accessible to the system. The notion of *accessibility* will be clarified as we focus on particular ensembles. In Example 4.2, an accessible state would correspond to one in which x and p had values such that the energy of the state was between E and $E + \delta E$.

Example 4.3 Returning to Example 4.1, we had 36 microstates overall. We could consider several possible ensembles:

- If we care about the roll of each individual die (perhaps you and a friend are each rolling a die and the high roller wins), then each microstate corresponds to a distinct macrostate, and our ensemble contains 36 macrostates.
- If we are interested in the sum of the two dice, then our ensemble contains 11 macrostates (with the same 36 microstates, so most of the macrostates correspond to more than one microstate).
- Finally, suppose we cared about the sum of the two dice, but only wanted to consider rolls of six, seven, or eight. Then we would have 16 microstates and three macrostates. In this case, not all possible microstates are *accessible* to the system, and we can see that Example 4.2 is analogous to this (where the "energy" of our dice is the outcome of a roll, with $E = 6$ and $\delta E = 2$ here).

Remember we are using the term "ensemble" in two different, but related, ways. Theoretically, as we said above, a **statistical ensemble** is a set of identical systems such that every possible state is included *once*.[2] Said another way, an ensemble is a just a *complete set of accessible states*, similar to a complete set of functions or vectors as discussed in linear algebra.

An **experimental ensemble**, discussed briefly in Sections 2.3 and 3.2 and relevant experimentally, is a large number M of identical systems upon which we perform a measurement. Technically, not all of the accessible states will be represented in the same way as in the probability distribution, however in the limit $M \to \infty$, we expect the experimental ensemble to correspond (probabilistically at least) to the statistical ensemble. That is, a given state appears m_r times out of M, so the *experimental probability* as we defined it in Eq. (2.5) corresponds to the expected probability of Eq. (2.1), meaning Eq. (2.6) holds.[3]

Example 4.4 You can visit the Dice Ensemble section of the companion site to simulate rolling two dice (with any number of sides) any number of times. For example, I "rolled" two six-sided dice 1000 times and obtained

2 More correctly, I should say that every state is included a number of times which is proportional to its probability.
3 The terminology here is not common, but I find it useful to think of these two different definitions in this way.

Sum of two dice	# of results	fraction of results	# of microstates with this sum	probability
2	31	0.031	1	0.027778
3	50	0.050	2	0.055556
4	86	0.086	3	0.083333
5	113	0.113	4	0.111111
6	155	0.155	5	0.138889
7	155	0.155	6	0.166667
8	129	0.129	5	0.138889
9	118	0.118	4	0.111111
10	82	0.082	3	0.083333
11	58	0.058	2	0.055556
12	23	0.023	1	0.027778

The second column shows the number of times the result in the first column was rolled, and the third is the fractional number of times it was rolled. The last two columns show the number of times the state appears in the ensemble, along with the theoretical probability. In the limit where the number of experiments goes to ∞, the third and fifth columns should agree precisely.

In statistical mechanics, there are three common and fairly general ensembles that can be applied to most systems. These are:

1. **Microcanonical ensemble**: Systems that we study using this ensemble are completely isolated from their surroundings. Thus, the energy of the system is constant and fixed within some resolution: The system is in a microstate r such that it has some total energy E_r between an energy E and another energy $E + \delta E$ with $\delta E \ll E$. Coffee in an insulated thermos could be approximately described using this ensemble, but in general it is difficult to construct such systems that are completely isolated. However, it is simpler to consider this ensemble first theoretically.[4]
2. **Canonical ensemble**: A system described by this ensemble is in equilibrium with another system much larger than it so that its temperature remains fixed.[5] Additionally, we will find that the *mean* energy of the system is fixed, but not the total energy. An example of this would be a system such as an uninsulated cup of coffee with a lid in a room—eventually the coffee will cool down (or warm up, depending on whether or not you like iced or hot coffee) to be at the temperature of the room. We will study this ensemble beginning in Chapter 8.
3. **Grand canonical ensemble**: In both of the first two ensembles above, we keep the total number of particles in the system fixed. This last ensemble is the same as the canonical ensemble, but with the number of particles in the system allowed to vary. This will be helpful when studying quantum statistics problems, so I'll first introduce this ensemble in Chapter 11.

As we will see, these ensembles are merely ways to set up our problem in statistical mechanics so that we can apply the ideas of probability and information theory to the system and thus determine physical quantities. We will see that under certain conditions, they can sometimes be used interchangeably. In Chapters 5–7, we will focus on the microcanonical ensemble and later

4 For systems with discrete energy levels, we could fix the energy precisely; however, to keep a consistent formulation I will use δE regardless.

5 I will be very clear on defining temperature later, but for now you can just think of it as you do in everyday life.

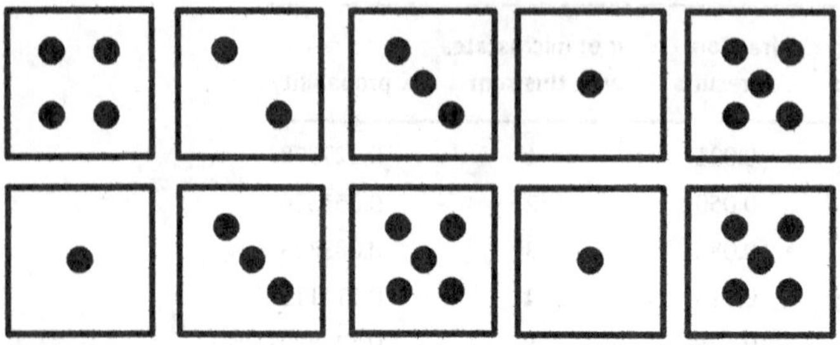

Figure 4.2 One of the $6^{10} = 60\,466\,176$ possible microstates for a system of 10 six-sided dice.

switch to the canonical ensemble (Chapter 8 and following). The grand canonical ensemble will be touched on as an easier approach to study quantum statistics (Chapters 11 and 12), but could also be used to study phase transitions and chemical reactions.

Let's examine several different systems, describe the different states accessible to them, and what the ensemble would correspond to.

Example 4.5 Before discussing physical systems, let's start with our dice, where we now consider rolling 10 six-sided dice. Again, the state of a die is whichever number (from one through six) is facing up after a roll. A microstate of the system is specified by denoting the roll of each individual die, where one such example is shown in Figure 4.2.

As each die has six outcomes, there are $6^{10} = 60\,466\,176$ total possible microstates, quite a bit more than with just two dice![6] In some cases, a microstate is unique (e.g., rolling 10 ones), but in other cases, there are many more. For the example in Figure 4.2, with three ones, one two, two threes, a four, and three fives, there are a total of 50 400 ways to achieve this outcome (if we don't care about which die has which value).[7]

We can calculate the number of bits of missing information in this system,

$$I = \log_2 6^{10} = 10 \log_2 6 = 10(\log_2 3 + 1) \approx 25.9 \text{ bits},$$

which means we would need 26 bits in computer memory to fully specify the system.

As always, the number of possible macrostates depends on what we are interested in. If we care only about the total sum of the 10 dice, then there are only 51 macrostates (because we could roll a 10, 11, 12, ..., 60). In some cases, there are very few microstates for a given macrostate (e.g., there is only one way to roll a 10 and 10 ways to roll an 11), but in many cases there are a lot of possible microstates for a macrostate. The sum of the dice in Figure 4.2 is 30, and as mentioned, there are 50 400 ways to roll this particular combination, but there are 2 930 455 ways in total to roll this sum. Calculating these numbers is possible using a formula found on pages 23–24 of Ref. [2]. If we have n dice with s-sides, then the number of ways N to roll a total sum S is given by

$$N = \sum_{k=0}^{\lfloor (S-n)/s \rfloor} \frac{(-1)^k n}{k!(n-k)!} \frac{(S-sk-1)!}{(S-sk-n)!}, \tag{4.2}$$

where $\lfloor x \rfloor$ is the **floor** of x: the integer closest to and less than or equal to x.

6 If I wanted to list all states, even doing so as briefly as possible (1111111111, 1111111112, 1111111121, etc.), it would take up just under 300 000 pages in this book! Perhaps not the best use of space.

7 This is found by realizing there are $10! = 3\,628\,800$ ways to arrange the different dice, but any time a number appears more than once, such as two threes, we have to correct for overcounting, so we divide by 2! for every double, 3! for every triple, and so forth, as discussed on page 9.

Example 4.6 Moving to a more physical example, imagine a system with three electrons in an external magnetic field, **H**. Electrons are intrinsically spin-1/2 particles, so even at rest, they have a magnetic dipole moment with magnitude

$$\mu = g\frac{e}{2m}\left(\frac{\hbar}{2}\right), \tag{4.3}$$

with $g \approx 2$. When measuring the spin of a spin-1/2 particle, we find that the spin can only point "up" or "down" along any given axis. Placing the electron in a background field **H** is akin to measuring the electron spin, as it can be aligned or anti-aligned with the field (e.g., the direction the field points in corresponds to the "up" direction of the electron spin). The potential energy of the system is

$$U = -\boldsymbol{\mu} \cdot \mathbf{H} = \mp\mu H, \tag{4.4}$$

so the lowest energy state is found when the electron is aligned with the field.

With three electrons in such a field, we can enumerate all of the possible states in Table 4.1. The first column is merely a label—an arbitrary choice so that we can easily refer to a given state. The second column is the microstate, where I show the orientation of each spin relative to the magnetic field (↑ = aligned with **H**, ↓ = anti-aligned with **H**), assuming I can identify each electron. With just three electrons we can write the states out explicitly and see that there are eight microstates, and that there are $I = \log_2 8 = 3$ bits of missing information.

The third and fourth columns are equivalent ways we might label macrostates, by the total magnetic moment (found by adding up the individual magnetic moments), or the total energy of the system. I'll discuss this in the next example.

Example 4.7 Now suppose we have more information: Not only are there three electrons in the field **H**, but we have measured the total energy of the system to be $U = -\mu H$. In this case, there are only three microstates that are accessible to the system in this ensemble, and we have $I = \log_2 3 \approx 1.58$ bits of missing information. As we have more information than in the previous case (we know the number of electrons as well as the total energy), the number of accessible states decreases, and thus of course there is less missing information.

Table 4.1 A list of all microstates corresponding to three electrons in a background magnetic field H, where an up arrow corresponds to the electron being aligned with the magnetic field, and with a down arrow, the electron is anti-aligned.

State label (r)	mMcrostate	Macrostate (μ_{total})	U
1	↑↑↑	$+3\mu$	$-3\mu H$
2	↓↑↑		
3	↑↓↑	$+\mu$	$-\mu H$
4	↑↑↓		
5	↓↓↑		
6	↓↑↓	$-\mu$	$+\mu H$
7	↑↓↓		
8	↓↓↓	-3μ	$+3\mu H$

I include the total magnetic moment (μ_{total}) of the system along with the total energy of the system in each state.

Example 4.8 In many cases we won't be able to list off every possible state explicitly, but rather we could give a general expression that denotes the microstates. Examples include:

- A quantum mechanical simple harmonic oscillator in one dimension has energies given by

$$E_n = \left(n + \frac{1}{2}\right)\hbar\omega, \; n = 0, 1, 2, \dots. \tag{4.5}$$

- A quantum mechanical particle of mass m constrained in a cube with sides L (a "particle in a box") has energies

$$E_{n_x,n_y,n_z} = \frac{\hbar^2\pi^2}{2mL^2}\left(n_x^2 + n_y^2 + n_z^2\right), \; n_{x,y,z} = 1, 2, 3, \dots. \tag{4.6}$$

- A classical one-dimensional simple harmonic oscillator of mass m and natural frequency ω, as discussed in Example 4.2, with position x and momentum p has energy

$$E_r = \frac{p^2}{2m} + \frac{1}{2}m\omega^2x^2. \tag{4.7}$$

Without any restrictions on the energies, we would have an ensemble with an infinite number of microstates. But if we use, say, the microcanonical ensemble, we only allow energies within some range, giving us a finite number of states. This is only strictly true for the quantum mechanical systems mentioned, as the energy levels are discrete; for the classical case there is still an infinite number of states in a finite energy range, but we will worry about this when the time comes.

4.4 From States to Information

Once we have enumerated the accessible microstates of our system, if we wish to apply our expression for the Shannon entropy in Eq. (3.21) to our system, we need to know the probability distribution of our microstates. In our non-physics examples, such as our various dice examples, if we assume every die is fair, or completely random, then this is simple. Any given roll of a single die, for example, is equally likely to be in any of its six possible states. If we roll two dice, and we only care about the sum of the dice, we just count the possible outcomes and calculate the probabilities. I show the outcomes in Table 4.2, where I am assuming the two dice are different colors, red and blue, so that we can distinguish them. Each *individual* roll has the same probability: 1/36, and the only reason a seven is more likely is only because there are six different ways to make a seven when rolling two dice. We now have the probabilities for every macrostate for this system, and as such the probability distribution.

If we know a system is truly random as in the dice example, then assigning these probabilities is straightforward, but how about for a physical system? There are equations of motion that govern how every molecule will interact with every other molecule, so how are we supposed to assign a probability to a given physical macrostate?

To come up with a solution, let us revisit our Shannon entropy, the missing information for an ensemble of systems. We will assume that we have enumerated all possible microstates, and we'll change our notation to be more standard: We'll say that the number of accessible microstates is denoted by Ω. If r labels a particular microstate,[8] then the Shannon entropy is of the form

$$S = -k_B \sum_{r=1}^{\Omega} P_r \ln P_r. \tag{4.8}$$

8 As above, this is just an arbitrary label for us to identify the microstate. Sometimes it will have some significance and other times it will just be a bookkeeping device.

Table 4.2 The possible sums and how to obtain them on two dice, which are assumed to be different colors (red and blue), along with the probability of each.

Total roll	Individual outcomes	Probability
2	11	1/36
3	12, 21	2/36 = 1/18
4	13, 22, 31	3/36 = 1/12
5	14, 23, 32, 41	4/36 = 1/9
6	15, 24, 33, 42, 51	5/36
7	16, 25, 34, 43, 52, 61	6/36 = 1/6
8	26, 35, 44, 53, 62	5/36
9	36, 45, 54, 63	4/36 = 1/9
10	46, 55, 64	3/36 = 1/12
11	56, 65	2/36 = 1/18
12	66	1/36

Note writing "23" implies I rolled a two on the blue die and a three on the red, while "32" is different: I rolled a three on the blue die and a two on the red.

Here we have set the constant $I(e) = k \equiv k_B$, the Boltzmann constant, which is discussed in Appendix A in our discussion of units, and has the value

$$k_B \equiv 1.380\,649 \times 10^{-23}\,\mathrm{J/K}. \tag{4.9}$$

This determines the units that we will use to measure entropy.

We are ultimately interested in bulk properties of the system, such that we only need to know something about the macrostate of the system, while knowing very little about the underlying microstate. Also, let me remind you of an important statement from before: **We are only interested in states that are in equilibrium**. That is, if there have been any disturbances to our system, we will wait a long enough time so the system returns to an equilibrium state before attempting to study it.

Given our restricted knowledge of the system, then it stands to reason that we assign the probabilities to the various microstates of the system such that the amount of missing information is maximized. That is, we don't know a lot about the underlying system, so we want S to be as large as possible given what we do know; we want to assume as little as possible about our system. As was shown in Problem 3.8, S is maximized when every microstate has equal probability, or

$$P_r \equiv \frac{1}{\Omega}. \tag{4.10}$$

This gives us

$$S = k_B \ln \Omega, \tag{4.11}$$

which was the first expression we derived for missing information, but we will just refer to this as **entropy** from now on. However, keeping in mind the origin of this expression from information theory will allow us to have a clearer idea of what entropy actually is, which we will discuss in detail later. Additionally, the meaning of the constant k_B will become clear later. Assuming the missing information is maximized is also known as the **fundamental postulate of statistical mechanics**: An isolated system in equilibrium is equally likely to be in any one of its accessible microstates.

Given our non-physics examples, this postulate is not unreasonable. In the case of rolling dice, yes, *technically* if I know the equations of motion describing the movement of the dice after they leave my hand and I know the initial conditions, I can predict exactly what the roll of each die will be. The point that I have made numerous times up to now is that this is highly impractical (and dare I say impossible) to accomplish. Thus, if we perform a large number of rolls (with randomly chosen initial conditions such that each initial condition is equally likely), we would find that our results would be consistent with the assumption that the dice are fair and every microstate is equally likely.

And so it follows for physical systems—without knowing the underlying equations of motion (or more importantly, the *solutions* to those equations), it makes sense to assume every microstate is equally likely for a system in equilibrium. Were we to find a system where this were not true, we would have to start over with a refinement of this assumption. This would require determining the true probabilities P_r of each state, then returning to Eq. (4.8), and working from there.

> As a brief aside, Claude Shannon was not the first to formulate an expression for entropy as given in Eq. (4.8), he merely did so from an information theory perspective. Josiah Gibbs (1839–1903) was the first to formulate this quantity in this form, from a different point of view (being one of the first to develop the field of statistical mechanics as distinct from thermodynamics). The form of entropy given in Eq. (4.11) is the original form used, from Ludwig Boltzmann (1844–1906), which of course is a special case of the more general Shannon or Gibbs entropy, only valid when every microstate is equally likely. I do not wish to omit the contributions of Boltzmann or Gibbs by referring to this only as the *Shannon entropy*, but rather am just sticking with the notion that these physical principles can actually be developed purely from an information theory point of view. This, in my mind, helps in understanding how we should understand entropy in general, which will be discussed shortly. You can read more on Gibbs (Shannon) vs. Boltzmann entropy and their relationships in Refs. [3, 4].

Entropy is an interesting quantity, as it is both crucial to the study of thermodynamics but also often terribly misunderstood. Many times (not just colloquially but in physics textbooks and articles) you will hear it described as the "disorder" of a system, which is a misleading term. To see why, we should remind ourselves of our origin of the expression for entropy from information theory. In Eq. (4.8), entropy is written in terms of the probabilities of the various microstates, and remember that when dealing with probabilistic systems, we never just consider a single system—we always consider an *ensemble* of systems. This is also clear with our more common expression for entropy, Eq. (4.11), as this shows specifically that the entropy is dependent on the number of states—a system is ever only in one state, although it can be in one of many possible states.

Thus, a system itself cannot have entropy; we should really say that an ensemble of identically prepared systems has entropy. This is rather long-winded, so it's usually just easier to talk about the entropy of a system. However, keeping the ensemble-system separation is important to remove this confusion of entropy as disorder. I am not really sure how to quantify disorder, but I can count the total number of states a system can be in.

Reference [5] discusses this distinction in an enlightening way. This article uses several models to argue why the notion of entropy as disorder is a flawed concept (and also considers another similarly flawed notion of entropy as "freedom"). One of the examples shown in this article is that of a lattice gas model, with two examples shown such that the comparison with disorder breaks down. I have reproduced similar examples in Figures 4.3 and 4.4, both of which show a system of

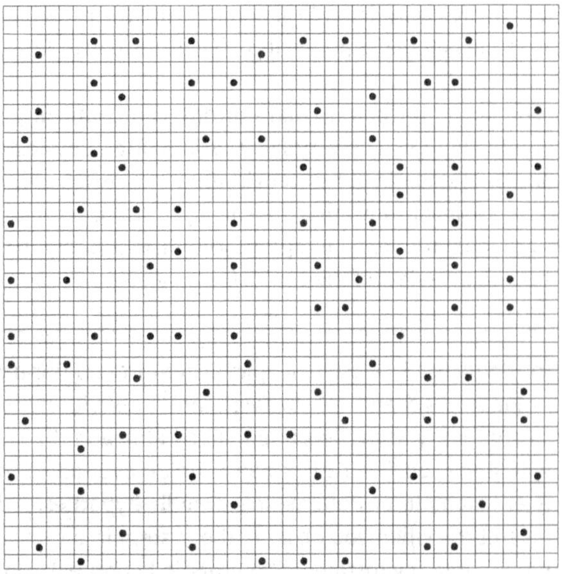

Figure 4.3 A system constructed from an ensemble with low entropy.

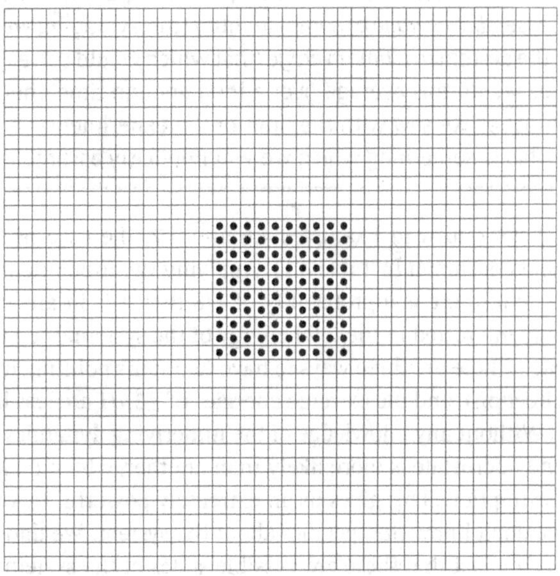

Figure 4.4 A system constructed from an ensemble with high entropy.

molecules arranged according to specific (and different) rules. One might say the former configuration looks "more disordered" than that in the latter (although that is also a subjective statement). However, the second figure was constructed using the rule that each particle could be anywhere (without overlapping) in the grid, while the former had an additional restriction that each particle couldn't be adjacent to any other particle. Thus, despite appearances, the system in Figure 4.4 is an example from an ensemble with a *higher* entropy.

Actually, to be clear, given that these figures consist of 100 particles filling a 40×40 grid, there are so many possible configurations that getting precisely what I show here is rather unlikely. For the higher entropy system, there are $\sim 10^{319}$ possible ways to put the particles into the grid, and even for the lower entropy system we have $\sim 10^{314}$ states (see Problem 4.1). I only state this because these figures were actually generated with the rules listed above in mind, but not randomly (I didn't have time before the publication deadline to wait for figures that displayed was I was interested in to show up).

The key point here is that entropy is a confusing concept, and there isn't really an easy way to understand it. The very fact that anything physical can be determined by merely counting the number of states in our system is quite surprising. This would normally not be something we would deem possible in any other area of physics, and yet this seems to be the case here. I would encourage you to look into Refs. [6, 7] for more insight into entropy through both computation and through a historical review of the confusing concept.

In reality, such a macroscopic system in equilibrium is not truly in a single state as shown in Figures 4.3 and 4.4. In fact, once in equilibrium, the system moves from one microstate to another (within the same macrostate), and the entropy (and other bulk properties that we will study soon) arise from an average over all of these allowed microstates. This averaging is an example of Boltzmann's **ergodic hypothesis** (discussed quite nicely in Ref. [8]). At any given time when you observe the system in a given macrostate, you will find it in any one of its possible microstates. The ergodic hypothesis states that over time, the system will move from microstate to microstate, sampling all of the possible microstates. We can connect the *ensemble average* of some quantity (the averaging we are discussing in a probabilistic sense) in the system to a *time average* (that which would correspond to an experimental average). So long as we do this average over a long enough time, then these two averages should be the same (just as we saw with the limit of the experimental probabilities and theoretical probabilities). This is something which we will implicitly assume throughout this text.

Let's return to Examples 4.6 and 4.7, the case of three electrons in a background magnetic field (as shown in Table 4.1). Applying the fundamental postulate of statistical mechanics to this system, there is a probability of 1/8 that it is in any of the individual microstates. But if we are only interested in the total energy of the system, then the probability of measuring $\pm\mu H$ is 3/8 for each (there are three ways to have two spins pointing in one direction and the third in the opposite direction), and the extreme energies of $\pm 3\mu H$ are less likely, each with a probability of 1/8.

Without any knowledge of the underlying dynamics of the system, there is no reason to assume one microstate is more likely than another, which makes the fundamental postulate a reasonable assumption. This gives us an extremely powerful theory, as it actually capitalizes on the fact that we don't know the specifics about the system. We don't need to know how the system came to be set up, just that these are possible states of the system. While we might need to know something about the underlying forces or interactions, we only need them in the most general sense—we will see we merely need to know how these lead to the possible energies of the system. As long as we have this information, we will find that there is quite a bit we can do with this theory. This gives a framework for solving a great many problems in physics, at least in the case of very large systems.

Of course everything changes if we have more information. For example, if we knew the 10 dice in Example 4.5 were weighted, so that rolling a six were more likely (as in Problem 3.9), then the fundamental postulate no longer applies. This additional information would require us to return to the expression in Eq. (3.21). However, while that would complicate the starting point of any solution to understanding a system, we will see that we can still work with many of the techniques we will discuss (for example, Problem 8.6). For now, however, we will stick with systems where the postulate does apply, and in Section 4.5 we will work out how to count the number of accessible states in the microcanonical ensemble.

4.5 Microcanonical Ensemble: Counting States

A system in the microcanonical ensemble is isolated from its surroundings and (as is always our preference) in equilibrium, and with this system, we will set up the statistical problem and derive many physical results of interest. The energy of an isolated system is constant, so a system in the microcanonical ensemble could be described as one that has a fixed energy.

For both practical reasons and a deeper theoretical reason, we don't want to say that the system has a precisely fixed energy in many cases. From a practical point of view, this cannot be done experimentally (you can never measure the energy of a system exactly, there is always some uncertainty). But as experimental limitations are not usually a consideration when we consider the theoretical development of a topic, there is another reason that will be relevant to us, and so we will always require the energy to be within some (small) range of energies.

Suppose we know the possible energies of each microstate of our system, and we'll label those as E_r, where again, r is just some label for a particular microstate. (Many different states might correspond to the same energy, or $E_r = E_{r'}$ for $r \neq r'$, which as we saw before just tells us that many microstates correspond to the same macrostate.) As stated above, we restrict our system such that the energies fall in range such that

$$E < E_r < E + \delta E, \tag{4.12}$$

where generally we will insist $\delta E \ll E$. That is, while we are requiring a range of energies, we still want the range to be small enough that making a comment such as, "the system has a fixed energy E," is fine.[9] In order to calculate the entropy S, we need to determine $\Omega(E)$, the number of states in this energy range, as the fundamental postulate states that they all are equally probable.

4.5.1 Discrete Systems

In some cases, some of which are in the problems (and some we saw in Chapter 3), calculating $\Omega(E)$ directly for a system with discrete energy levels is feasible by brute force (or using combinatorics). However, in general this is a difficult problem, so let us first look at a simpler quantity to calculate: the number of states with energy *less than* E,

$$\phi(E) \equiv \text{number of states with } E_r < E. \tag{4.13}$$

For small systems, or single-particle systems, ϕ can be fairly easy to evaluate. For example if we care about how many states a single one-dimensional quantum oscillator can have with energy less than $25\hbar\omega$, it is simply a matter of counting the states. Here we can use $E_n = (n + 1/2)\hbar\omega$ (we have changed our label r for n as the latter label is more conventional for this situation). The allowed values of n range from 0 to 24 (inclusive), so $\phi\,(25\hbar\omega) = 25$. Of course we see our choice of E does not have to be an allowed energy of the system.

It quickly becomes difficult to determine ϕ even if the energy levels are discrete (as in most quantum systems) as we increase the energy, especially when we consider more than one particle.

---

9 And honestly much easier to say than always referring to this range of energies.

Exercise 4.2 Consider a system of five quantum oscillators that are weakly interacting so you can treat them as independent (the energy of each oscillator doesn't affect the others), but they each could have an energy given by $E_n = (n + 1/2)\hbar\omega$. For this system, determine $\phi(5\hbar\omega)$ and show that it is 21.

Hint: It's simplest, although tedious, to make a table of the possible values and calculate the many possible energies.

Consider $\phi(100\hbar\omega)$ and specifically how much more challenging this becomes even for five oscillators.

As you can see from this exercise, just having a few additional particles (that is, when we increase the size of the system), enumerating the states quickly becomes difficult. Consider a system with roughly a mole of quantum oscillators, so $N \sim 10^{24}$, and let's assume a relevant scale of energy to be $\hbar\omega \sim 0.01$ eV. With this many oscillators, the energy scale has to be around $(N/2)\hbar\omega \sim 10^{22}$ eV or greater. If we were interested in $E \sim 10^{25}$ eV, then determining $\phi(E)$ using the same method of Exercise 4.2 would not be recommended. Enumerating all possible combinations of the N oscillators (and then counting the number of states) would take more time than the universe has been around! Not only would this be a waste of time, it would not give us how ϕ depends on the energy, which is (as we shall see) something we will be interested in.

At this point I would like take a step back and try to remember how to count, as silly as this sounds. That is, when we count, we actually are just "adding 1" an appropriate number of times, which we could write $\sum 1$. The restrictions appear in the limits of the sum, which are often hard to write down, but in our case, we can write it as

$$\phi(E) = \sum_{E_r < E} 1, \tag{4.14}$$

where this means we include all values of r such that $E_r < E$.

Let's see how abstracting something we know how to do actually helps. In this particular case, and as we'll see is often true, the energy E is large enough that it can be treated as a continuous variable. If we consider the case where our N oscillators above were all in the ground state, so $E = (N/2)\hbar\omega \sim 10^{22}$ eV, then as we increment n in the energies of each oscillator, E changes by ~ 0.01 eV for each change in n. This is small enough compared with E that E is essentially continuous.

Example 4.9 If this is the case, we can use Eq. (4.14) to determine ϕ for a single simple harmonic oscillator. First we realize that the difference between neighboring integers is one, or

$$\Delta n = 1,$$

which implies that, for a simple harmonic oscillator,

$$\Delta E = (\Delta n)\hbar\omega.$$

Then we can write $E = \left(n_{\max} + 1/2\right)\hbar\omega$ and

$$\phi(E) = \sum_{E_n < E} 1 = \sum_{n=0}^{n_{\max}} \Delta n = \sum_{E_n = \hbar\omega/2}^{E} \frac{\Delta E}{\hbar\omega}.$$

Treating E as continuous is equivalent to replacing this sum by an integral (and since $\hbar\omega \ll E$, we can replace the lower limit with 0), so that

$$\phi(E) = \sum_{E_n = \hbar\omega/2}^{E} \frac{\Delta E}{\hbar\omega} \to \int_0^E \frac{dE}{\hbar\omega} = \frac{E}{\hbar\omega}. \tag{4.15}$$

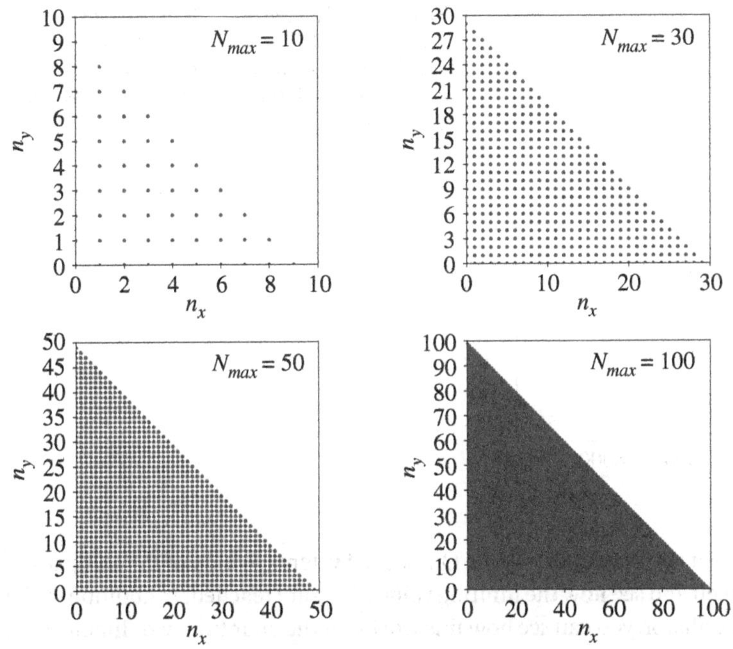

Figure 4.5 Plots of n_y vs. n_x for $N_{\text{max}} = 10$, 30, 50, and 100.

To be concrete, suppose we have $\hbar\omega = 0.01$ eV for our oscillator, and we want to calculate $\phi(E)$ for an energy $E = 7 \times 10^{22}$ eV. In this case, $n_{\text{max}} \sim 7 \times 10^{24}$, and Eq. (4.15) gives us $\phi(E) = 7 \times 10^{24}$. While we keep this as an exact equality, it generally is off by a few (more precisely $\phi = 7 \times 10^{24} \pm 1$), but that will not affect our results. The numbers we will deal with here will be so large that any errors introduced in this case will be completely negligible.

Exercise 4.3 Let's consider a two-dimensional quantum oscillator, so that the energy of a state is labeled by two integers, n_x and n_y, and we have

$$E_{n_x,n_y} = \left(n_x + n_y + 1\right)\hbar\omega, \quad n_x, n_y = 0, 1, 2, \ldots \tag{4.16}$$

1. Calculate $\phi(E)$ for $E = N_{\text{max}}\hbar\omega$ for $1 \leq N_{\text{max}} \leq 10$. (Do this by "brute force," by just enumerating and then counting the possible states.)
2. For $N_{\text{max}} \gg 1$, use the method above to convert the sum into a two-dimensional integral to show that $\phi(E) = \frac{1}{2}N_{\text{max}}^2 = E^2/(2\hbar^2\omega^2)$.

The best way to see how this integral approximation works is to make a plot of n_y vs. n_x. Each point on such a plot corresponds to one of the possible states, such that (for the two-dimensional simple harmonic oscillator) $n_x + n_x \leq N_{\text{max}} - 1$. As N_{max} increases, the density of points increases so much that $\phi(E)$ merely corresponds to the area of a triangle, as shown in Figure 4.5, hence the result of Exercise 4.3.

If we compare the exact number of states for the two-dimensional quantum oscillator in Exercise 4.3 for large N_{max} and the approximate value $\frac{1}{2}N_{\text{max}}^2$, we find the percent difference between the two decreases rapidly, as shown in Figure 4.6. For $N_{\text{max}} = 10\,000$, this percent difference is less than 0.01%. Given that statistical systems will have $N_{\text{max}} \sim 10^{24}$ or higher, our continuous approximation will clearly be sufficient!

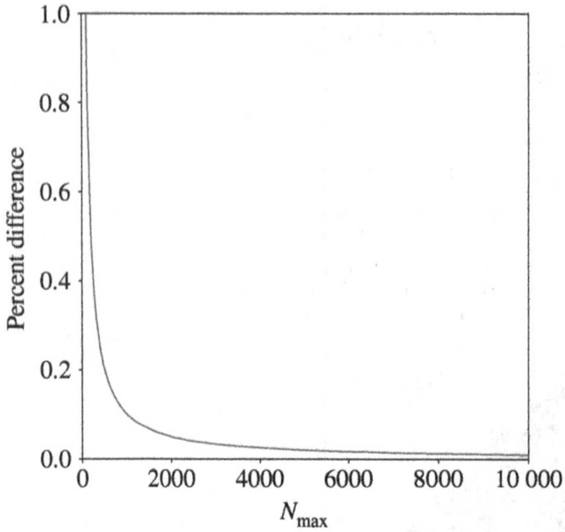

Figure 4.6 The percent difference between the exact number of states for the two-dimensional quantum oscillator in Exercise 4.3 for large N_{max} and the approximate value $\frac{1}{2}N_{max}^2$.

Computer Exercise 4.1 Visit the Two-dimensional Quantum Systems section of the companion site to create these plots yourself and see how the limiting values for ϕ are reached. In addition to the two-dimensional quantum oscillator, you can see how this works for the quantum two-dimensional particle in a box. This code directly calculates the number of states below a certain value and compares the exact result to the integral approximation.

How large would N_{max} have to be for the difference in either case to be around 1%?

4.5.2 Continuous Systems

For a classical system, we already have a continuous energy spectrum (and thus an infinite number of states with energy less than some value), so we have to treat this with a bit more care. In general, for a three-dimensional system, we describe our system by an appropriate set of coordinates,

$$\{\mathbf{r}_i, \mathbf{p}_i\},\ i = 1, \ldots, N; \tag{4.17}$$

and the total energy of a system is given by a function of these degrees of freedom,

$$E_r \equiv E(\mathbf{r}_i, \mathbf{p}_i).$$

As above, r is a label that refers to the set of coordinates in Eq. (4.17). With such continuous variables, then when we calculate $\phi(E)$ via Eq. (4.13), we get an infinite number of states with $E_r < E$. However, this is where we can consider a real system that we were attempting to study: even though *in principle* we can measure all of the positions and momenta exactly, we know that in practice there is always some uncertainty.[10] We can incorporate this resolution into our counting of states by looking at the available **phase space**.

First let's look at a single particle (say the $i = 1$ particle) and its position and momentum in a given direction (say the x-direction). The phase space, or allowed values of x_1 and p_{x1}, available to this particle is commonly represented in a plot of the momentum vs. position. For example, suppose we have a single one-dimensional classical simple harmonic oscillator with mass m and natural frequency ω, then the phase space plot would be an ellipse given by the equation

10 Of course we know that there is a quantum mechanical limitation as well, but right now we'll ignore that and just think about the classical case.

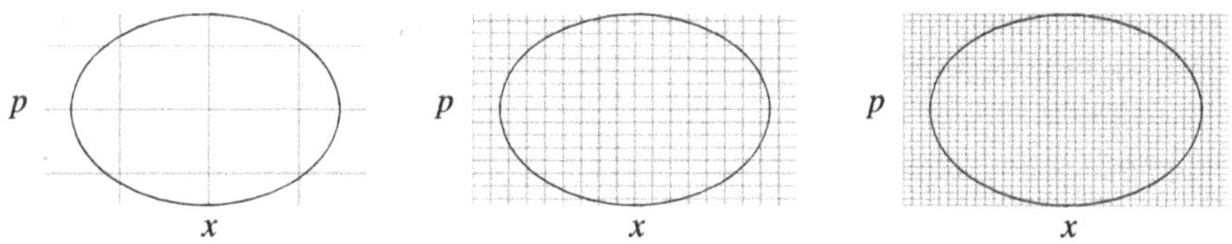

Figure 4.7 Phase space for a single classical oscillator in one dimension. From left to right, we show this for decreasing values of h_0.

$p_{x1}^2/(2m) + m\omega^2 x_1^2/2 = E$. To count the number of states and obtain a finite result, we divide phase space into boxes of area (often called a *volume* even though it's two-dimensional) h_0, so that

$$\delta x_1 \delta p_{x1} = h_0, \tag{4.18}$$

with δx_1 and δp_{x1} the uncertainties in measuring the position and momentum, respectively of our particle, and consider each of these boxes a single state. We additionally define Eq. (4.18) for the y- and z-directions of this particle, as well as for every other particle in our system. While we could choose a different h_0 for each direction and particle, we'll keep them the same.[11] This turns our continuous system back into a discrete system, so that our sum for ϕ in Eq. (4.13) is applicable to the classical case and it will result in a finite number of states.

I show this for a classical oscillator for three different choices of h_0 in Figure 4.7. To determine ϕ, we would just count the distinct boxes inside the ellipse, not including those boxes that the curve passes through (we only count states with energy *less than* E). As we decrease h_0, we find of course more states (from left to right, we obtain four, 140, and somewhere around 600 states).

As a quick aside, there is an important theorem in classical mechanics which is satisfied by the density of states in classical phase space, known as **Liouville's theorem**, which was first published by Joseph Liouville (1809–1882) in 1838. In a simple form, this theorem states that the density of states in phase space is constant in time. To see the derivation of this and how it could be applied here, see Section 7.12 of Ref. [1] for example. While this theorem is important for a rigorous treatment of statistical mechanics, I will let you explore this on your own (or in later courses if you delve deeper into the subject in more advanced approaches).

From this we find in the classical case that Eq. (4.13) can be written as

$$\phi(E) = \sum_{E_r<E} \underbrace{\left(\frac{\delta x_1 \delta p_{x1}}{h_0}\right)\left(\frac{\delta y_1 \delta p_{y1}}{h_0}\right)\left(\frac{\delta z_1 \delta p_{z1}}{h_0}\right)\cdots\left(\frac{\delta z_N \delta p_{zN}}{h_0}\right)}_{3N \text{ factors}},$$

and if we assume that δx_1, δp_{x1}, etc. are small enough so that the positions and momenta can be treated as continuous, this sum becomes an integral as before,

$$\phi(E) = \int_{E_r<E} \prod_{i=1}^{N} \left(\frac{d^3\mathbf{r}_i d^3\mathbf{p}_i}{h_0^3}\right). \tag{4.19}$$

11 This both won't make a difference for most quantities we are interested in, and also we shall see when studying quantum statistics in Chapter 11 that this is the correct choice.

Besides the sheer number of integrations needed here (6N of them!), the real difficulty arises from the limits of integration. In the case $N = 1$, we have

$$E_r = \frac{p^2}{2m} + \frac{1}{2}m\omega^2 x^2, \tag{4.20}$$

so ϕ becomes

$$\phi(E) = \int_{E_r<E} \frac{dxdp}{h_0}, \tag{4.21}$$

and x and p are restricted such that

$$\frac{p^2}{2m} + \frac{1}{2}m\omega^2 x^2 \leq E \Rightarrow \frac{p^2}{2mE} + \frac{1}{2E}m\omega^2 x^2 \leq 1. \tag{4.22}$$

In this one-dimensional phase space, x and p form an ellipse with the semi-major and semi-minor axes $\sqrt{2mE}$ and $\sqrt{2E/(m\omega^2)}$.[12] The integral in Eq. (4.21) is just the area of the ellipse defined in Eq. (4.22), π times the product of the semi-major and semi-minor axes, so

$$\phi(E) = \frac{2\pi E}{h_0\omega}. \tag{4.23}$$

For this particular case, extending this calculation to higher dimensions is not too difficult (see Appendix D), although it gets trickier for more complicated systems of course.

4.5.3 From $\phi \to \Omega$

In Section 4.5.2 we saw how to calculate the number of states *less than* some energy E, but remember, for the microcanonical ensemble, we are interested in $\Omega(E)$, the number of states *in the range* E to $E + \delta E$. If we know $\phi(E)$, we can easily calculate this as

$$\Omega(E) = \phi(E + \delta E) - \phi(E). \tag{4.24}$$

As discussed previously, we only consider cases such that $\delta E \ll E$. And because both classical and quantum systems can be treated such that $\phi(E)$ is a continuous function of E, we can Taylor expand ϕ to get

$$\Omega(E) \approx \frac{\partial \phi(E)}{\partial E}\delta E + \cdots. \tag{4.25}$$

For macroscopic systems, we might consider $\delta E \sim 10^{-6}$ J, while $E \sim 10^5$ J, so the condition where $\delta E \ll E$ is valid, so we can drop the approximation and treat Eq. (4.25) as an exact equality.

Example 4.10 Consider the two-dimensional simple harmonic oscillator from Exercise 4.3, where (for large energy) $\phi(E) = \frac{1}{2}N_{\max}^2$, with $N_{\max} = E/\hbar\omega$, so $\phi(E) = \frac{1}{2}\frac{E^2}{(\hbar\omega)^2}$. Differentiating we see that for this case

$$\Omega(E) = \frac{E}{(\hbar\omega)^2}\delta E.$$

Exercise 4.4 Using Eq. (4.23), show that $\Omega = 2\pi\delta E/(h_0\omega)$ for the one-dimensional classical oscillator.

12 I won't specify which is the semi-major or semi-minor axis; as these have different dimensions, we cannot compare their sizes to each other. We could correct this by considering dimensionally equivalent quantities $p/\sqrt{2m}$ and $\sqrt{m\omega^2/2}x$, but that's not really essential for our purposes.

One important point to make after these examples is that we can see it is often simpler to calculate ϕ instead of Ω. Thus, by considering our system's energy to be within a small range instead of a precise value, we allow ourselves an easier method to determine the number of states by simple differentiation. This is the second reason I alluded to earlier as to why we are considering our system to be allowed a range of energies, even if it is not required by systems with discrete energy levels.

As another quick aside before continuing: You will sometimes see in these problems the **density of states**, which is defined as

$$g(E) \equiv \frac{\partial \phi(E)}{\partial E} = \text{number of states per unit energy.} \tag{4.26}$$

As this does not depend upon the arbitrary δE, it can be useful to consider for many systems. I won't generally use this for our problems; however, you will likely come across it in other sources.

Our expression for Ω makes sense from some simple reasoning: For one, if we increase the energy range δE, then the number of accessible states should increase. While δE is an arbitrary value, we shall see that it will not affect (most) physical observables. Also, we expect that as the energy we consider increases, so should the number of states, as we saw in the two-dimensional oscillator example above. This happens because the number of ways to get a larger energy usually increases as the energy increases.[13]

Recall in Section 4.2 we defined the number of **degrees of freedom** f for a general system. As applied here, this corresponds to the number of variables that can change in order to change the energy (or more specifically, the state of the system). For a system of N quantum simple harmonic oscillators, there are as many degrees of freedom as the number of dimensions times the number of particles, or $f = N$, $2N$, or $3N$ for one, two, or three dimensions, respectively. These degrees of freedom correspond to the integers n_x, n_y, and n_z that can vary to change the energy. For a classical system of N particles, there are $2N$ degrees of freedom for each dimension (because we have a position and a momentum for each dimension and particle), so $f = 6N$ for a three-dimensional classical system.

In either case, f is of order N, which is generally very large for our systems. We will often refer to f instead of N, as a system with N particles in one dimension has the same number of degrees of freedom as a system with $N/2$ particles in two dimensions, and so on. Thus, f is a useful variable to consider systems more generally.

We can estimate the dependence Ω should have on energy, at least roughly for general systems. Suppose our system has f degrees of freedom, and for this estimate we will assume each degree of freedom is independent. On average, each degree of freedom has some energy ε, and let's focus on one of these for now. For this single degree of freedom, if each energy level is spaced on average $\Delta\varepsilon$ apart, then the number of states with energy less than ε can be approximately given by

$$\phi(\varepsilon) \sim \frac{\varepsilon}{\Delta\varepsilon},$$

as shown in Figure 4.8. This is precisely true for the quantum simple harmonic oscillator (see Eq. (4.15)), but usually the spacing between energies is not exactly constant. Often, the spacing between energies will depend on the energy itself, but for now we'll ignore this for a simple

13 If you're thinking carefully while reading this, you might have found (at least) two cases where this is not true. First, this is not the case in the one-dimensional example, which is fine for our work here, because we will never study a single particle in one dimension. Second, this is not true for a finite number of magnetic dipoles in a magnetic field as in Table 4.1. In this case, we see that after a certain point, Ω will *decrease* with increasing energy. Keep that exception in the back of your mind, as we will return to a discussion of this later.

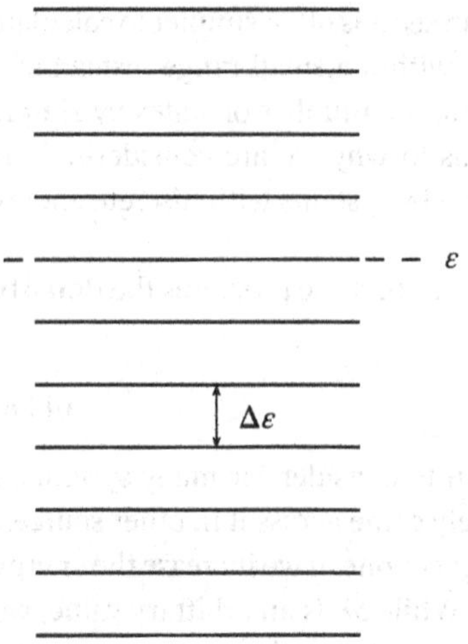

Figure 4.8 A schematic of equally spaced energy levels for a single particle in a general system.

estimate. As we increase ε, ϕ will generally increase, and while for the quantum simple harmonic oscillator or Figure 4.8 it would do so linearly, we could be a bit more general by stating

$$\phi(\varepsilon) \sim \left(\frac{\varepsilon}{\Delta\varepsilon}\right)^{a}, \quad a > 0. \tag{4.27}$$

The power a is likely of order 1, but we won't require this to be the case.

For each of our f degrees of freedom, let's suppose the number of states they could occupy corresponds to Eq. (4.27). Additionally we will assume the interactions among these degrees of freedom are weak.[14] The entire system has total energy E, which for this rough estimate, we will assume to be equally distributed over the degrees of freedom, so we can write

$$\varepsilon \approx \frac{E}{f}.$$

If each degree of freedom has $\phi(\varepsilon)$ states with energy less than ε, then the total number of states in the system with energy less than E can be found by multiplying ϕ for each degree of freedom together,[15]

$$\Phi(E) \sim \phi(\varepsilon)^{f} = \phi(E/f)^{f},$$

or

$$\Phi(E) \propto \varepsilon^{af} \propto E^{af},$$

where we are dropping any multiplicative constant as we are only interested in the energy dependence. Using Eq. (4.25), we have[16]

$$\Omega(E) \propto afE^{af-1}\delta E = (\text{const})E^{af}\left(\frac{\delta E}{E}\right). \tag{4.28}$$

14 This will be a common theme here: The interactions are weak but they must be nonzero; otherwise, the system is really a set of f completely independent systems and the system could never actually approach equilibrium unless it was already in equilibrium.

15 Here we are using lower case ϕ for a single degree of freedom, and Φ for the entire system. This won't always be the case, so beware of the context!

16 We can also (and will do this later) use the fact that $f \gg 1$ so we can assume $af - 1 \approx af$ to simplify such expressions.

With this estimate of the number of states, assuming they are equally likely, we can calculate the entropy,

$$S = k_B \ln \Omega(E) = k_B \left[af \ln E + \ln \left(\frac{\delta E}{E} \right) + \text{const} \right]. \tag{4.29}$$

For a macroscopic system the first term will be $\sim 10^{24}$ while the other two terms will generally be much smaller than that (order one, or perhaps 10 or 100, but much smaller than 10^{24}), so they are negligible. Additionally, they will not be relevant for most physical quantities, as we shall see those tend to depend on derivatives of the entropy. This is another reason why the precise choice of δE is irrelevant for our calculations.

As is clear from this estimate, $\Omega(E)$ grows extremely rapidly, especially for $f \sim 10^{24}$ degrees of freedom. Consider a system with $f \sim 10^{24}$ and $E \sim 3000$ J, then $\Omega \sim (3 \times 10^3)^{3\times 10^{24}} \sim 10^{10^{25}}$. The entropy is a much more reasonable number to consider because of the logarithm ($S \sim 2 \times 10^{25}$ is still very large, but won't make your calculator have an overflow error). Keep in mind that Eqs. (4.28) and (4.29) are not true for all systems in general, but rather they are meant to show our expectation that $\Omega(E)$ will be a rapidly increasing function of E for systems we're interested in here.

Example 4.11 Consider our system just discussed with $a = 1, f = 10^{24}$, and for a macroscopic energy $E = 10^7$ J. We find that our choice for δE can matter quite a bit, if we were to choose $\delta E = 10^{-10}$ J. This seems very small because of course $\delta E/E = 10^{-17}$, but if we compare Ω and ϕ, we find

$$\frac{\Omega}{\phi} = 10^7,$$

that is, $\Omega > \phi$! How can this be, given that from our definition in Eq. (4.24), Ω should be less than ϕ? Clearly the value of δE we chose is not small enough, and as such our truncation of the expansion of ϕ that led to Eq. (4.25) was not valid. To determine what value of δE is "small enough," we consider

$$\frac{\Omega}{\phi} = af \frac{\delta E}{E},$$

and require

$$\frac{\delta E}{E} \ll \frac{1}{af}.$$

For a system with $af = 10^{24}$, then we would want to choose $\delta E < 10^{-24} E$ so that Ω is truly smaller than ϕ.

The origin of this bizarre result is that most of the states that contribute to ϕ reside in the region of width δE around E, which is discussed quite nicely (with the origin of this issue) in Section 15.5 of Ref. [9]. You can explore this in more detail in Problem 4.9. We will assume that any δE we choose will be small enough such that our expansion above is valid.

4.5.4 Classical Ideal Gas

The most well-known and simplest example of a thermodynamic system is the **classical ideal gas**. We have a gas of N molecules with mass m confined to region of volume V that we describe classically. The energy of the system in state r is

$$E_r = K + U + E_{\text{int}}, \tag{4.30}$$

where as usual, r is our label for a classical microstate and includes the positions $\mathbf{r}_i$ and momenta $\mathbf{p}_i$ of the N molecules. The kinetic energy of the system is given by

$$K = \frac{1}{2m}\sum_{i=1}^{N}\mathbf{p}_i^2, \tag{4.31}$$

and we assume the potential energy to be independent of the momenta, so

$$U = U(\mathbf{r}_1, \mathbf{r}_2, \ldots, \mathbf{r}_N). \tag{4.32}$$

E_{int} is the **intramolecular energy**, which is nonzero only for **polyatomic molecules** (molecules composed of more than one atom). This is generally a function of internal positions R_j and momenta P_j ($j = 1, \ldots, M$, where M is the number of relevant internal coordinates) of the atoms that compose the molecule. These positions include, for example, relative distances between atoms and angles relative to various axes connecting atoms, and the momenta are those defined in terms of these variables (see Section 11.7 where we consider the explicit case of the diatomic molecule). The energy in Eq. (4.30) is valid for most classical systems and is quite general. However, we cannot make any headway without knowing the potential energy or intramolecular energy, and so we will make a significant simplification to move forward with it.

It would be simplest to consider the case $U = 0$; however, this would be completely uninteresting, because it corresponds to a non-interacting system. If we have a non-interacting system that initially is not in equilibrium, it would never be allowed to reach equilibrium (because the molecules wouldn't interact to be able to transfer energy to each other to make this happen), and thus we could not use the techniques here to learn anything about the system. Instead, we will assume the molecules interact very weakly, so that $U \ll K$ and E_{int} for any r, so that we can consider it to be zero to a good approximation.[17]

The expression for the number of states with energy less than E is given by

$$\phi(E) = \frac{1}{h_0^{3N+M}}\int_{E_f<E} d^3\mathbf{r}_1\cdots d^3\mathbf{r}_N d^3\mathbf{p}_1\cdots d^3\mathbf{p}_N dR_1\cdots dR_M dP_1\cdots dP_M, \tag{4.33}$$

where we included h_0, the phase space area for each pair of position–momentum degrees of freedom. This is a $(6N + 2M)$-dimensional integral, which seems completely intractable, especially given the constraint that we require $E_r < E$! First though, if we can neglect the interactions between molecules, U, then there is no position dependence in E_r, so the $\mathbf{r}_i$ integrals are actually quite simple (the internal energy only depends upon the R_i's and P_i's). The molecules can be anywhere in the region (of volume V), so we have

$$\int d^3\mathbf{r}_i = V,\ i = 1, \ldots, N, \tag{4.34}$$

for each of our N molecules.[18] This immediately simplifies our expression for $\phi(E)$ to

$$\phi(E) = \frac{V^N}{h_0^{3N+M}}\int_{E_f<E} d^3\mathbf{p}_1\cdots d^3\mathbf{p}_N dR_1\cdots dR_M dP_1\cdots dP_M, = V^N f(E), \tag{4.35}$$

so ϕ has two factors: one that only depends upon the volume and another that only depends on the energy E.

17 This seems like a purely academic consideration, and like we're now splitting hairs. However, we always want to be sure to be as precise as possible so that we know every approximation that is made. This way if there are ever deviations from experimental observations, we need to know precisely where we started to then modify our assumptions.

18 In class, I often like to say, "Look, we just performed $3N$-integrals instantly!" My students are not impressed.

If we simplify to the case of the **monatomic ideal gas**, there are no internal degrees of freedom, and we can drop the R_j's and P_j's (along with the M factors of h_0 that came with them), to obtain

$$\phi(E) = \frac{V^N}{h_0^{3N}} \int_{E_f<E} d^3\mathbf{p}_1 \cdots d^3\mathbf{p}_N. \tag{4.36}$$

Despite appearances, this is actually simpler to evaluate than you might think. First note that if $E_r < E$, then we have

$$\sum_{i=1}^{N} \mathbf{p}_i^2 = \sum_{j=1}^{f} p_j^2 < 2mE, \tag{4.37}$$

where $f = 3N$, and we use a different label j just to make it clear that because we are integrating over all of them, there is not really any difference between the different momentum components.

This integral is merely the volume of a $3N$-dimensional sphere of radius $\sqrt{2mE}$ in momentum space and is calculated in Appendix D. We can determine the result of this integral exactly using the results in that appendix, but we really only need Eq. (D.10), which when applied here is $b(2mE)^{3N/2}$. b is an irrelevant constant, and we get

$$\phi(E) = AV^N E^{3N/2}. \tag{4.38}$$

Here, A includes all factors (such as b, m, and h_0) that do not depend upon the energy or volume. As we will see, this multiplicative constant will have no effect on (most) physical observables, so leaving it in this form will be enough for our purposes.

From this we can determine the number of states with energy between E and $E + \delta E$ from Eq. (4.25),

$$\Omega = A\left(\frac{3N}{2}\right) V^N E^{3N/2-1}\delta E = BV^N E^{3N/2}, \tag{4.39}$$

where $B = A(3N/2)\delta E$ is independent of the energy, and with $N \gg 1$, we can approximate $3N/2 - 1 \approx 3N/2$. Then the entropy for the classical monatomic ideal gas is

$$S = k_B \ln \Omega = Nk_B\left(\ln V + \frac{3}{2}\ln E\right) + k_B \ln B. \tag{4.40}$$

The last term is an irrelevant constant. The constant B can be determined exactly (see Exercise 4.5), but for now this is a good example of calculating the number of states. We see that Ω is proportional to E^{af}, with f related to the number of degrees of freedom (here $f = 3N$), and thus $a = 1/2$ from Eq. (4.28), so it scales as we generally expect.

Exercise 4.5 Use the results of Appendix D to determine an exact form of S for the classical monatomic ideal gas.

As a quick aside, we can also calculate $\Omega(E)$ for a general classical ideal gas, from Eq. (4.35), and we obtain

$$\Omega = V^N f'(E)\delta E = V^N \chi(E). \tag{4.41}$$

The most important feature here again is that the volume and energy dependence factorizes; that Ω is the product of a function only of V and one only of E, which we define as $\chi(E)$.

In the problems at the end of the chapter, we work through several other examples, including different ways to determine Ω instead of always starting from ϕ, as there are some cases where there are simpler approaches. Now that we have the way to calculate the number of states, we will next discuss how systems interact so that we can see how to calculate physical quantities from Ω (or equivalently S).

4.6 Interactions Between Systems

Now that we have figured out how to describe a single system, the next step is to determine how different systems interact. There are two primary interactions we will consider here, **thermal interactions** and **mechanical interactions**. A third type, **diffusive** or **chemical interactions**, will be discussed in Chapter 10.

4.6.1 Thermal Interaction

In addition to the total energy E, a system can be described by a set of **external parameters**—those parameters that we can control which change the state of the system. For example, the volume V of a system, an external magnetic field H, and an electric field $\mathcal{E}$. As the most common external parameter we will consider is volume, I will use V as the only relevant parameter. Generalizing to systems with other external parameters, or with more than one parameter, is straightforward and will be the topic of some of the problems. For us, we can write the energy of a state generally as a function of this external parameter, or

$$E_r = E_r(V),$$

as it may often depend upon the external parameter (such as the quantum mechanical particle in a box in Example 4.8, where the box has a volume $V = L^3$). Generally the energies of our system can fluctuate, so given the probabilistic nature of the system, we will be interested in the mean energy, $\overline{E} = \sum_r P_r E_r$, which averages over all accessible microstates (in the microcanonical ensemble, $\overline{E} = E$, but we'll keep the bar over the E for now to remind us that the mean energy is what is significant when we consider other ensembles in the future).

Let's put two systems, labeled A and A', in contact with each other to form a combined system $A^{(0)} = A + A'$. The energies of each system are $\overline{E}$ and $\overline{E}'$, respectively, and we assume this combined system is completely isolated from the outside world, but A and A' can interact while the external parameters remain fixed. In this case, as shown in Figure 4.9, the partition which divides the two systems cannot move but energy can flow between the two systems, which is known as a **thermal interaction**:

> A purely thermal interaction is one in which energy flows between two systems while *all* of their external parameters remain unchanged.

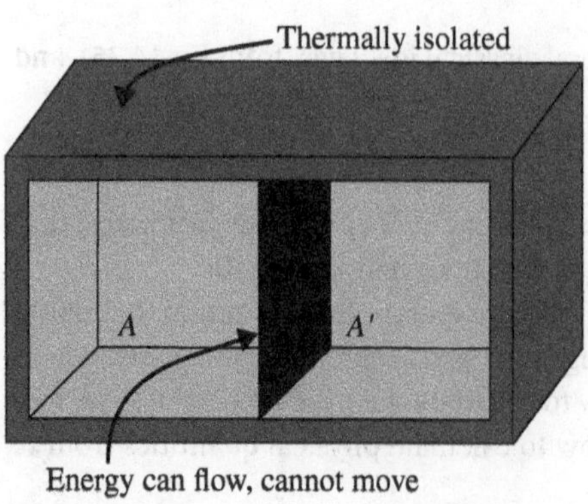

Figure 4.9 Two systems, A and A', in thermal contact with each other. The partition between the two cannot move but energy can flow between the systems.

For example, when you pour a hot cup of coffee and put it on the kitchen counter without a lid, after a short time, the coffee will get colder. Energy has left the coffee, but the volume of the coffee is unchanged. This is a purely thermal interaction (well, *approximately* a purely thermal interaction, as some of the coffee evaporates so it actually loses molecules and its volume decreases).

The energy absorbed by a system (leading to an energy *addition*) is called the **heat** added to the system, Q. The total energy $\overline{E}^{(0)}$ of the combined system will remain unchanged because it is isolated, so that

$$\Delta\overline{E}^{(0)} = \Delta\overline{E} + \Delta\overline{E}' = Q + Q' = 0. \tag{4.42}$$

Thus we have

$$Q' = -Q,$$

or, the heat *emitted* by A' is *absorbed* by A. The convention for Q is such that if $Q > 0$, energy is *added* to the system A and if $Q < 0$, energy leaves the system (and similarly for A').

Example 4.12 Let's revisit Example 4.6. Volume is not a relevant parameter here, as the spins are considered to be fixed in place and only their orientations (up or down) relative to the external field matter. In this case, the external field is a relevant external parameter, but suppose it remains fixed, and we find that the energy of the system has increased from $-\mu H \rightarrow 3\mu H$. Energy in the form of heat has transferred to this system from some outside source, specifically an amount $Q = 4\mu H$.

When two systems interact, we say they undergo a **thermodynamic process**: The system will transition from one (equilibrium) state to another. We will start classifying processes that our system undergoes in Section 4.6.2, according to which parameters may (or may not) change. In this section we focused only on a process which allows the energy of the system to change while the external parameters are fixed, and in Section 4.6.2 we will consider processes where the energy changes when the external parameters change.

4.6.2 Mechanical Interaction

Now suppose we have two systems in contact but are **thermally isolated** from each other so that heat cannot flow between the two. Such a process is known as an **adiabatic process**. As long as the external parameters can change, the energies of the systems can change, and this type of interaction is a purely **mechanical interaction**.

Imagine a setup as shown in Figure 4.10, where a gas is in a thermally isolating container with a lid of mass M that is allowed to move, and it is such that if we release it, the lid will fall a distance h, compressing the gas. The lid has a weight Mg, with g the acceleration due to gravity, and it does an amount of **work** on the gas given by $W_{\text{on}\,A} = Mgh$; as such the energy of the system will change. As we want to focus on variables that only depend upon our system of interest, then we are interested in the work done *by* the system A. The work done *by* the system is denoted as W (without any subscript) and is given by

$$W = -W_{\text{on}\,A}.$$

When we refer to work, we always refer the work that is *done by the system*. **Be careful with the minus signs!** If $W > 0$ then the energy of the system is *decreasing* because it is doing work on another system.

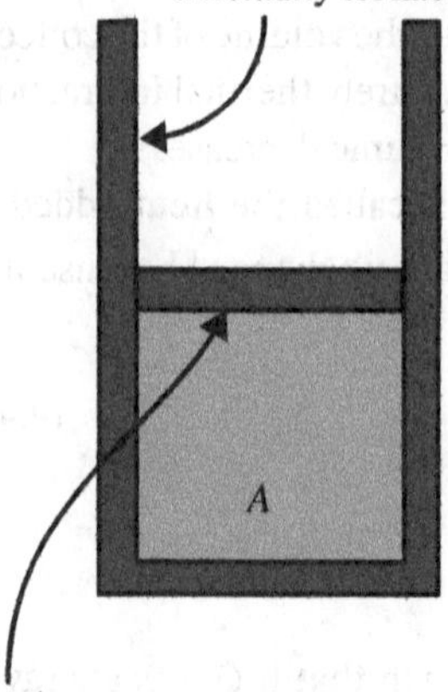

Figure 4.10 A system with a lid which is allowed to move, and under the influence of gravity perhaps it can fall, changing the volume of the system.

Again, since the combined system is isolated (in Figure 4.10 the system A' is implied to be surrounding the system A), the total energy is conserved so that

$$\Delta\overline{E}^{(0)} = \Delta\overline{E} + \Delta\overline{E}' = -W - W' = 0, \tag{4.43}$$

or

$$W' = -W.$$

Thus the work done on system A must be equal to the work done by system A' (and vice versa). In our example above, W' is the work done by the lid as it falls and is given in that case by Mgh.

Example 4.13 Again we revisit Example 4.6. The external field is our external parameter, and if we increase it, we can change the energy of the system while it is thermally isolated, thus doing work on the system.

In general, two systems can interact both thermally and mechanically. The total energy change of our system A is then

$$\Delta\overline{E} \equiv Q - W. \tag{4.44}$$

This expression can be used to define precisely what heat is, as generally it is a difficult concept to quantify (or measure directly). The heat flow into a system is the change of the internal energy that is *not* from a mechanical interaction:

$$Q \equiv \Delta\overline{E} + W. \tag{4.45}$$

The quantities Q, W, and $\Delta\overline{E}$ are all *finite* quantities. What if the system does only an infinitesimal amount of work $đW$, or has an infinitesimal amount of heat $đQ$ flowing into it, such that in either case, there is an infinitesimal change in the energy $d\overline{E}$? We can write Eq. (4.44) as

$$d\overline{E} = đQ - đW. \tag{4.46}$$

We use the symbol $đ$ for infinitesimals that are not changes in those quantities, but just for infinitesimal amounts. While $d\overline{E}$ corresponds to an infinitesimal change in the energy, $đQ$ is an infinitesimal amount of heat added to the system and $đW$ is an infinitesimal amount of work done by the system. We are never *changing* work or heat, but we are changing the energy. This difference arises because the energy is a property of our system depending on its state, while heat and work are not: They correspond to ways in which the energy can change and are not properties of the system.[19] This notation will remind us of the difference between these types of quantities, and we will discuss them in more detail in Section 4.7.1.

The only problem with talking about these different sorts of interactions is that if they undergo such interactions, then clearly our systems are not in equilibrium, which is what we have been requiring this entire time! We need a way to allow our systems to change, and it needs to be consistent with our requirement for systems remaining in equilibrium, which we discuss in Section 4.7.

4.7 Quasistatic Processes

Suppose we have a system in equilibrium, so we can use our techniques to study its properties.[20] If we allow it to interact with another system (say we start to push down on the lid covering the system in Figure 4.10), then it will no longer be in the same equilibrium state as it started in.

In most cases, we will consider any process that allows a system to change from one equilibrium state to another equilibrium state to be **quasistatic**. Such processes occur slowly enough (on whatever time scale is relevant to the problem) that between each change in the system, there is enough time for the system to return to a new equilibrium state. In other words, we can assume that at every moment during the change, the system is always in equilibrium even though it is changing from one equilibrium state to another.

A nice analogy is as follows. Suppose you're balancing on a (large enough) ball, so you're in equilibrium at a particular location, but you want to move to the other side of the room. You will use your feet to cause the ball to roll, but each time you move a bit you pause so that you remain balanced. The changes are slow, and you wait until you return to equilibrium before you move forward a bit more [10].

For the case of pushing down on the lid in Figure 4.10, you would do so by first pushing the lid a distance dy in some time dt that is much larger than the time it would take the gas to return to an equilibrium state. Once that happens, you push it down another distance dy in a time dt until eventually you have moved the lid down a total distance h. Of course this is merely an approximation, as processes in Nature are not quasistatic, but again, many can be treated as such to an excellent approximation. (And as usual, if we make a prediction with this assumption and it no longer matches our empirical observation, then we will revisit this.)

Let's first consider the thermally isolated system in Figure 4.11. You push on the lid while it's in the initial position (on the left), and you push slowly enough so we can consider this a quasistatic mechanical interaction. The lid has a cross-sectional area A, you push with a constant force F, and the lid moves down a distance dy. The small amount of work you do is given by

$$đW_{\text{on}\,A} = Fdy = \left(\frac{F}{A}\right)(Ady) = -pdV, \tag{4.47}$$

so the work done is given by the **pressure** p times the change in volume dV. The negative sign arises because we are considering the case where we are pushing *down*, which implies that if $dy > 0$, then the volume of the system decreases, or $dV < 0$.

The energy of an individual state of the system in general changes by a small amount

$$dE_r = \left(\frac{\partial E_r}{\partial V}\right) dV, \tag{4.48}$$

19 We wouldn't say a system has heat or work, but it *gains* heat or *does* work.

20 Okay, so we haven't actually gotten to calculating actual observables, but that will come soon!

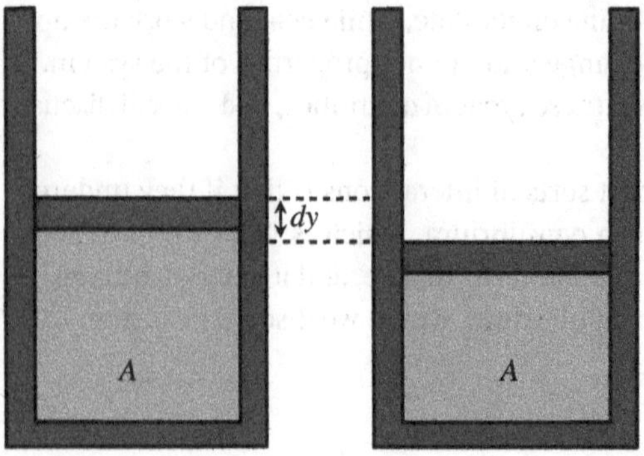

Figure 4.11 Our system above before (on the left) and after (on the right) you pushed the lid down a distance *dy*.

and the work done *by* the system is then defined by

$$đ W_r \equiv -dE_r = p_r dV, \tag{4.49}$$

where we have used Eq. (4.47) to relate the pressure of a state to the change in the energy,[21]

$$p_r \equiv -\frac{\partial E_r}{\partial V}. \tag{4.50}$$

For different external parameters, Eqs. (4.49) and (4.50) will differ in two ways: The external parameter (thus what E_r is being differentiated with respect to) and the physical interpretation of that derivative. In this case, when the external parameter to change is volume, then the pressure arises naturally in the equations above. In other cases it could be a force (when the external parameter is a length), or even the magnetization (if the external parameter is a magnetic field). However, the expression for the work done by the system will be of the same form. We will see examples of other types of systems in the problems.

Now let's consider the macroscopic (albeit infinitesimal) work done ***due to a quasistatic change*** on an ensemble of such systems, which will arise from the mean energy change,

$$đ W = \overline{đ W_r} = -\overline{\left(\frac{\partial E_r}{\partial V}\right)} dV = \overline{p} dV. \tag{4.51}$$

The mean pressure of the system comes from averaging over the individual microstates in our ensemble. For a finite amount of work, we merely integrate this expression from some initial state i to a final state f,

$$W = \int_{V_i}^{V_f} \overline{p} dV, \tag{4.52}$$

and this, as we'll discuss next, is dependent on the integration path. But for now we can see that if we can plot $\overline{p}$ as a function of V, then the total work done can be found as the area under the curve. Additionally, even though we looked at this with the specific example above, this holds for any system with any volume change.

4.7.1 Exact vs. Inexact Differentials

Earlier we used two different notations for infinitesimal quantities, such as $đ W$ and $d\overline{E}$. In this section I will clarify the difference between the two.

21 Note that Eq. (4.47) is the work done *on* the system, hence the minus signs in Eqs. (4.49) and (4.50).

Let's first look at a differential equation that can be written in the form

$$A(x,y)dx + B(x,y)dy = 0. \tag{4.53}$$

Such a differential equation is called **exact** if there exists a function $f(x,y)$ such that we can write this equation as

$$df = 0. \tag{4.54}$$

If this is possible, then one can show that

$$df = \left(\frac{\partial f}{\partial x}\right)_y dx + \left(\frac{\partial f}{\partial y}\right)_x dy = A(x,y)dx + B(x,y)dy, \tag{4.55}$$

where the notation

$$\left(\frac{\partial f}{\partial x}\right)_y$$

makes it explicit that when we differentiate f with respect to x, we hold y constant. Usually this is implied in partial differentiation, but in thermodynamics, we will see plenty of cases where being more precise is essential. If this is the case, then we will call df an **exact differential**.

What matters most for our purposes is that if we look at a finite difference of this function between two points (x_i, y_i) and (x_f, y_f),

$$\Delta f = f(x_f, y_f) - f(x_i, y_i) = \int_i^f \left[A(x,y)dx + B(x,y)dy\right], \tag{4.56}$$

it only depends on the endpoints, not on the particular path we integrate over to get from $(x_i, y_i) \rightarrow (x_f, y_f)$. This is familiar in classical physics when looking at conservative forces—work done by a conservative force is independent of the path, so we can define a potential energy U such that the force is minus the gradient of the potential energy. dU in this case would be called an exact differential. We call the function f a **state function** (commonly used in thermodynamics): It is a function that only depends upon the state our system is in.

Example 4.14 Suppose we are given the expression

$$(1 + y^2)dx + 2xydy. \tag{4.57}$$

Is there a function f that would give

$$\left(\frac{\partial f}{\partial x}\right)_y = 1 + y^2, \text{ and}$$

$$\left(\frac{\partial f}{\partial y}\right)_x = 2xy?$$

One (brute force) way to approach this is to just integrate the first of these with respect to x and the second with respect to y, to get

$$f(x,y) = x + xy^2 + h_1(y), \text{ and}$$

$$f(x,y) = xy^2 + h_2(x),$$

where the functions h_1 and h_2 are "constants" of integration (insofar as h_1 is independent of x and h_2 is independent of y) that are determined to make these two expressions for f the same. If we choose $h_1(y) = c$ where c is independent of x and y, and $h_2(x) = x + c$, then

$$f(x,y) = xy^2 + x + c \tag{4.58}$$

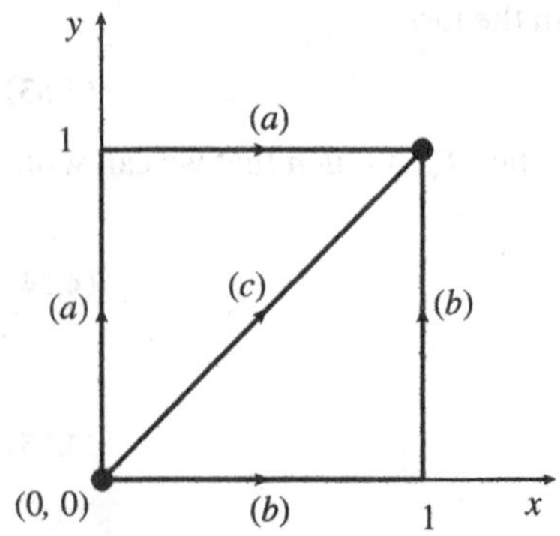

Figure 4.12 The three paths to perform the integral $\int_{0,0}^{1,1} df$ for Exercise 4.6.

is the function we were looking for. Because we can find this function, then Eq. (4.57) is an exact differential, df. We will discuss a simpler way to check if such a function exists shortly.

Exercise 4.6 Verify that the integral of $f(x, y)$ as defined by Eq. (4.58), $\int_{0,0}^{1,1} df$ is independent of the path for the following three paths (shown in Figure 4.12):

a. $(0, 0) \to (0, 1) \to (1, 1)$ along straight lines,
b. $(0, 0) \to (1, 0) \to (1, 1)$ along straight lines,
c. $(0, 0) \to (1, 1)$ such that $y = x$ along the path.

In each case you should find that it is equal to $\Delta f = f(1, 1) - f(0, 0) = 2$. Note that the constant c is irrelevant here.

Suppose now we have an inexact differential equation

$$C(x, y)dx + D(x, y)dy = 0, \tag{4.59}$$

that cannot be written as the differential of some function g. This might still represent something physical, so we define an **inexact differential** $đg \equiv Cdx + Ddy$. In this case, the integral from one point to another,

$$g = \int_i^f [C(x, y)dx + D(x, y)dy], \tag{4.60}$$

is dependent upon the integration path, just like the work done by a non-conservative force such as friction. Thus it cannot be written as $\Delta g = g(x_f, y_f) - g(x_i, y_i)$. $đg$ is a perfectly well-defined infinitesimal quantity, but it is not strictly speaking a differential[22] and g, unlike f, is not a state function.

In order to check if we have an exact or inexact differential, we can check the second derivatives of our function. If you have an exact differential, then you know that Eq. (4.55) holds, and for a continuous function with continuous first derivatives, then the order in which we take the mixed second derivatives shouldn't matter. That is,

$$\left(\frac{\partial}{\partial y}\left(\frac{\partial f}{\partial x}\right)_y\right)_x = \left(\frac{\partial}{\partial x}\left(\frac{\partial f}{\partial y}\right)_x\right)_y, \tag{4.61}$$

or

$$\left(\frac{\partial A}{\partial y}\right)_x = \left(\frac{\partial B}{\partial x}\right)_y. \tag{4.62}$$

22 If we were to be more precise, we would just refer to exact differentials as differentials and give another name to inexact differentials. This terminology is conventional in thermodynamics, so we'll stick with it.

If Eq. (4.62) does not hold, then we can often use the methods of ordinary differential equations to come up with an **integrating factor** to turn an inexact differential into an exact one. We can usually find a function $\mu(x,y)$ such that

$$df = \mu(x,y)\, đg = \mu(x,y)C(x,y)dx + \mu(x,y)D(x,y)dy \tag{4.63}$$

is an exact differential even if $đg$ is not.

Exercise 4.7 Show that df in Eq. (4.57) is an exact differential by calculating the mixed second derivatives.

Example 4.15 Suppose you are given the following differential,

$$đg = y^2dx + xydy. \tag{4.64}$$

We can easily evaluate the mixed partial derivatives,

$$\left(\frac{\partial(y^2)}{\partial y}\right)_x = 2y, \text{ and}$$
$$\left(\frac{\partial(xy)}{\partial x}\right)_y = y,$$

and see they are not equal.

Exercise 4.8 Show that if we multiply Eq. (4.64) by x, then $df = x\, đg$ (with $đg$ from the previous example) is in fact an exact differential. Additionally, show that $\int_{(0,0)}^{(1,1)} đg$ is different for the three paths shown in Figure 4.12, but $\int_{(0,0)}^{(1,1)} df$ is not. You should get for each path for $\int đg$: (a) 1, (b) 1/2, and (c) 2/3, while for any of them, $\int df = 1/2$.

4.7.2 Physical Examples

Everything in Section 4.7.1 was purely mathematical, so let's consider some physical examples of exact and inexact differentials.

A system in a certain macrostate r has some energy E_r, and an ensemble of similar systems has some well-defined mean energy $\overline{E} = \sum_r P_r E_r$.[23] Energy, like the mean pressure $\overline{p}$, is a state function: It only depends on the state that it is in, not how it got there. Thus, the change in the mean energy of a system is the integral of an exact differential,

$$\Delta\overline{E} = \overline{E}_f - \overline{E}_i = \int_i^f d\overline{E}. \tag{4.65}$$

How the state moved from the initial to the final state is irrelevant, we just need to measure the energy at each point and take the difference. The different ways the energy of the system can change from $\overline{E}_i$ to $\overline{E}_f$ is a different story.

A system can do an infinitesimal amount of work $đW$ (via a mechanical interaction) or absorb an infinitesimal amount of heat $đQ$ (via a thermal interaction), so the infinitesimal energy change is

$$d\overline{E} = đQ - đW. \tag{4.66}$$

23 This comment is confusing, because in the microcanonical ensemble, we are essentially fixing the energy to a value E, so this averaging is not needed. However, when we extend to different ensembles later, we will need to keep this in mind.

$đW$ and $đQ$ are inexact differentials, because they correspond to the work done and heat flow as a system goes from the initial to final state. A small energy change $d\overline{E}$ can arise from either only work done or only heat absorbed, or any combination of each.

We can convert these inexact differentials into exact ones if we can find an appropriate integrating factor, which we have already found for work. If volume is the relevant external parameter, then $đW = \overline{p}dV$. Volume is clearly a state function, since it is something we can measure for a system and it only depends on the size of the system. As such, dV must be an exact differential, and $1/\overline{p}$ is the integrating factor so that $(1/\overline{p})đW$ an exact differential. Later we will determine the integrating factor that we can multiply by $đQ$ to obtain an exact differential.

The total work done as we go from $\overline{E}_i \to \overline{E}_f$ is

$$W = \int_i^f đW, \tag{4.67}$$

and the total heat absorbed is[24]

$$Q = \int_i^f đQ, \tag{4.68}$$

so that for a finite amount of heat absorbed and work done,

$$\Delta\overline{E} = Q - W. \tag{4.69}$$

The different processes we can take to go from the initial energy to the final energy will dictate what Q and W will be. For example, the system could undergo an adiabatic process, so that $Q = 0$, and thus $\Delta\overline{E} = -W$. Alternatively, the system could undergo a process where the external parameters are fixed, such that $W = 0$ and thus $\Delta\overline{E} = Q$. If the initial and final states are the same, then $\Delta\overline{E}$ is the same for either process but Q and W would be different for each. An important thing to note here is that in the future we will often find it easy to calculate the work done by a system as well as the energy change of the system, but not the heat flow. Luckily with Eq. (4.69) we can find Q with W and ΔE.

Equation. (4.69), and the infinitesimal form in Eq. (4.66), is an important enough equation that we give it a name:

The First Law of Thermodynamics

The energy of a thermally isolated system is constant. If the system is not isolated, then the energy changes by an amount given by the heat absorbed minus the work done by the system. For infinitesimal quantities we write this as

$$d\overline{E} = đQ - đW,$$

and for finite quantities,

$$\Delta\overline{E} = Q - W.$$

The first law is really nothing more than a statement of the law of the conservation of energy as applied to thermodynamics, but this law is one of the most important principles in physics.

24 Note you might see in some textbooks ΔW and ΔQ written for the work done or heat absorbed, but I will avoid this. This creates some confusion as it implies that work and heat are quantities which can change, but they are what *cause* something else (the energy) to change.

Exercise 4.9 We'll see later that the mean pressure of a thermally isolated amount of a classical ideal gas varies with the volume according to

$$\bar{p}V^{\gamma} = K, \tag{4.70}$$

where γ and K are positive constants, and specifically $\gamma > 1$. What is the work done in this quasistatic (and adiabatic) process only in terms of the initial and final pressures (p_i and p_f) and volumes (V_i and V_f)? You should find

$$W = \frac{p_i V_i - p_f V_f}{\gamma - 1}. \tag{4.71}$$

With the setup of this chapter, we are almost ready to tackle most problems in statistical mechanics. We need to first discuss equilibrium in some detail and introduce (finally!) the concept of temperature, which is the topic of Chapter 5.

4.8 Summary

- Once we have identified our system, then we can enumerate the microstates in the system by either the relevant positions and momenta (for a classical system), or a set of quantum numbers (for a quantum system).
- For most systems of interest here, the macrostates we are interested in are those with a specific energy for now, with the microcanonical ensemble. In this ensemble, we restrict our systems such that $E \leq E_r \leq E + \delta E$ with $\delta E \ll E$. From there we can determine the number of states with energy in this range, $\Omega(E)$. Determining $\Omega(E)$ can be difficult but is an essential part of setting up such problems in statistical mechanics.
- The fundamental postulate of statistical mechanics states that all accessible microstates for a system in equilibrium are equally likely, which essentially takes the idea that we would like to maximize the missing information in the system, or now as we call it, the entropy, $S = k_B \ln \Omega$.
- Energy can be exchanged between two systems through the flow of heat or by work being done by the systems.
- If no work is done by a system during an interaction, it is a purely thermal interaction. If no heat flows into the system, it is an adiabatic process.
- As we are best set up to understand systems in equilibrium, most processes we can work with here will be quasistatic processes; those that occur on such a slow timescale that we can assume that the systems are in thermal equilibrium at every stage.

Problems

4.1 Consider the systems discussed that correspond to Figures 4.3 and 4.4. Recall the (high entropy) system in Figure 4.4 was constructed allowing the 100 particles to be anywhere in the grid (without sharing a space), while the (low entropy) system in Figure 4.3 restricted them to be separated by at least one space in between each other.

(a) For the high entropy case, where each of the 100 particles could be anywhere in the 40×40 grid, show that there are $\sim 10^{319}$ possible states.

(b) For the low entropy case, where each of the 100 particles could be anywhere in the 40×40 grid *except* orthogonally adjacent to any other particle, show that there are $\sim 10^{314}$ possible states.

Hint: Start by "choosing" any of the 100 particles, and put it in a box; there are 1600 options. Then, for one of the remaining 99 particles, you can place them in any one of the remaining boxes (1599 options for part (a), or fewer options for part (b)).

4.2 Return to the two-dimensional quantum oscillator of Example 4.3. Show that the exact number of states with $E_r < E = (N_{\max} + 1)\hbar\omega$ is given by $(N_{\max}^2 + N_{\max})/2$.
Hint: Consider the plots in Figure 4.5—and relate them to symmetric $N_{\max} \times N_{\max}$ matrices. How many independent elements are there in such a matrix?

4.3 Consider a three-dimensional classical simple harmonic oscillator with mass m and natural frequency ω, where the energy of a state is given by

$$\varepsilon_r = \frac{p_x^2 + p_y^2 + p_z^2}{2m} + \frac{1}{2}m\omega^2(x^2 + y^2 + z^2).$$

(a) Convert the position variables to a system of units such that

$$x' = \frac{\omega x}{\sqrt{m}},\ y' = \frac{\omega y}{\sqrt{m}},\ z' = \frac{\omega z}{\sqrt{m}},$$

and write ε_r in terms of these variables. Use this expression and the method of Section 4.5.4 to calculate $\phi(\varepsilon)$, the number of states with energy less than ε. You can use $\int d^n\mathbf{r} = br^n$ for an integral over an n-dimensional "sphere," and you can keep your answer in terms of b.

(b) With the same units, now calculate $\Phi(E)$, the number of states with energy less than E for a system of N such molecules.

(c) With $\delta E \ll E$, calculate the number of states $\Omega(E)$ between E and $E + \delta E$.

4.4 (a) Show that the number of states $\phi(\varepsilon)$ of energy less than ε for a quantum mechanical particle of mass m in a cubical box of side L is

$$\phi(\varepsilon) = \frac{\pi}{6}\left(\frac{L}{\pi\hbar}\right)^3 (2m\varepsilon)^{3/2}.$$

Use the energy levels

$$E_{n_x,n_y,n_z} = \frac{\hbar^2\pi^2}{2mL^2}\left(n_x^2 + n_y^2 + n_z^2\right).$$

(b) Calculate $\Omega(\varepsilon)$, the number of states between ε and $\varepsilon + \delta\varepsilon$, for such a particle.

(c) An argon atom at room temperature has a typical energy of 6×10^{-21} J. Calculate $\phi(\varepsilon)$ for such a molecule in a box 0.1 m on a side.

(d) Choose $\delta\varepsilon = 10^{-31}$ J (safely smaller than ε), and show that about 10^{19} states lie between ε and $\varepsilon + \delta\varepsilon$.

(e) Make a *very rough* numerical estimate of $\Omega(E)$ for a liter of argon gas at room temperature if $E = N\varepsilon$ (N is the number of atoms in the liter), and $\delta E = 10^{-31}$ J (safely less than E considering the constraint on large systems in Example 4.11). Perhaps this will convince you that $\ln\Omega$ is a more tractable quantity than Ω.

Hint: First estimate $\Phi(E)$ for the gas by assuming that each atom has any energy up to ε and can be in any one of the states found in part (c).

4.5 (a) Repeat parts (a) and (b) of Problem 4.4 for an ultra-relativistic gas, where the energy levels are

$$E_{n_x,n_y,n_z} = \frac{\pi \hbar c}{L}\sqrt{n_x^2 + n_y^2 + n_z^2},$$

where c is the speed of light.
Hint: You shouldn't have to recalculate the integral.

(b) Show that $\ln \Omega(\varepsilon)$ has the form

$$\ln \Omega(\varepsilon) = \ln V + 2\ln \varepsilon + \text{const.},$$

where V is the volume of the box, and determine the constant in terms of $\hbar$, c, $\delta\varepsilon$, and pure numbers.

4.6 Suppose you have an isolated system with N spin-1/2 molecules, with $N \gg 1$. Each molecule has a magnetic moment μ which can point parallel or anti-parallel to an applied magnetic field H, and each molecule will have energy $U = \pm\mu H$, where the minus (plus) sign is for a spin (anti-)parallel to H.

(a) If you have a state such that n_1 spins are aligned parallel to H and n_2 spins are aligned anti-parallel to H (where $N = n_1 + n_2$), what is the total energy of the system in this state?

(b) How many such microstates will have the same energy as you calculated in part (a)?
Hint: Recall the discussion of combinatorics on page 9.

(c) Treat this system using the microcanonical ensemble, where we restrict the energy to be between E and $E + \delta E$ ($\delta E \ll E$), but we also assume $\delta E \gg \mu H$. What is the total number of states $\Omega(E)$ lying in this energy range in terms of N and E (not n_1 and n_2)?
Hint: Assume that for this energy range, every energy has the same number of microstates.

(d) Write down an expression for $\ln \Omega(E)$ as a function of E. Simplify this expression by applying Stirling's formula in its simplest form.

4.7 Suppose you have a system of N quantum harmonic oscillators, and the system has some total energy $E = \frac{1}{2}N\hbar\omega + M\hbar\omega$.

(a) Show that the number of states with exactly this energy equals the number of ways of distributing M identical objects among N compartments and that this number is

$$\frac{(M+N-1)!}{M!(N-1)!}.$$

Hint: Consider the number of distinct arrangements of a set of M objects and $N-1$ partitions.

(b) Assuming $\hbar\omega \ll \delta E \ll E$, write down an expression for $\Omega(E)$, the number of states with energy between E and $E + \delta E$.
Hint: Assume that the number calculated in part (a) doesn't change appreciably in the small range between E and $E + \delta E$.

(c) Use Stirling's formula to show that when $N \gg 1$ and $M \gg 1$,

$$\ln \Omega \sim (M+N)\ln(M+N) - M\ln M - N\ln N + \ln(\delta E/\hbar\omega).$$

If N is of the order of Avogadro's number, can you ignore the last term? Explain.

(d) Show that when $M \gg N$, the above result leads to a dependence of $\ln \Omega$ on E which is compatible with the general result $\ln \Omega \sim f \ln(E) + \text{const.}$, where f is the number of degrees of freedom of the system and "const." is independent of E.

4.8 Consider a non-relativistic particle confined in a rectangular box of side L_x, L_y, and L_z, whose energy levels are given, as usual, by

$$\epsilon_{n_x,n_y,n_z} = \frac{\hbar^2\pi^2}{2m}\left(\frac{n_x^2}{L_x^2} + \frac{n_y^2}{L_y^2} + \frac{n_z^2}{L_z^2}\right).$$

(This is the same as in Problem 4.4, but with the three sides of the box not necessarily equal).

(a) Let the particle be in a given state r specified by particular values of the three integers n_x, n_y, and n_z. Find the force on the wall perpendicular to the x-axis

$$F_{r,x} = -\frac{\partial E_r}{\partial L_x}$$

and calculate from this the pressure p_x (force per unit area) on this wall.

(b) Express the sum $p_x + p_y + p_z$ for any state in terms of the volume $V = L_x L_y L_z$ and the energy E_r of that state.

(c) Show that the mean pressure $\overline{p}$ exerted by the particle in equilibrium is related to its mean energy $\overline{E}$ by

$$\overline{p} = \frac{2\overline{E}}{3V}.$$

Hint: Use the result of part (b) and the fact that, in equilibrium, the mean pressure on every wall is the same.

(d) Show that if the length of the box is increased adiabatically and quasistatically from L_x to $8L_x$, the work W done by the particle is 3/4 of the initial mean energy $\overline{E}_{\text{init}}$.
Hint: Convert $đW = \overline{p}dV$ to a differential equation involving only $\overline{E}$ and V.

(e) Show that an attempt to calculate the work done by averaging the *finite* energy change in each state resulting from an increase of length from L_x to $8L_x$ over the initial distribution of values of n_x, n_y, and n_z leads to the incorrect result $W = (21/64)\overline{E}_{\text{init}}$. Explain clearly why this method is not an appropriate description of a quasistatic process.

4.9 There is another approach to calculating the number of states in the microcanonical ensemble to determine the entropy. It hinges on the issue discussed in Example 4.11, where if δE was not small enough, we found that $\phi > \Omega$. That is, the number of states with energy less than E was *greater* than the number of states with energy between E and $E + \delta E$ as calculated in Eq. (4.25).

To study this, we use a slightly different, but equivalent, definition of Ω in terms of ϕ, which makes the problem when $\phi > \Omega$ more obvious,

$$\Omega(E) = \phi(E) - \phi(E - \delta E).$$

(a) Consider our expression for ϕ for a general system where

$$\phi(E) = CE^f,$$

with C some constant and f the number of degrees of freedom. Show that Ω can be written as

$$\Omega(E) = \phi(E)\left[1 - \left(1 - \frac{\delta E}{E}\right)^f\right]. \tag{4.72}$$

(b) Given that $\delta E/E < 1$ (maybe not *much less than* one), Taylor-expand the second term and show that

$$\Omega(E) = f\frac{\delta E}{E}\phi(E)\left(1 - \frac{f-1}{2}\frac{\delta E}{E} + \cdots\right). \tag{4.73}$$

(c) Our approximation for Ω neglects the second term in parentheses (and all subsequent terms). How must δE compare with E for this to be valid? With the result you obtain, is Ω greater or less than ϕ?

(d) Another approximation can be made starting from your result in part (a). Given that $\delta E/E < 1$, then $(1 - \delta E/E)^f$ is negligible for values of f that are relevant to us. In fact, for $\delta E/E = 0.1$, calculate $(1 - \delta E/E)^f$ for $f = 10, 100, 1000$, and $10\,000$ and see that it very quickly becomes zero.

This allows us to use an alternate approximation for Ω,

$$\Omega(E) \approx \phi(E), \tag{4.74}$$

which allows us to skip one step when calculating the entropy in the microcanonical ensemble (no derivatives need to be taken!). I will stick with using the expression in Eq. (4.25), but I encourage you to use both methods when calculating physical quantities to see how they give equivalent physical results.

4.10 Consider the infinitesimal quantity

$$(x^2 - y)dx + xdy. \tag{4.75}$$

(a) Is this an exact differential?

(b) Evaluate the integral of this differential between the points (1,1) and (2,2) along three straight-line paths:

i) $(1,1) \rightarrow (1,2) \rightarrow (2,2)$

ii) $(1,1) \rightarrow (2,1) \rightarrow (2,2)$

iii) $(1,1) \rightarrow (2,2)$

(c) Suppose that Eq. (4.75) is divided by x^2. Show that the resulting differential is exact, so you can write

$$df = \frac{x^2 - y}{x^2}dx + \frac{dy}{x}.$$

(d) Evaluate $\int df$ along the paths in part (b), and show that it is the same as what you would obtain if you calculated $f(2,2) - f(1,1)$.

4.11 A system undergoes a **cyclic** quasistatic process, during with the mean pressure $\overline{p}$ changes with volume V as a closed curve as shown. This is a cyclic process because the system ends up in a final macrostate which is identical to its initial macrostate. Show that the work done by the system is given by the shaded area contained within the closed curve.

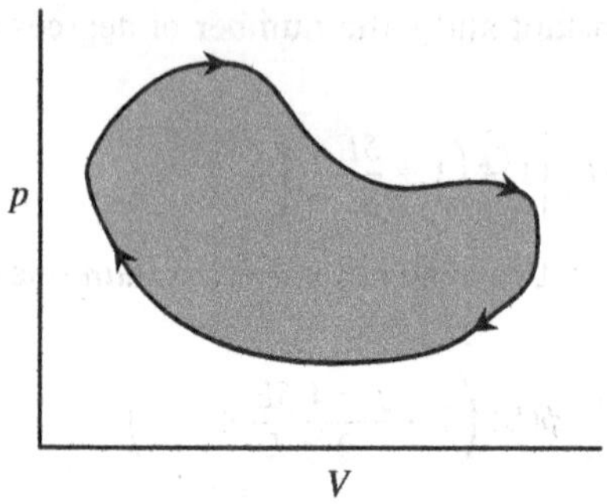

4.12 A classical ideal gas undergoes a quasistatic and adiabatic expansion from $A \to B$ as shown in the diagram. The mean pressure $\overline{p}$ of the gas changes with its volume V (shown by the dotted curve, path (d)) according to the relation

$$\overline{p} = KV^{-5/3},$$

where K is a constant. Consider the following three different processes which take the system from macrostate A to macrostate B, and find the quasistatic work done W and the net heat Q absorbed by the system in each case. All processes (including the adiabatic process) decrease the pressure from 3.2 MPa to 0.1 MPa and increase the volume from 10^{-3} m ^{3} to 8×10^{-3} m ^{3}.

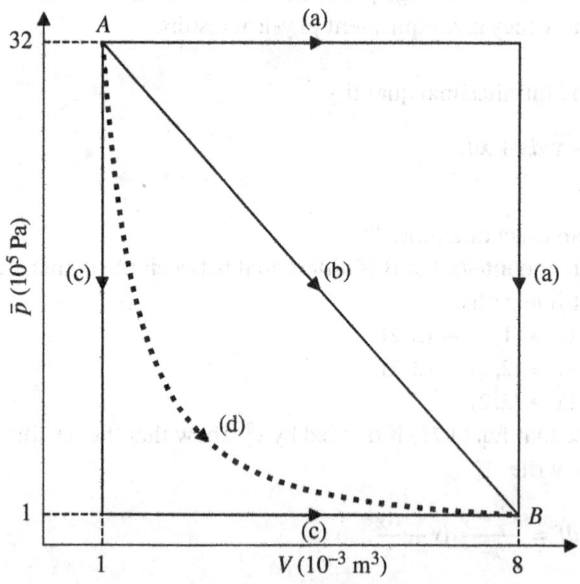

(a) The system expands from its initial to its final volume, while heat is added to maintain constant (initial) pressure. The volume is then fixed, while heat is extracted to reduce the pressure to its final value.
(b) The volume is increased and heat is supplied in such a way so that the pressure decreases linearly with the volume.
(c) The two steps of process (a) are performed in the opposite order.

References

1 J. B. Marion and S. T. Thornton. *Classical Dynamics of Particles and Systems*. Cengage Learning, 5th edition, 2003.

2 J. V. Uspensky. *Introduction to Mathematical Probability*. McGraw-Hill, 1937.

3 Gibbs vs Boltzmann entropies. *American Journal of Physics*, 33(5):391–398, 1965.

4 P. županović and D. Kuić. Relation between Boltzmann and Gibbs entropy and example with multinomial distribution. *Journal of Physics Communications*, 2(4):045002, 2018.

5 D. F. Styer. Insight into entropy. *American Journal of Physics*, 68(12):1090–1096, 2000. Erratum: *America Journal of Physics*, 90(5):396, 2022.

6 T. Salagaram and N. Chetty. Enhancing the understanding of entropy through computation. *American Journal of Physics*, 79(11):1127–1132, 2011.

7 D. Styer. Entropy as disorder: history of a misconception. *The Physics Teacher*, 57(7):454–458, 2019.

8 J. von Plato. Boltzmann's ergodic hypothesis. *Archive for History of Exact Sciences*, 42(1):71–89, 1991.

9 H. B. Callen. *Thermodynamics and An Introduction to Thermostatistics*. Wiley, New York, NY, 2nd edition, 1985.

10 T. Weideman. Thermodynamic processes. https://phys.libretexts.org/@go/page/18474?pdf. Accessed: 20 February 2023.

5

Equilibrium and Temperature

We are almost at the point where we can study thermodynamics with everything we have discussed up to now. However, there are two more topics that need to be covered before we are ready to truly dive in. The first of these is the concept of thermal equilibrium, which we have mentioned before (as only systems in equilibrium are of interest to us). Equilibrium is familiar in physics generally; however, now we would like to go more into detail about not just the thermal equilibrium of a system, but also how a system *approaches* equilibrium, what happens when a system in equilibrium is changed, and how to determine if two systems are in equilibrium with each other.

Once this is taken care of, we can finally introduce the concept of temperature, initially from a fundamental point of view. From this theoretical vantage point, we will then connect the abstract idea of temperature to what we are all familiar with from everyday life. Doing so will not only allow us to easily connect Chapters 2 through 4 to temperature, but also to see some new and strange ideas appear.

After finishing this chapter, you should be able to

- understand thermodynamic equilibrium and the approach to it in more detail,
- understand the notion of irreversibility, and
- understand temperature and the properties of it.

5.1 Equilibrium and the Approach to it

In statistical mechanics, we focus on a system in equilibrium, where we have calculated the number of accessible states in a system, given by Ω. Generally this is a function of many different parameters in the system and more importantly different constraints. The question we first pose is: "What happens to Ω when you remove one of these constraints?" When a constraint is removed, the system will undergo a **spontaneous process**: a process which brings the system from one equilibrium state (with a constraint) to another equilibrium state (without that constraint).

5.1.1 Equilibrium

Just vaguely referring to "constraints" is not useful when trying to understand this notion of a spontaneous process. Therefore, to be concrete, let's imagine a container that is divided in half, as shown in Figure 5.1, where the left side contains some amount of oxygen gas in equilibrium. We will consider the right side to either be filled with some gas or empty. The divider in the center

Statistical Thermodynamics: An Information Theory Approach, First Edition. Christopher Aubin.

Companion website: www.wiley.com/go/Aubin/StatisticalThermodynamics

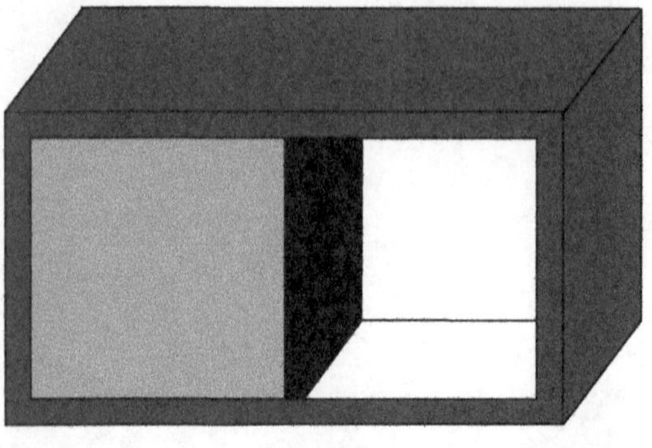

Figure 5.1 A container with oxygen gas in the left half, and either vacuum or another gas in the right half. The two halves are separated by a removable divider.

is the constraint that we can easily remove at any time, and we will assume that the system is in equilibrium, with some initial number of states given by Ω_i.

In the microcanonical ensemble, the fundamental postulate states that all accessible microstates are equally likely, so the probability our system is in a state i is given by a constant, which we will write as

$$P_i = \frac{\Omega_i}{\Omega}, \tag{5.1}$$

where we define $\Omega \equiv \sum_i \Omega_i$ and see that $\sum_i P_i = 1$. Ω is the total number of states accessible to the system, subject to whatever constraint(s) we require, while Ω_i is the number of ways to arrange a system to be in a particular state, which will be more clear in the following discussion.

Remember, for macroscopic systems, such as the container of oxygen in Figure 5.1, we will generally have a number of particles so large (again, to be concrete we'll consider $N = 10^{24}$) that enumerating all of these states is difficult, if not impossible. However, most of the accessible states are not important, because the system will be extremely unlikely to be in them. Considering only the portion of our container above with the oxygen gas, suppose we label the various states of each molecule as being in the front half or rear half of the left side—if this is all we care about, then there are only two states accessible to each oxygen molecule. We can start considering what Ω_i would be if we had zero, one, two, etc. molecules in the front half. For now, we will use the subscript on Ω_i to denote the number of molecules in the front half for clarity, so Ω_0 is the number of states with zero molecules in the front, Ω_1 is the number with one molecule in the front, and so forth.

For zero molecules in the front half, then all 10^{24} molecules are in the rear half. There are $\Omega = 2^{10^{24}}$ total states and only $\Omega_0 = 1$ way to have all of them in the front, so

$$P_0 = \frac{1}{2^{10^{24}}} \approx 10^{-10^{24}}. \tag{5.2}$$

To have one molecule in the front, there are $\Omega_1 = 10^{24}$ ways for that to happen, as we can choose any of the oxygen molecules to put there, so in this case there are quite a few more states. However, the probability would still be Eq. (5.2) to any reasonable precision, so we would say that it is just as likely to have zero or one particle in the front half (in either case, it is not likely at all!).

Exercise 5.1 Perform the calculation to argue the probability that one oxygen molecule is in the front half is $P_1 = P_0$ to an excellent approximation.

As we allow more molecules to move to the front, P_i would increase. Given how large N is, we would need a substantial number of molecules, close to half, on one side before we see any noticeable change to P_i. In other words, the most likely state is when half of our molecules are in the front and the other half are in the back—or rather, they are equally distributed throughout the left side of the container.

Additionally, combining the fundamental postulate of statistical mechanics with Eq. (5.1), we can say

> The thermodynamic equilibrium state is the one where the number of accessible states is maximum.

This in and of itself is not a new statement, as we have seen it in other systems in Chapters 3 and 4. However, it will give us a starting point for how to *determine* the equilibrium state for a system. But now that we have a single, well-defined, equilibrium state, we will revert to the subscript i referring to the single initial equilibrium state our system is in.

5.1.2 Irreversible and Reversible Processes

As mentioned before, only looking at systems in equilibrium would make for a rather boring subject. We have already discussed interactions among systems, along with quasistatic processes that we can use to describe the transition between two equilibrium states without losing the tools at our disposal.

More generally we would like to discuss what would happen when we have a spontaneous process—that is, what would happen to our system when we remove a constraint that has been placed on it? For example, when we remove the divider in Figure 5.1, what is going to happen? The answer depends on what lies on the other side of the partition.

First we assume the other half of the container is empty. Initially the number of states accessible to our gas of N oxygen molecules is Ω_i, and given the previous discussion, these N molecules are evenly spread throughout the entire left half of the container. When we remove the divider, now the number of states accessible to our gas is Ω_f. Given that the gas is now able to be in a much larger volume, we have $\Omega_f > \Omega_i$.

Exercise 5.2 Quantify this by assuming the oxygen constituted a classical ideal gas. Use Eq. (4.41) to calculate the ratio Ω_f/Ω_i and show explicitly that if the volume accessible to the gas is doubled, this ratio is greater than one.

However, suppose we had a slightly different setup: Assume $N/2$ of the oxygen molecules are in the left half of the container, and $N/2$ in the right half. In this case we have Ω_i states accessible to the gas,[1] and now we remove the divider, and now the gas has Ω_f accessible states. As we argued above, the equilibrium state is such that the gas will be distributed roughly equally throughout the volume. As the gas already was equally distributed throughout the container before the divider was removed, then we expect that in this case $\Omega_f = \Omega_i$.

Is there a case where $\Omega_f < \Omega_i$? As a spontaneous process occurs generally when a constraint is *removed*, then this is not a possibility. When you remove a constraint, you are always going to allow the system to be in more states than before (except in that special case where the number of states is the same). As such, we can say:

1 We use the same notation, but this Ω_i is not going to be the same as our previous number of states with all N molecules on the left side.

In any spontaneous process of a system that takes it from some initial (equilibrium) state i with a number of states given by Ω_i to some final (equilibrium) state f with a number of states given by Ω_f, the number of states must never decrease, or

$$\Omega_f \geq \Omega_i. \tag{5.3}$$

This can be understood in the context of information theory—if we remove the constraint, then we now have *less information* (or at the very least not more information) about the system, or rather the amount of missing information has *increased* (or stayed the same), so that

$$S_f = k_B \ln \Omega_f \geq k_B \ln \Omega_i = S_i. \tag{5.4}$$

Those spontaneous processes where $\Omega_f > \Omega_i$ are called **irreversible**, while those with $\Omega_f = \Omega_i$ are **reversible**. A reversible process is one in which we could replace the constraint (the divider in our container above) and return the system to the original state without any other action. For the irreversible case, where the right half of the container was empty in our example, we cannot simply replace the divider—half of the gas is now on the right side. We would have to do work on the gas to push it all back to the left. In that case, by doing work on the system we are changing its energy, and thus we will have a different number of states than Ω_i even though all of the gas is back on the left.

5.1.3 Two Systems in Equilibrium

Let's return to our two systems A and A' from before (such as in Figure 4.9), where the combined system $A^{(0)} = A + A'$ is completely isolated, and these two systems can thermally interact with each other. Both systems are in equilibrium: system A has $\Omega(E)$ states with energy in the range E and $E + \delta E$, while system A' has $\Omega'(E')$ states with energy in the range E' and $E' + \delta E'$. While in contact, the energies E and E' may change; however, the total energy $E^{(0)} = E + E'$ will remain constant given that the combined system is isolated. In this case, we can write everything in terms of the energy E of system A, by using $E' = E^{(0)} - E$. Thus as the energy of system A increases, the energy of system A' must decrease.

Similarly, the total number of states of the combined system $\Omega^{(0)}$ can be thought of as a function only of E, or

$$\Omega^{(0)} = \Omega(E)\Omega'(E') = \Omega(E)\Omega'(E^{(0)} - E), \tag{5.5}$$

where as usual we can determine the total number of states of a combined system by taking the product of the number of states of the individual systems.

Using Eq. (5.1), the probability of the combined system being in a state such that system A has energy E is

$$P(E) = C\Omega^{(0)} = C\Omega(E)\Omega'(E^{(0)} - E), \tag{5.6}$$

where C is a constant that is independent of E. We could determine it if we knew specifically how Ω and Ω' depended on E, by normalizing the probability.

What we would like to know is what $P(E)$ looks like for this combined system. Specifically, we have already argued that the equilibrium state occurs when $P(E)$ is maximized, but at what value of E will this happen?

Example 5.1 Let's consider a simple model of two small systems: one with five degrees of freedom and the other with seven degrees of freedom, and assume that they behave according to our general expectation of Eq. (4.28), with $a = 1$. From that expression, we write

$$\Omega(E) = CE^5, \text{ and } \Omega'(E') = C'E'^7. \tag{5.7}$$

Additionally, we will use some arbitrary units for the energy so $0 \leq E, E' \leq 16$, and the total energy of the combined system is fixed to $E^{(0)} = 16$. $\Omega(E)$ and $\Omega'(E^{(0)} - E)$ would be as shown in Figure 5.2, where we have scaled each vertical axis by the maximum value of the number of states on each side. The solid curve is $\Omega(E)$ scaled by its maximum value, while the dashed curve is a plot of $\Omega'(E^{(0)} - E)$ scaled by its maximum value (only the behaviors of each curve are relevant). In Figure 5.3, we show the probability (divided by its maximum value, P_0) from Eq. (5.6) for this system.

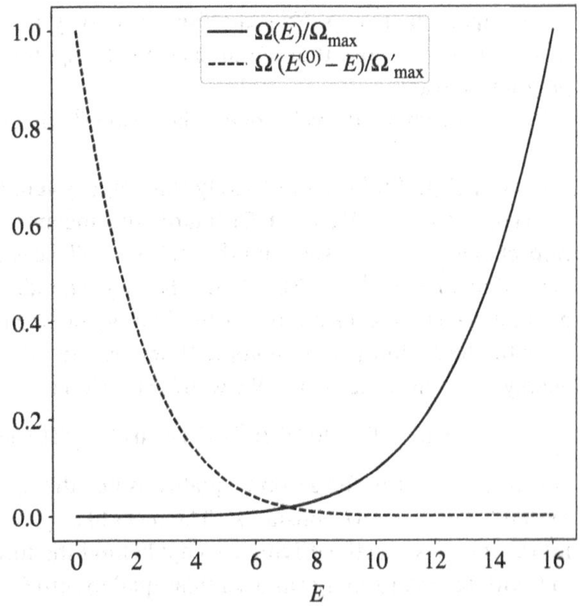

Figure 5.2 The number of states as a function of the energy E for our systems described by Eq. (5.7). The solid curve is a plot of $\Omega(E)$ scaled by the maximum value of Ω, while the dashed curve is a plot of $\Omega'(E^{(0)} - E)$, scaled by the maximum value of Ω'.

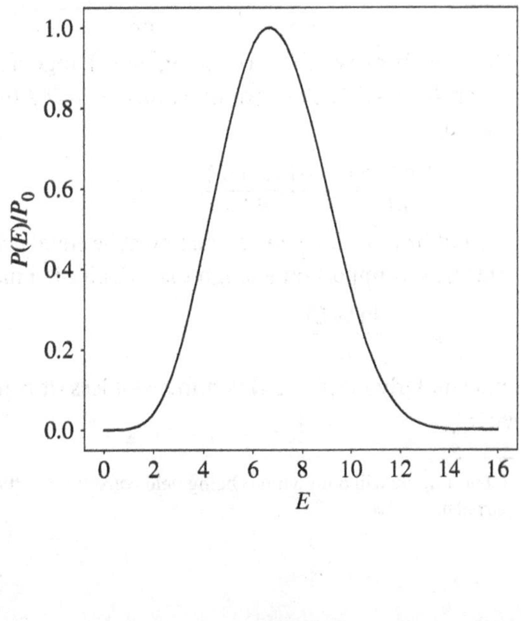

Figure 5.3 The product of $\Omega(E)$ and $\Omega'(E^{(0)} - E)$ as a function of the energy E for our systems described by Eq. (5.7), where we scaled the vertical axis by the maximum value of $\Omega\Omega'$.

Since Ω is a rapidly increasing function of E, we find that Ω' rapidly *decreases* as a function of E. The product of the two, which gives us $P(E)$, will peak at some point in-between. For our simple example here, it turns out to be precisely at $E = 20/3 \approx 6.67$ in our units. Notice that it is not exactly halfway between the allowed energies ($E = 8$), but is skewed a little below that due to the different functions describing the number of states accessible to each system.

Exercise 5.3 Plot $\Omega(E)$, $\Omega'(E^{(0)} - E)$, and $P(E)$ for various numbers of degrees of freedom, such as

1. $f = f' = 5$
2. $f = 1, f' = 10$
3. $f = 9, f' = 2$

and compare the results. You can visit the Two Systems in Contact section of the companion site to use the Jupyter code that I used to create the figures in the previous example, or use some other plotting software.

In general, what is the value of E (relative to $E^{(0)}$) that maximizes the probability for such systems?

If we wish to find the most likely state of a system (and thus the equilibrium state), we want to maximize $P(E)$ in Eq. (5.6). Ordinarily this means we would like to set the derivative of P with respect to E to zero. However for the systems we'll be working with, with a large number of degrees of freedom ($f \sim 10^{24} \gg 5,7$), $P(E)$ will be such a rapidly changing function around the equilibrium point, it is nicer to consider $\ln P$ when finding the maximum.

As the logarithm is a monotonic function, then the maximum of $\ln P$ will occur at the same energy as the maximum of P. We want to maximize

$$\ln P = \ln \Omega(E) + \ln \Omega'(E^{(0)} - E) + \ln C = \left[S(E) + S'(E')\right]/k_B + \ln C, \tag{5.8}$$

which, as is clear in the second equality, is just the total entropy of our system (up to some additive and multiplicative constants). The most likely state is just that which maximizes the missing information, as was the driving concept behind the fundamental postulate! So let us differentiate $\ln P$ with respect to energy and set this equal to zero,[2]

$$\frac{\partial \ln P}{\partial E} = \frac{\partial \ln \Omega(E)}{\partial E} + \frac{\partial \ln \Omega'(E^{(0)} - E)}{\partial E} = 0. \tag{5.9}$$

However, because $E^{(0)}$ is constant, any change in E must be equal and opposite to any change in E', or $dE = -dE'$. Thus we must have $\partial \ln \Omega'/\partial E = -\partial \ln \Omega'/\partial E'$, and putting this into Eq. (5.9) gives us

$$\frac{\partial \ln \Omega(E)}{\partial E} = \frac{\partial \ln \Omega'(E')}{\partial E'} \tag{5.10}$$

in equilibrium. Since this is what must be equal between two systems at equilibrium, the derivative $\partial \ln \Omega/\partial E$ is important enough that we give it a name and its own variable,

$$\beta \equiv \frac{\partial \ln \Omega(E)}{\partial E}, \tag{5.11}$$

and similarly for β'. As Ω is dimensionless (it is just the number of states), β has units of inverse energy.

2 For now we will omit what is being held constant, but there could be other parameters that are, hence the use of partial derivatives.

Exercise 5.4 You might wonder if you would obtain Eq. (5.10) without considering the maximum of $\ln P$. Show that if you maximize P directly, you still arrive at this same condition.

If we multiply β and β' by k_B, we can write the above expressions in terms of the entropy (because $S = k_B \ln \Omega$) to get

$$k_B\beta = \frac{\partial S}{\partial E} = \frac{\partial S'}{\partial E'} = k_B\beta'. \tag{5.12}$$

In everyday usage, β is not very common, but rather we define

$$\beta \equiv \frac{1}{k_B T}, \tag{5.13}$$

where T is the **absolute temperature** of the system. Thus, our condition that two systems are in **thermal equilibrium** is that their temperatures are equal,

$$T = T' \text{ or } \beta = \beta'.$$

We can also rewrite Eq. (5.12) as

$$\frac{1}{T} = \frac{\partial S}{\partial E}. \tag{5.14}$$

Exercise 5.5 Show that the energy for a general classical system with f degrees of freedom as a function of T is given by $E = afk_BT$. Use the expression from Eq. (4.28).

Exercise 5.6 Calculate T for a classical monatomic ideal gas using Ω from Eq. (4.39). Solve this to obtain E as a function of T. You should get $E = \frac{3}{2}Nk_BT$, so that the energy is independent of the volume for a classical monatomic ideal gas.

Exercise 5.7 Calculate T for a general classical ideal gas using Ω from Eq. (4.41) and use this to show that the energy is independent of the volume for *any* classical ideal gas.

While not obvious yet, temperature as defined in this manner is (almost) precisely what you know about from everyday life. From a statistical mechanics point of view, temperature is nothing more than a quantity related to a system such that when two systems are in thermal equilibrium, their temperatures (or β's) are equal. Of course, from an everyday perspective, temperature is related to how hot or cold you might feel when you step outside of a building. I want to dissuade you from thinking this way too much yet, as we will come up with a more precise definition of hot vs. cold in Section 5.1.4.

5.1.4 Approaching Thermal Equilibrium

Suppose we have two systems that we wish to put into contact with each other so they can interact thermally (via a spontaneous process), and we study how they approach equilibrium. They each have initial energies $\overline{E}_i$ and $\overline{E}'_i$ (where we refer to the mean energies again here to be more general for later ensembles) with the total energy $\overline{E}^{(0)} = \overline{E}_i + \overline{E}'_i$ constant. While interacting, their energies may change to $\overline{E}_f$ and $\overline{E}'_f$, and the total energy $\overline{E}^{(0)}$ is of course unchanged. Initially they may have different temperatures, T_i and T'_i (or equivalently β_i and β'_i); however, once they reach their final equilibrium states, $T_f = T'_f$ ($\beta_f = \beta'_f$).

Initially the combined system has some number of states $\Omega_i^{(0)}$, and after interacting and returning to equilibrium, it has some number of states $\Omega_f^{(0)}$ which is not less than the initial number of states. We can also say

$$\ln \Omega_f^{(0)} \geq \ln \Omega_i^{(0)},$$

or

$$S_f^{(0)} \geq S_i^{(0)}. \tag{5.15}$$

In terms of the initial and final entropies of the individual systems,

$$S_f' + S_f \geq S_i' + S_i,$$

or

$$\Delta S^{(0)} = \Delta S + \Delta S' \geq 0.$$

This is our first glimpse of the

> **Second Law of Thermodynamics:** The total entropy of an isolated system can never decrease. We will discuss this in greater detail in Chapter 6.

Exercise 5.8 Argue that putting two systems into thermal contact constitutes a spontaneous process.

For a purely thermal interaction, the change in energy arises solely from the heat absorbed by each system, so

$$Q = \overline{E}_f - \overline{E}_i$$

and

$$Q' = \overline{E}_f' - \overline{E}_i'.$$

Because $\Delta E^{(0)} = 0$, we know that $Q + Q' = 0$, which we saw before.

At this point we should discuss the notion of a *hot* or *cold* system. The problem with this idea is that we cannot say that a system is hot or cold, but rather we can compare it to other systems: It can be **hotter** or **colder** than some other system. We define this specifically as follows:

> If system A is **hotter** than system A', then when we put the two systems into thermal contact, we would find $Q < 0$. That is, A loses energy in the form of heat to system A'.

Notice that our definition of "hotter" or "colder" has nothing to do with temperature!

5.2 Temperature

We are all familiar with temperature in an everyday sense, but let's first discuss the more abstract concept of absolute temperature, $T = 1/(k_B\beta)$. Suppose we have two systems that are not in thermal contact with each other but are separately in equilibrium, then we bring them into thermal contact. One of two things will happen:

1. If they initially have the same value of β, when brought into contact, they will remain in their original equilibrium states with the same value of this parameter. (That is, nothing happens.)
2. If they initially have different values of β, then they will *not* remain in equilibrium when brought into thermal contact. Instead they will exchange energy in the form of heat until they do have the same β (different from either of the two initial β's).

More interestingly, suppose we have three systems, A_1, A_2, and A_3, that all have temperatures β_1, β_2, and β_3 respectively. If the first two are in thermal equilibrium, then we know that

$$\beta_1 = \beta_2.$$

Additionally, suppose A_1 and A_3 are also in thermal equilibrium, so we have

$$\beta_1 = \beta_3.$$

Thus, we must also have $\beta_2 = \beta_3$, or systems A_2 and A_3 are also in thermal equilibrium with each other. This is stated as the

Zeroth Law of Thermodynamics: If two systems are in equilibrium with a third system, then they must be in equilibrium with each other.

This might seem like a somewhat obvious statement,[3] but this law specifically allows for the construction of **thermometers**, which is where our everyday understanding of temperature comes from. Thermometers use a **thermometric parameter** that varies in a known way with the absolute temperature of the system. When we think of temperature, we really are thinking of some thermometric parameter.

Some common thermometers and their thermometric parameters are:

1. Old-school mercury or alcohol thermometers which use the height of mercury or alcohol in a tube as its thermometric parameter.
2. Some gas thermometers can operate at constant volume where the thermometric parameter is the mean pressure of the gas. Similarly you could have a gas thermometer operating at constant pressure, where the thermometric parameter is the volume of the gas.
3. Many modern digital thermometers use the electrical resistance of a wire as the thermometric parameter.
4. Even more recently, infrared thermometers measure the energy (in the infrared part of the spectrum) radiated from a body and relate that to the temperature.

Different thermometers may yield different values of the temperature because they use different expressions to relate the thermometric parameter to absolute temperature. Effectively we are measuring the temperature in different units. If β as defined above is used as the thermometric

3 In a sense it is the transitive property of mathematics applied to a physical system!

parameter, then any thermometer would result in the same value of the temperature. So while absolute temperature is an abstract concept, it acts as a useful parameter that can calibrate other thermometers effectively.

The connection between the abstract concept of absolute temperature and our everyday notion will come about as we study various systems in Chapters 6 and 7. There are two truly absolute temperature scales (Kelvin and Rankine) and two common non-absolute scales (Celsius and Fahrenheit). We will discuss these (and other) scales, along with the SI definition of temperature in Section 5.3.2.

5.3 Properties of Temperature

Let us examine several interesting properties of temperature that we can determine from its definition,

$$\frac{1}{k_B T} = \beta = \frac{\partial \ln \Omega}{\partial E}.$$

We saw earlier that, in general, the number of states Ω is a rapidly increasing function of E. That means that we would usually expect

$$\beta > 0 \quad \text{or} \quad T > 0.$$

For example, from Eq. (4.28), we argued that

$$\Omega = (\text{const})\overline{E}^f$$
$$\Rightarrow \ln \Omega = f \ln \overline{E} + (\text{const}).$$

Differentiating this we have

$$\beta = \frac{f}{\overline{E}} \Rightarrow k_B T = \frac{\overline{E}}{f}. \tag{5.16}$$

In other words, we find that temperature is a rough measure of *the mean energy per degree of freedom* (this is an example of a much more general theorem that we will return to in the future). Often it is stated that temperature measures the kinetic energy in a system but that is not entirely true: It is a measure of how much thermal energy there is (on average) in a system. This thermal energy comes from both kinetic and potential energy sources.

Let's return to our two systems A and A' which have fixed external parameters and suppose they are in nearby (but different) equilibrium states. After they are put into thermal contact, the number of states of the combined system, or equivalently the probability, must *increase* as they approach equilibrium with each other. Since we can treat the probability P as a function of $\overline{E}$ and $\overline{E}'$, we can write the total differential dP as

$$dP = \frac{\partial \ln \Omega(\overline{E})}{\partial E} d\overline{E} + \frac{\partial \ln \Omega(\overline{E}')}{\partial E'} d\overline{E}' \geq 0. \tag{5.17}$$

From our definition of β, we can write this as

$$\beta d\overline{E} + \beta' d\overline{E}' \geq 0.$$

With the external parameters fixed, the only way the energy will change is through the transfer of heat, so $d\overline{E} = đQ$ and $d\overline{E}' = đQ'$,

$$\beta đQ + \beta' đQ' \geq 0$$
$$(\beta - \beta') đQ \geq 0, \tag{5.18}$$

where we used $đQ = -đQ'$. Now suppose system A' is *hotter* than A. From our definition of this earlier, that means energy flows *from A' to A*, so $đQ > 0$, and

$$\beta \geq \beta', \Rightarrow T \leq T'. \tag{5.19}$$

Thus, heat flows from a system with a *higher* temperature to a system with a *lower* temperature, which matches our every day intuition. When we want to determine which system is hotter among several, we don't have to put them into thermal contact to see which way the heat flows, but rather can just measure their temperatures to make this determination.

This all assumes that the temperatures of the systems are positive, which is usually true—as long as we are looking at absolute temperature scales, and so long as $\Omega(E)$ grows with the energy. If the latter is not true, then the absolute temperature can actually be negative!

5.3.1 Negative Absolute Temperature

There are some cases where that absolute temperature of a system can actually become negative. When can this happen? Primarily this can only occur when the number of states decreases when the energy of a system increases, as can be seen by looking at the definition of T.

Ordinary (classical) systems take into account the kinetic energy of the particles. Thus, even though there is a lower bound to the energy due to our systems having a ground state, there is no upper bound—this implies that as the energy increases so must the number of states, or the entropy. However, suppose we have a system, such as a system of spins, where the kinetic energy is fixed (or not relevant), then there exists not only a lower bound, but an upper bound as well (see Problem 5.5). Such a system would have an entropy that could look like Figure 5.4. Note for energies to the right of the dashed line, $\partial S/\partial E < 0$, so the system has a negative absolute temperature. Any system where there is a region where the number of states accessible to the system decreases as we increase the energy will have a negative temperature.

Let's go back to our systems which are in nearby (different) equilibrium states and brought into thermal contact. This time, assume system A has a *negative* absolute temperature, so we will write $\beta = -|\beta|$, and A' still has a positive absolute temperature. Then, Eq. (5.18) can be written as

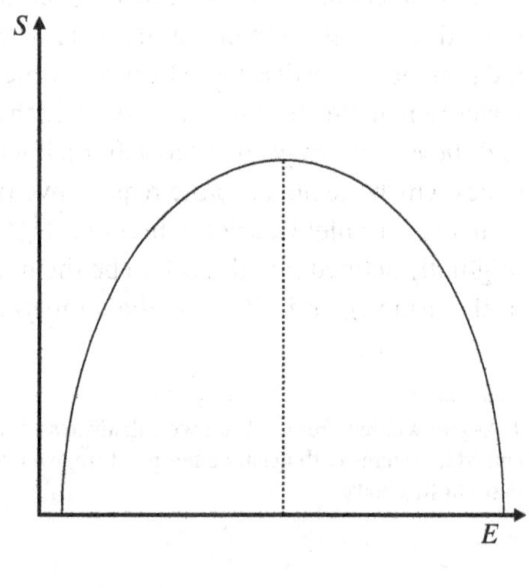

Figure 5.4 The entropy as a function of energy for a sample system that can have a negative temperature (for energies to the right of the dotted line).

$$(\beta - \beta')đQ \geq 0 \Rightarrow (|\beta| + \beta')đQ \leq 0. \tag{5.20}$$

Because we required $\beta' > 0$, then it must be true that $đQ < 0$. In other words,

Systems with negative temperatures are *hotter* that those with positive temperatures!

Note that this does not strictly apply to non-absolute temperature scales, like the Celsius scale which we'll discuss in Section 5.3.2. Those have negative temperatures that are colder than positive temperatures, but all absolute scales have an actual *zero point*, known as **absolute zero**. In all systems, temperature can never drop below absolute zero—that is, nothing is *colder* than a system at this lowest temperature. However, as we just saw, a system with negative absolute temperature is actually hotter than those with positive temperatures, so heat will flow from the system with negative temperature to a system with positive temperature.

If we were to order temperatures from coldest to hottest, we would do so as

$$T = 0, \ldots, +\infty, -\infty, \ldots, -0; \tag{5.21}$$

which is rather confusing given that $+\infty$ and $-\infty$ are right next to each other. This actually makes more sense when we write this in terms of β (again from coldest to hottest),

$$\beta = +\infty, \ldots, +0, -0, \ldots, -\infty; \tag{5.22}$$

as it makes sense that we can go from slightly above zero (positive numbers) to slightly below zero (negative numbers) quite easily. So from a theoretical standpoint, β is more useful than T, while T is a more useful parameter from a practical standpoint.

Such systems have been observed experimentally, as discussed in Refs. [1, 2]. In these references, the authors review the various observations made of negative absolute temperature of nuclear spin systems as well as the changes needed when studying these systems in statistical mechanics.[4] More recently, negative absolute temperatures for systems with motional degrees of freedom (that is, when classical kinetic energy cannot be neglected) have been observed in Ref. [3].

5.3.2 Temperature Scales

We have spent all of this time talking about temperature from an abstract perspective, so let's take some time to discuss temperature as it is used. There are four main temperature scales that are used today, which I will discuss in historical order. These can be divided into two categories: temperature scales which are absolute and those which are not. Absolute temperature scales use the definition of β above and a single point to define a linear scale (the second point is absolute zero). Temperature scales which are not absolute require two points of reference to set a linear scale.

In 1724, Daniel Gabriel Fahrenheit (1686–1736) came up with the first widely used scale. He originally defined it with 96°F to be the body temperature of a healthy human body (as measured in the armpit), and 0°F to be the temperature of a brine of ice, water, and ammonium chloride.

4 As you will see, from now on we will always assume $T > 0$ so that we can be concrete with many of our claims. To consider systems with negative temperature, we would have to start over to some degree, so we won't deal with them in this book.

Not the most obvious or practical choice, but it has remained in use in the United States, Liberia, as well as a few small island nations around the world.

Currently this scale is such that 32°F is the freezing point of water and 212°F its boiling point (at a pressure of 1 atm), and is not an absolute temperature scale, as it uses two points to set the scale. The current standard definition of this scale is in terms of the Celsius scale which we'll discuss next.

Originally in 1742, Anders Celsius (1701–1744) defined his scale such that 100°C was the freezing point of water while 0°C was the boiling point, so it went in the "wrong" direction![5] In 1744 Carl Linnaeus (1707–1778) reversed the scale to what we know today, making 100°C the boiling point and 0°C the freezing point of water. This too is not an absolute temperature scale, and while these values are still used to define the scale colloquially (such as the values quoted for the Fahrenheit scale above), it is more precisely defined in terms of the first developed absolute temperature scale, the **kelvin**.

Over a century after the Celsius scale entered the picture, Lord William Thomson Kelvin (1824–1907) defined the first absolute temperature scale in 1848. This scale was defined such that the **triple point of water** (the temperature and pressure such that water can exist as liquid, solid, and gas) is 273.16 K (at a pressure of about 0.006 atm), and one kelvin is equal to 1/273.16 of this, to ensure the size of 1 K is that of 1°C.

Finally, in 1859, William John Macquorn Rankine (1820–1872) decided there should be an absolute scale that had a unit which corresponded to the same size as a degree fahrenheit. This scale sets the triple point of water as 491.688 R, which is $\frac{9}{5}$ times the triple point in kelvin, and is used in the United States for some engineering applications.

As of May 2019 in the SI system, the kelvin is defined more precisely so that

> the kelvin, symbol K, is the SI unit of thermodynamic temperature; its magnitude is set by fixing the numerical value of the Boltzmann constant to be equal to exactly
>
> $$k_B = 1.380\,649 \times 10^{-23}\ \text{J/K}. \tag{5.23}$$

For everyday use, the "original" definition is sufficient, relating to the triple point of water; however, this is more precise and thus useful for high precision experiments. It is useful to keep in mind some "ordinary" temperatures, as they can be helpful for calculations. While 0°C = 273.15 K and 100°C = 373.15 K, we also use **room temperature** as a standard temperature, even though it is not a precise value. Generally a "comfortable" temperature for most people is around 25°C (77°F), which is close to 300 K. From now on, if you are asked to consider a system at room temperature, you can safely assume it to be 300 K.

One last note: with the modern definition of temperature in the SI system, the kelvin is the actual unit of measurement, without saying "degree kelvin," as we do for Celsius or Fahrenheit temperatures. This is a convention since 1968, when the 13th General Conference on Weights and Measures dropped "degree" from this unit. As such, you might still see °K mentioned in some texts, especially older ones; however, we will stick with the modern usage.[6]

5 But this is only wrong because nowadays we are so used to higher temperatures being hotter; there is no reason it couldn't be this way.

6 And as the kelvin is the proper SI unit, still other texts will use "deg" or "degree" for the symbol K.

Many other temperature scales have been proposed throughout history, which I have plotted in Figure 5.5, each as a function of temperature in kelvin. Those that are still in use today are shown with solid lines, while those no longer used are shown with dashed lines. The two vertical lines show the freezing and boiling points of water.

Rømer: In 1701, Ole Christensen Rømer (1644–1710) defined zero degrees as the freezing temperature of a brine, and later updated this lower point such that the freezing point of water is set to be 7.5 degrees. For the other point to define his scale, he set the boiling point of water to be 60 degrees.

Newton: Also in 1701, Sir Isaac Newton (1643–1727) came up with his own temperature scale, although there is very little evidence of the exact numerical values he used. He published anonymously that "the heat of air in winter where water begins to freeze" was to have "0 degrees heat" while there was a range of other references for the second point. These were rather ambiguous and difficult to standardize (such as "the heat of midday in July"). As such, the Newton line in Figure 5.5 should be treated as largely conjectural.

Réaumur: In 1730, René-Antoine Ferchault de Réaumur (1683–1757) chose zero degrees as the freezing point of water and 80 degrees as the boiling point of water at normal atmospheric pressure.

Delisle: In 1732, Joseph-Nicolas Delisle (1688–1768) set the freezing point of water to be 150 degrees, and the boiling point of water to be zero. Like the original Celsius scale, Delisle chose an inverted temperature scale.

There is a nice blog post about this with other sources cited in Ref. [4].

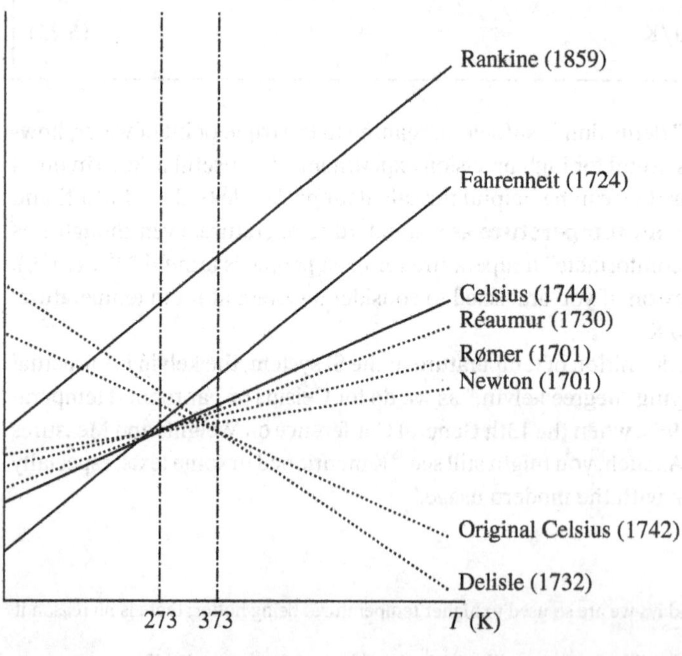

Figure 5.5 Various temperature scales that have been defined in the history of the subject. The solid lines are those temperature scales still in use today, while the dashed lines are not.

5.4 Summary

- A system is in thermal equilibrium if the number of states accessible to it, given any external constraints, is maximized.
- Spontaneous interactions never decrease the number of accessible states to a system, so $\Omega_f \geq \Omega_i$. If, after such a process, $\Omega_f = \Omega_i$ then the process is **reversible** and if $\Omega_f > \Omega_i$ then the process is **irreversible**. This is known as the **second law of thermodynamics**, that for an isolated system, $\Delta S \geq 0$ for any spontaneous process, as $S = k_B \ln \Omega$.
- Temperature, from a fundamental point of view, is defined by considering two systems in thermal equilibrium. The temperatures of two such systems are equal, and this arises from the requirement that the probability of a state must be maximized in equilibrium. It is defined as $T = 1/(k_B\beta)$, with $\beta = \partial \ln \Omega/\partial E$.
- The **zeroth law of thermodynamics** states that if two systems are each separately in thermal equilibrium with a third system, then they are in thermal equilibrium with each other.
- If two systems are in thermal contact, the *hotter* system is defined as the one which loses energy to the other (so $Q < 0$ for the hotter system). For positive absolute temperatures, this corresponds to the temperature of the hotter system being higher than that of the colder system.

Problems

5.1 Consider a macroscopic system at room temperature. Estimate by what *factor* $\Omega(E)$ increases for such a system when it absorbs a single quantum of visible light with energy $\Delta E \ll E$. Do this by calculating

$$\ln\left[\frac{\Omega(E+\Delta E)}{\Omega(E)}\right] = \ln \Omega(E+\Delta E) - \ln \Omega(E) \approx \frac{\partial \ln \Omega}{\partial E}\Delta E,$$

where the approximation is valid for $\Delta E \ll E$.

5.2 Suppose that a substance existed for which $\partial^2 \ln \Omega/\partial E^2 = \partial\beta/\partial E$ was **positive** for some region, such as A to B in the plot shown. If the substance, with an energy corresponding

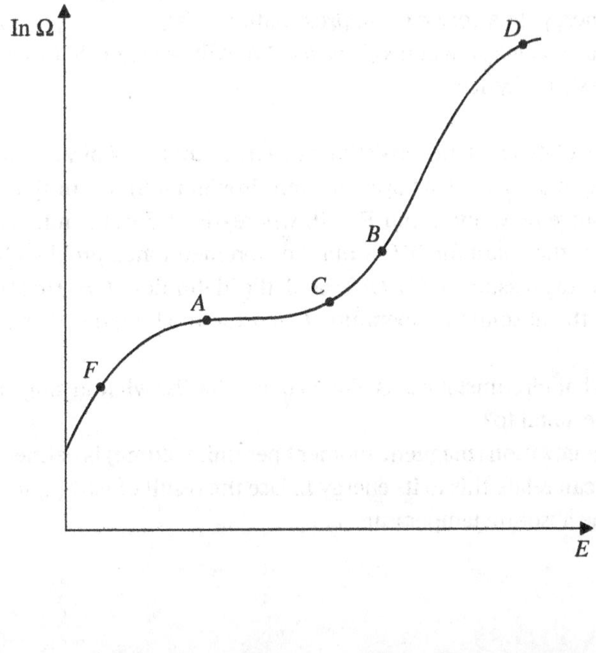

to point C on the plot, were brought into thermal contact with a **heat reservoir** at a very slightly higher temperature, what would happen? At which of the points A, B, C, D, or F, would the substance most likely be after coming to equilibrium with the heat reservoir? Assume both the substance and the reservoir are at a positive absolute temperature. Explain clearly. [Note: we will define a heat reservoir in Chapter 6, but for this problem, you just need to know that such a system can absorb or emit any amount of heat while its temperature does not change.]

5.3 In Problem 4.3, we had a set of N classical oscillators with total energy

$$\frac{p_x^2 + p_y^2 + p_z^2}{2m} + \frac{1}{2}m\omega^2(x^2 + y^2 + z^2),$$

and found that

$$\Omega(E) = (\text{const})E^{3N}.$$

Determine T from this, and solve your expression to obtain E as a function of T.

5.4 (a) In Problem 4.7, we had a set of N quantum mechanical oscillators with total energy

$$E = \frac{1}{2}N\hbar\omega + M\hbar\omega,$$

and you found the number of states with energy between E and $E + \delta E$ to be

$$\ln\Omega \sim (M + N)\ln(M + N) - M\ln M - N\ln N + \ln(\delta E/\hbar\omega).$$

Write this in terms of E (not M) and show that a collection of N harmonic oscillators at temperature T has energy

$$E = \frac{N\hbar\omega}{2}\left(\frac{e^{\hbar\omega/k_BT} + 1}{e^{\hbar\omega/k_BT} - 1}\right) = N\left(\frac{\hbar\omega}{2} + \frac{\hbar\omega}{e^{\hbar\omega/k_BT} - 1}\right).$$

(b) Show that when $k_BT \ll \hbar\omega$, the oscillators are all sitting in their ground states (state of lowest energy) to a very good approximation.

(c) Show that $E \approx Nk_BT$ when $k_BT \gg \hbar\omega$. We will see later that this is the answer one gets from classical physics.

5.5 Consider an isolated system consisting of a large number N of very weakly interacting localized particles of spin 1/2 (the system from Problem 4.6). As in that problem, we consider the energy range between E and $E + \delta E$ where $\delta E \ll E$ but is microscopically large so that $\delta E \gg \mu H$. Use the result for $\Omega(E)$ from the aforementioned problem here.

(a) With the expression for $\ln\Omega(E)$ and the definition $\beta = \partial\ln\Omega/\partial E$, find the relation between the absolute temperature T and the total energy E of this system, solving for $E(T)$.

(b) Under what circumstances is T negative? That is, what arrangement of the spins does that correspond to?

(c) The magnetization (magnetic moment per unit volume) is defined as $M \equiv (n_1 - n_2)\mu/V$, and you can relate this to its energy E. Use the result of part (a) to find M as a function of H and the absolute temperature T.

References

1 A. S. Oja and O. V. Lounasmaa. Nuclear magnetic ordering in simple metals at positive and negative nanokelvin temperatures. *Reviews of Modern Physics*, 69:1–136, Jan 1997.

2 M. Baldovin, S. Iubini, R. Livi, and A. Vulpiani. Statistical mechanics of systems with negative temperature. *Physics Reports*, 923:1–50, 2021. Statistical mechanics of systems with negative temperature.

3 S. Braun, J. P. Ronzheimer, M. Schreiber, S. S. Hodgman, T. Rom, I. Bloch, and U. Schneider. Negative absolute temperature for motional degrees of freedom. *Science*, 339(6115):52–55, 2013.

4 S. Matthiesen. A short history of temperature measurements and temperature scales. https://www.stephan-matthiesen.de/en/texts-blog/only-english-texts/138-a-short-history-of-temperature-measurements-and-temperature-scales.html. Accessed: 5 June 2023.

6

Thermodynamics: The Laws and the Mathematics

At this point we have worked through the fundamentals of statistical mechanics and made our first connections to thermodynamics. Mostly this has been from a rather abstract point of view, so now let's start considering the physical applications of what we have set up. We saw in Chapter 5 how the condition for two systems to be in equilibrium allows us to obtain the temperature of a system from its entropy, and now we will see how the entropy can lead to the other important thermodynamic quantities. While our results will follow from studying systems using the microcanonical ensemble, we will later see that they are quite general.

After finishing this chapter, you should be able to

- understand the basic interactions between systems,
- apply the laws of thermodynamics to real systems,
- learn how to calculate thermodynamic observables and relate "first derivatives" that are not easily measurable to a select few that are, and
- both be able to derive thermodynamic properties from first principles while also determining statistical quantities from an empirical point of view.

6.1 Interactions Between Systems

As discussed in Section 4.6, we aren't only interested in systems by themselves, but in how systems interact with each other. We have already discussed the first two types of interactions we will focus on: thermal interactions (where energy is transferred between systems in the form of heat) and mechanical interactions (where we allow the external parameters to change, so energy is transferred in the form of work). In both cases we will consider the processes that take place to be quasistatic, and we begin with the former of these two.

6.1.1 Quasistatic Thermal Interaction

We begin with two systems A and A' in thermal contact, but otherwise isolated from the surroundings, and held at fixed volumes (or any other external parameters), and as such the energy change of each system is given solely by heat transfer. We will also assume that the systems are in nearby

Statistical Thermodynamics: An Information Theory Approach, First Edition. Christopher Aubin.

Companion website: www.wiley.com/go/Aubin/StatisticalThermodynamics

equilibrium states so that the process is both quasistatic and infinitesimal, which means we can write $d\bar{E} = đQ$ for our system.[1]

We would like to relate this transfer of heat to entropy and temperature to then connect statistical mechanics more completely to thermodynamics. When our system absorbs some energy $đQ$, the entropy, which is a function of the mean energy, changes from $S(\bar{E}) \to S(\bar{E} + đQ)$. If $đQ$ is infinitesimal, we can expand $S(\bar{E} + đQ)$ to linear order in $đQ$, so the corresponding infinitesimal change in the entropy is

$$\begin{aligned} dS = S(\bar{E} + đQ) - S(\bar{E}) &= \left(\frac{\partial S}{\partial \bar{E}}\right) đQ + \frac{1}{2}\left(\frac{\partial^2 S}{\partial \bar{E}^2}\right)(đQ)^2 + \cdots \\ &\approx \frac{đQ}{T}. \end{aligned}$$

The second line arises from the definition of temperature along with neglecting the second-order term. If this is truly infinitesimal, we can drop the approximation symbol to get

$$dS = \frac{đQ}{T}. \tag{6.1}$$

Thus we have found an integrating factor for $đQ$: multiplying $đQ$ by $1/T$ turns this inexact differential into an exact differential, dS.

Exercise 6.1 Argue clearly why dS must be an exact differential.

The expression in Eq. (6.1) is true for all quasistatic processes, and if we are interested in the finite entropy change when going from an initial state i to a final state f, we just integrate it to obtain

$$\Delta S = \int_i^f \frac{đQ}{T}. \tag{6.2}$$

While the heat flow to go from one state to another, $Q = \int_i^f đQ$, depends on the particular process (i.e., whether or not work is also performed by the system), the entropy change only depends upon the initial and final states because S is a state function.

6.1.2 The Heat Reservoir

There's a special system we need to discuss that we have seen briefly in Problem 5.2: a **heat reservoir** (or heat bath). Such a system is much larger than other systems that might come into contact with it, and is quite common, both as a theoretical tool and in Nature. As with anything that has to do with sizes, this is a relative concept: A system which may be a heat reservoir for one may not be so for another. I will give examples of this later, but first let's define precisely what we mean by this concept.

Yet again, suppose we have two systems, A and A', in thermal contact with each other, such that A' is much larger than A, as shown schematically in Figure 6.1 (for now assume the walls of the container that holds A are fixed, so only thermal interactions are possible). Each system is at a temperature T and T', respectively, when we first put them into contact. As A' is much larger than A, we consider it a heat reservoir: It can absorb or emit any amount of heat and still remain at the same temperature, T'. This means that the entropy does not change significantly (it still changes, but not by much compared to its original entropy) when the internal energy of A' changes, or more

1 We will focus on one of the two systems for now, but if we were interested in A', we could then also write $d\bar{E}' = đQ'$.

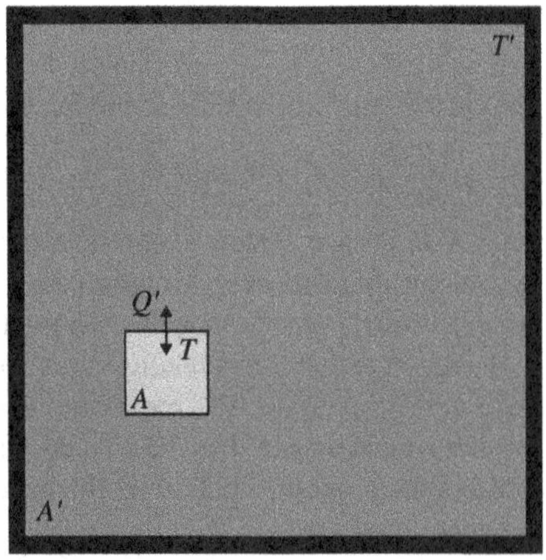

Figure 6.1 A system A (with rigid boundaries) in thermal contact with a heat reservoir, A'. The combined system is thermally isolated from the outside world.

fundamentally, the number of accessible states in our heat reservoir doesn't change significantly with the energy.

We can immediately guess what the change in the entropy of the heat reservoir should be by using Eq. (6.1). If T' remains constant even if a finite amount of heat Q' is absorbed, we can just say that the *finite* change in the entropy of the reservoir is given by

$$\Delta S' = \frac{Q'}{T'}. \tag{6.3}$$

While this is a good guess (it is correct, in fact), it would be good to see how this comes about formally.[2]

To verify our guess, Eq. (6.3), we must quantify what "fixed temperature" actually means, because it can't truly be fixed if the heat reservoir is not initially in thermal equilibrium with another system. We start by considering β', because even though T' is easier to understand from a practical perspective, β' is more useful for theoretical calculations. Mathematically, we can consider β' a function of the energy of the system $\bar{E}'$, so we can write, if Q' is small (compared to $\bar{E}'$),

$$\beta'\left(\bar{E}' + Q'\right) = \beta'(\bar{E}') + \left(\frac{\partial \beta'}{\partial \bar{E}'}\right) Q' + \cdots. \tag{6.4}$$

For a heat reservoir, the change in temperature should be negligible no matter how much energy is added, so we would expect $\beta'(\bar{E}' + Q') = \beta'(\bar{E}')$ to any reasonable precision. More specifically, we require

$$\left|\frac{\partial \beta'}{\partial \bar{E}'} Q'\right| \ll \beta', \tag{6.5}$$

where I take the absolute value since energy may be added or removed. As usual, of course, we are only interested in positive absolute temperatures (so $\beta' > 0$).

On the surface, Eq. (6.5) defines the condition for a heat reservoir irrespective of any other system. However, the energy our reservoir absorbs must come from somewhere, so even if not mentioned, there is a second system implied. With this in mind, recall that for a general system with f degrees of freedom, we saw from Exercise 5.5 that (applying this to our reservoir) $\bar{E}' = afk_BT'$ where a is a constant dependent on the system. We can write this as $\beta' = af/\bar{E}'$, and thus

$$\left|\frac{\partial \beta'}{\partial \bar{E}'}\right| = \frac{af}{\bar{E}'^2} = \frac{\beta'}{\bar{E}'}.$$

2 You can skip the rest of this section if you are interested in going right into applications.

When in contact with system A, the most energy that can be absorbed by A' would be $\bar{E}$, so we can say that $Q'_{max} = \bar{E}$, or we can say that generally Q' is at most on the order of the energy of the second system. Putting $Q' \sim \bar{E}$ into Eq. (6.5), we have

$$\left|\frac{\beta'}{\bar{E}'}\bar{E}\right| \ll \beta' \Rightarrow \bar{E} \ll \bar{E}'. \tag{6.6}$$

Thus A' is a heat reservoir relative to A if its energy is much larger than that of A. These are general statements that will work for most cases, but keep in mind that Eq. (6.6) is not our *definition* of a heat reservoir. Even so, we will often use this because it is simpler to conceptualize than Eq. (6.5).

Never forget that this is relative! Imagine the room you're sitting in while reading this and drinking a hot cup of coffee. The room can be considered a heat reservoir when compared to the coffee, but not when compared to the outside. Usually we won't have to worry about this as it'll be clear which system in our problems is the heat reservoir, but it's good to remember that it is a relative concept.

The concept of a heat reservoir is familiar in other contexts. For example, when working with electric circuits, the concept of **electrical ground** is important and considered an "infinite source or sink of electrons." This could be phrased in a similar way to Eq. (6.5) if one wished to be more rigorous.

I like to consider the ocean as another example—in this case, this is an infinite source (or sink) of water. You won't make a significant dent in the amount of water in the ocean if you take a cup (or several gallons) of water out of it.

These two examples allow us to consider our heat reservoir as an "infinite source or sink of heat," at least in a qualitative sense.

Now let's examine the entropy change of the heat reservoir, using the condition in Eq. (6.5). After the reservoir absorbs some heat Q', the entropy change is

$$\Delta S' = k_B \left[\ln \Omega'(\bar{E}' + Q') - \ln \Omega'(\bar{E}')\right].$$

Because $Q' \ll \bar{E}'$, we can expand this (for clarity, we'll assume that Ω' without an argument is $\Omega'(\bar{E}')$),

$$\begin{aligned}\Delta S' &= k_B \left(\ln \Omega' + \frac{\partial \ln \Omega'}{\partial \bar{E}'} Q' + \frac{1}{2}\frac{\partial^2 \ln \Omega'}{\partial \bar{E}'^2} Q'^2 - \ln \Omega'\right) + \cdots \\ &= k_B Q' \left(\beta' + \frac{1}{2}\frac{\partial \beta'}{\partial \bar{E}'} Q'\right) + \cdots . \end{aligned} \tag{6.7}$$

Exercise 6.2 Use the condition in Eq. (6.5) to argue that Eq. (6.7) leads directly to Eq. (6.3), our original guess for the entropy change of a heat reservoir.

6.1.3 General Interactions Between Systems

We have already seen that if we are given the number of states Ω in a system as a function of the energy, then we are able to not only determine (quite trivially) the entropy of a system, but also the temperature. This was shown assuming only thermal interactions, but let's now allow for a mechanical interaction so that the external parameter(s) can change as well. As before, we will specialize to the case where the only relevant external parameter is the volume, so that the work done is as shown in Eq. (4.51). We have two goals here:

1. First to determine the mean pressure $\bar{p}$ from the number of states, and
2. then to determine the requirements for mechanical equilibrium.

There are two ways to go about this: One way is easier to understand conceptually (but requires a few more assumptions), while the other is much more abstract (and thus is more general). I could just cover the first of these (again because it is simpler), but the second method allows us to see how we can gain a physical understanding about a system from a fundamental statistical approach. In either case, from now on, we will label the number of states to be explicitly a function of the energy as well as the volume: $\Omega(\bar{E}, V)$.[3]

6.1.3.1 Obtaining $\bar{p}$ from Ω

The first approach we will take requires us to examine the combined system $A^{(0)} = A + A'$ (where A' is not necessarily a heat reservoir) in Figure 6.2, and determine the requirements for them to be in equilibrium, as we did in Section 5.1.3. The difference here is that we will require them to be in both thermal and mechanical equilibrium. We isolate the combined system so that as before, the total energy $\bar{E}^{(0)} = \bar{E} + \bar{E}'$ is fixed, and the external walls are rigid, so the total volume $V^{(0)} = V + V'$ is constant as well. But unlike a purely thermal interaction, we allow the partition between the systems to move, so V and V' can each change.

In this case, we can write the total number of states as

$$\Omega^{(0)} = \Omega(\bar{E}, V)\Omega'(\bar{E}', V') = \Omega(\bar{E}, V)\Omega'(\bar{E}^{(0)} - \bar{E}, V^{(0)} - V), \tag{6.8}$$

where as before, in the second equality it is explicit that this is really only a function of the parameters of one system, $\bar{E}$ and V. The combined system is in equilibrium when the probability (or equivalently the logarithm of the probability) of that state is maximized, and this now requires us to include variations in the volume as well as the energy. We get for a total change in $\ln P = \ln \Omega^{(0)} + \text{constant}$ (now dropping the arguments in Ω and Ω' to avoid clutter),[4]

$$\begin{aligned} d\ln\Omega^{(0)} &= \left(\frac{\partial \ln \Omega^{(0)}}{\partial \bar{E}}\right)_V d\bar{E} + \left(\frac{\partial \ln \Omega^{(0)}}{\partial V}\right)_E dV \\ &= \left[\left(\frac{\partial \ln \Omega}{\partial \bar{E}}\right)_V + \left(\frac{\partial \ln \Omega'}{\partial \bar{E}}\right)_V\right] d\bar{E} + \left[\left(\frac{\partial \ln \Omega}{\partial V}\right)_E + \left(\frac{\partial \ln \Omega'}{\partial V}\right)_E\right] dV. \end{aligned}$$

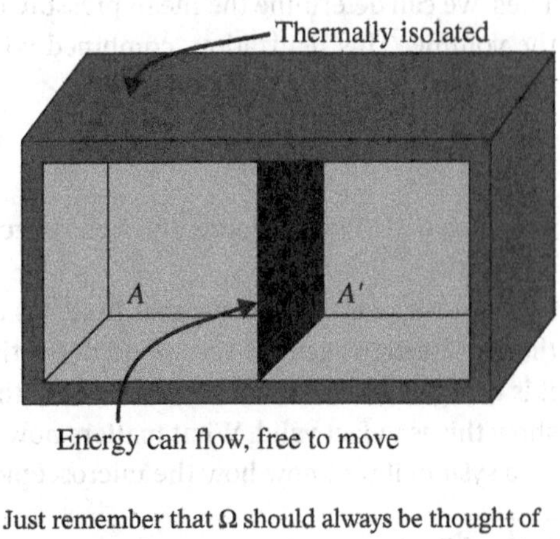

Figure 6.2 Two systems in thermal contact with each other, where the partition between them can move. The combined system is isolated from the outside world.

3 But sometimes I will omit the arguments as above for clarity: Just remember that Ω should always be thought of as $\Omega(\bar{E}, V)$ in this section.

4 A quick notation comment: I do not put the bars on $\bar{E}$ or $\bar{p}$ if those are variables being held constant, just because it looks a bit messier. Keep this in mind from now on.

Because of the dependence of Ω' on $\bar{E}$ and V, we have[5]

$$\left(\frac{\partial \ln \Omega'}{\partial \bar{E}}\right)_V = -\left(\frac{\partial \ln \Omega'}{\partial \bar{E}'}\right)_V, \quad \left(\frac{\partial \ln \Omega'}{\partial V}\right)_E = -\left(\frac{\partial \ln \Omega'}{\partial V'}\right)_E, \tag{6.9}$$

so that

$$d\ln\Omega^{(0)} = \left[\left(\frac{\partial \ln \Omega}{\partial \bar{E}}\right)_V - \left(\frac{\partial \ln \Omega'}{\partial \bar{E}'}\right)_V\right] d\bar{E} + \left[\left(\frac{\partial \ln \Omega}{\partial V}\right)_E - \left(\frac{\partial \ln \Omega'}{\partial V'}\right)_E\right] dV.$$

For this to vanish, the factors in square brackets have to separately vanish (as $\bar{E}$ and V are independent), so we again have

$$\left(\frac{\partial \ln \Omega}{\partial \bar{E}}\right)_V = \left(\frac{\partial \ln \Omega'}{\partial \bar{E}'}\right)_V,$$

or $\beta = \beta' \Rightarrow T = T'$—the temperatures of the two systems are equal. This is nothing new, although now it is more explicit that volume (or in more general systems, any external parameters) is held constant in our definition of temperature.

The second condition is that

$$\left(\frac{\partial \ln \Omega}{\partial V}\right)_E = \left(\frac{\partial \ln \Omega'}{\partial V'}\right)_E. \tag{6.10}$$

But what does this quantity physically correspond to? To figure this out, we look at the differential of $\ln \Omega$ for a single system,

$$d\ln\Omega = \left(\frac{\partial \ln \Omega}{\partial \bar{E}}\right)_V d\bar{E} + \left(\frac{\partial \ln \Omega}{\partial V}\right)_E dV, \tag{6.11}$$

and a couple of steps (see Exercise 6.3) allow us to write this in terms of β and S, so

$$dS = \frac{d\bar{E}}{T} + \left(\frac{\partial S}{\partial V}\right)_E dV. \tag{6.12}$$

Given that we know $d\bar{E} = đQ - đW$, and for a quasistatic process we have $đQ = TdS$ and $đW = \bar{p}dV$, we can write

$$\frac{\bar{p}}{T} = \left(\frac{\partial S}{\partial V}\right)_E \Rightarrow \beta\bar{p} = \left(\frac{\partial \ln \Omega}{\partial V}\right)_E. \tag{6.13}$$

Thus, we can determine the mean pressure of a system if we know how the entropy depends upon the volume. This realization, combined with Eq. (6.10), allows us to see that our condition for mechanical equilibrium is that

$$\bar{p} = \bar{p}'. \tag{6.14}$$

Exercise 6.3 Work through the steps to get from Eq. (6.11) to Eq. (6.12) and finally to Eq. (6.13).

Of course, this only shows that $\ln \Omega^{(0)}$ is an *extremum*, and to be certain that it is a *maximum*, then we need to examine its second derivative and ensure it is negative. For now let's just assume it is true (we have seen some cases where this is clearly the case), and later in Chapter 10 we will show this is in fact valid. What matters now is that we have a way to determine the mean pressure on a system if we know how the microscopic number of states depends upon the volume.

5 Also, following up on the previous footnote, for simplicity we also do not bother with the primes on variables that are being held constant; generally it will be clear which variables we are referring to. Additionally, here it does not matter given the relationship between them: If $\bar{E}$ is constant, so is $\bar{E}'$ and the same with V and V'.

Equation (6.13) allows us to obtain $\overline{p}$ as a function of other thermodynamic variables, known as an **equation of state**. Similarly, Eq. (5.14) allows us to obtain the energy as a function of other thermodynamic variables. These two equations, we will see, allow us to determine everything we will need to know about a given system. While ideally we would determine these from the microscopic theory (by first calculating S), we will find that we can also determine these through other, empirical, means. This will be useful later when we consider more complicated systems.

6.1.3.2 An Alternative Derivation of the Relationship Between $\overline{p}$ and Ω

Another way to derive this relationship from a more statistical point of view comes from chapter 3 of Ref. [1], which we will work through here. You can skip this section if you would prefer to continue with the study of thermodynamics, but this gives a marvelous (and deeper) insight to how we can connect statistical mechanics to thermodynamics.

In general, the energy of a given microstate r is dependent on the volume, $E_r(V)$. For example, we can write for a three-dimensional quantum mechanical particle in a cubical box with side L,

$$E_r = \frac{\hbar^2\pi^2}{2mL^2}(n_x^2 + n_y^2 + n_z^2) = \frac{\hbar^2\pi^2}{2mV^{2/3}}(n_x^2 + n_y^2 + n_z^2). \tag{6.15}$$

By changing the volume from V to $V + dV$, the energy of each microstate $E_r(V)$ will change by an amount

$$dE_r = \left(\frac{\partial E_r}{\partial V}\right) dV = -p_r dV. \tag{6.16}$$

When this happens, the number of states in the range E to $E + \delta E$ can change as well. Imagine, for example, we have a state r with energy $E_r < E$ initially, and after changing the volume by an amount dV, the new energy $E_r - p_r\,dV$ could lie within the range E to $E + \delta E$. The key is determining how to quantify this, and this is where we take an abstract turn.

Let's denote $\Omega_y(E, V)$ as the number of states with energy in the range $\{E, E + \delta E\}$ with volume V, *and* such that their derivative $(\partial E_r/\partial V)$ is given in some range $\{y, y + \delta y\}$. y here has the dimensions of pressure, and we will relate these two quantities shortly. The total number of states can be obtained by summing over all allowed values of y,

$$\Omega(E, V) = \sum_y \Omega_y(E, V). \tag{6.17}$$

Now consider some particular energy E. After changing the volume, some states with $E_r < E$ might gain energy and thus enter the $\{E, E + \delta E\}$ region and some states with $E < E_r < E + \delta E$ will gain energy and then have an energy above $E + \delta E$, leaving this region. Obviously some states' energies will change but remain within the region. To quantify these changes, we define

$\omega(E)$ = # of states whose energy increases from an amount
less than E to between $\{E, E + \delta E\}$,

to give us precisely how many states have been *added* to the energy range from below, thereby increasing Ω. As we did for Ω, we can also consider $\omega_y(E)$, the number of states that are in the region $\{E - ydV, E\}$ (and thus *below* the energy E), and that have an energy change given by

$$\left(\frac{\partial E_r}{\partial V}\right) dV,$$

which is sufficient to increase E_r to be in the range $\{E, E + \delta E\}$. The density of states (number of states per unit energy) in this region of width ydV is given by $\Omega_y(E, V)/\delta E$, and so the total number of states with energy in the range $\{E - ydV, E\}$ can be written as

$$\omega_y(E) = \frac{\Omega_y(E, V)}{\delta E} ydV. \tag{6.18}$$

We then sum over all possible values of y to get $\omega(E)$,

$$\begin{aligned}\omega(E) &= \sum_y \frac{\Omega_y(E, V)}{\delta E} ydV \\ &= \sum_y \frac{\Omega(E, V)}{\delta E} \frac{\Omega_y(E, V)y}{\Omega(E, V)} dV.\end{aligned} \tag{6.19}$$

In the second line I multiplied by $1 = \Omega(E, V)/\Omega(E, V)$ to be able to write this in terms of the mean value of y,

$$\sum_y y \frac{\Omega_y(E, V)}{\Omega(E, V)} = \sum_y yP_y = \bar{y},$$

allowing us to write

$$\omega(E) = \left[\frac{\Omega(E, V)}{\delta E}\right] \bar{y} dV. \tag{6.20}$$

We noted above that y has dimensions of pressure, but to be more precise, recalling Eq. (4.50),

$$\bar{y} = \overline{\left(\frac{\partial E_r}{\partial V}\right)} \equiv -\bar{p}.$$

$\omega(E)$ is the number of states with energy less than E such that they have an energy in the range $\{E, E + \delta E\}$ after the volume changes. But in the microcanonical ensemble, we are more interested in the *total* number of states in that range, Ω. Specifically we are asking, "How does the number of states in the range E to $E + \delta E$ change when we change the volume by an amount dV at fixed energy?" Mathematically, we wish to determine

$$\left(\frac{\partial \Omega(E, V)}{\partial V}\right)_E dV.$$

We can get this by considering the number of states *entering* the region and then subtracting the number of states *leaving* it, or

$$\left(\frac{\partial \Omega(E, V)}{\partial V}\right)_E = \omega(E) - \omega(E + \delta E) = -\frac{\partial \omega(E)}{\partial E} \delta E. \tag{6.21}$$

The second equality is technically only approximately true, when $\delta E \ll E$, but as we have seen, for our purposes this can be treated as an exact equality.

Example 6.1 Consider the energy levels shown schematically in Figure 6.3 for a single particle in a three-dimensional box as described by Eq. (6.15). The solid lines are energy levels in the region of interest before we change the volume (there are 45 energy levels), while the dashed lines are not. We can read off $\omega_y(E)$ and $\omega_y(E + \delta E)$ (c.f. Eq. (6.18)) by counting the states between the thick solid line and dash-dotted line: there are six states that will enter from below energy E and seven states that will leave from just below energy $E + \delta E$. Thus, after changing the volume, there is a net loss of one state in the region of interest. Of course, there are other values of y that can exist, hence the need to sum over all possible values of y that would change the number of states as in Eq. (6.19). Keep in mind that we cannot apply Eq. (6.21) to this figure, because we don't know how ω depends on the energy; this is meant as a way to visualize what we are doing here.

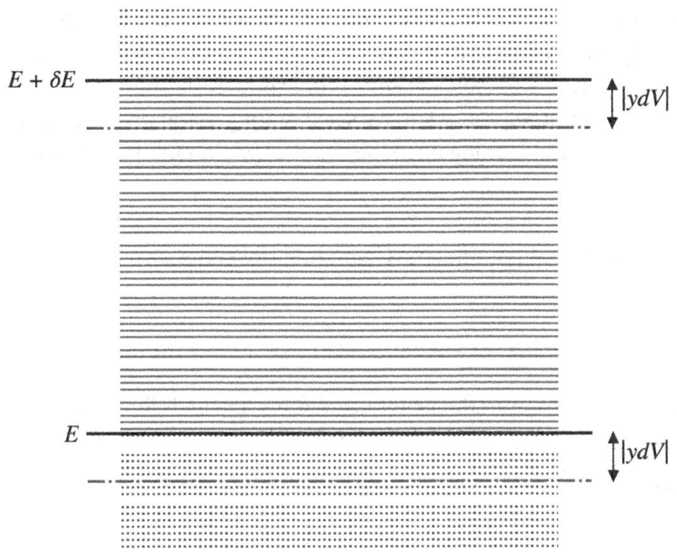

Figure 6.3 A graphical representation of the motion of states as we change the volume of the system. Those states just below $E + \delta E$ in the region of size $|ydV|$ will leave the region when we change the volume and those states just below E in the region of size $|ydV|$ will enter the region.

We can work more with Eq. (6.21) because we have an expression for $\omega(E)$. Putting Eq. (6.20) into Eq. (6.21), we obtain

$$\begin{aligned}\left(\frac{\partial\Omega(E,V)}{\partial V}\right)_E dV &= -\left[\frac{\partial(\Omega\overline{y})}{\partial E}\frac{1}{\delta E}dV\right]\delta E\\ &= -\frac{\partial(\Omega\overline{y})}{\partial E}dV,\end{aligned} \tag{6.22}$$

and using this we can show that

$$\left(\frac{\partial\Omega}{\partial V}\right)_E = \frac{\partial\overline{p}}{\partial E} + \frac{\partial\ln\Omega}{\partial E}\overline{p}. \tag{6.23}$$

Exercise 6.4 Fill in the steps to go from Eq. (6.22) to Eq. (6.23). You will want to multiply by Ω/Ω in the appropriate place and use the fact that $\overline{y} = -\overline{p}$.

The first term on the right-hand side in Eq. (6.23) is generally of order $\overline{p}/E$ for large systems, while the second term we can use the result in Eq. (4.28) we derived in Section 4.5.3, where $\Omega \sim (\text{const})E^{af}$, so

$$\frac{\partial\ln\Omega}{\partial E}\overline{p} \sim \frac{\partial\ln E^{af}}{\partial E}\overline{p} \sim \frac{af\overline{p}}{E} \gg \frac{\overline{p}}{E}. \tag{6.24}$$

Therefore, for macroscopic systems (with $f \sim 10^{24}$ or more) we can neglect the first term, and to an excellent approximation,

$$\frac{\partial\ln\Omega}{\partial V} = \beta\overline{p} \Rightarrow \overline{p} = T\left(\frac{\partial S}{\partial V}\right)_E, \tag{6.25}$$

which is just what we had in Eq. (6.13).

We could generalize this if we had different external parameters,

$$\left(\frac{\partial\ln\Omega}{\partial x_i}\right)_E = \beta\overline{X}_i, \tag{6.26}$$

where X_l is the **generalized force conjugate to the external parameter** x_l (for volume, the "force" is pressure). We can also easily include more than one external parameter—for each case, the energy and all the other external parameters are held constant when differentiating with respect to a given parameter. Comparing this to our other approach, we see that for general problems with external problems other than volume, the forces conjugate to the external parameter must be equal in thermodynamic equilibrium.

Exercise 6.5 Consider the following external parameters. Given that the generalized force conjugate to the external parameter is

$$X = -\frac{\partial E_r}{\partial x},$$

determine the physical quantity that corresponds to the generalized force for the following:

1. The length of a rubber band, ℓ.
2. The magnetic field H of a region.
3. The electric field $\mathcal{E}$ of a region.

6.1.3.3 The Classical Ideal Gas Revisited

Let's determine the mean pressure for a classical ideal gas. The number of microstates with energy from E to $E + \delta E$ is given by Eq. (4.41) for any ideal gas (not necessarily monatomic),

$$\Omega(E, V) = V^N \chi(E),$$

where χ is independent of V. We can write $\ln \Omega = N \ln V + \ln \chi(E)$, and using Eq. (6.25), we find

$$\bar{p} = \frac{1}{\beta}\frac{N}{V},$$

or

$$\bar{p}V = Nk_BT, \tag{6.27}$$

which is just the ideal gas equation: the equation of state for the ideal gas. This was originally determined from Boyle's law ($\bar{p} \propto 1/V$ at constant temperature),[6] Charles' law ($V \propto T$ at fixed pressure),[7] and Avogadro's law ($V \propto N$ at fixed temperature and pressure), all of which were determined empirically. In our case, however, using a statistical approach, we have derived this from a first-principles, microscopic understanding of this simple system. Note it is not dependent on the *type* of gas, because Ω factorizes into two functions: one of which depends only on the energy (which would be dependent on the type of gas) and the other a known function of the volume, which is the same for all ideal gases.

There are several other forms of the ideal gas equation we will use at different times. The first requires the **number density of particles,**

$$n \equiv \frac{N}{V},$$

and in terms of n the ideal gas equation becomes

$$\bar{p} = nk_BT. \tag{6.28}$$

---

6 Robert Boyle (1627–1691).
7 Jacques Charles (1746–1823).

You have likely seen the version of this equation in terms of the gas constant $R = N_A k_B$, or

$$\bar{p}V = \nu RT. \tag{6.29}$$

In this equation, $\nu = N/N_A$ is the number of moles and N_A is Avogadro's number (the number of molecules in a mole, given by Eq. (A.1)). All of these forms are useful in their own ways, depending on the situation.

We can also calculate the mean energy from Ω, which we did in Chapter 5 but it's useful to review this. With $\beta = \left(\partial \ln \Omega / \partial \bar{E}\right)_V$, we have

$$\frac{1}{k_B T} = \frac{d \ln \chi}{d\bar{E}} \Rightarrow \bar{E}(T, V) = f(\bar{E}), \tag{6.30}$$

or rather, the mean energy of an ideal gas is *independent of the volume*, which we saw in Exercise 5.7. For the specific case of the monatomic ideal gas, we found $\chi(\bar{E}) = B\bar{E}^{3N/2}$, so in this case,

$$\frac{d \ln \chi}{d\bar{E}} = \frac{3N}{2\bar{E}},$$

and thus

$$\bar{E} = \frac{3}{2} N k_B T = \frac{3}{2} \nu RT. \tag{6.31}$$

This was the result of Exercise 5.6.

Note the tremendous power of statistical mechanics—we have found two equations that constrain the five variables $S, \bar{E}, V, T$, and $\bar{p}$. Remember that each of these equations above (for $\bar{E}(T, V)$ and $\bar{p}(T, V)$) is called an equation of state (although sometimes only that involving the mean pressure is referred to as such). From a thermodynamic point of view they are often obtained empirically (and we will do this later), which is an important way to understand how systems work; however, it is rather exciting to be able to derive them by just knowing how the fundamental system behaves.

Extensive vs. Intensive Parameters

Before continuing with more physics discussions, let's take a brief pause to discuss a way to classify the large number of macroscopic bulk parameters that we have developed at this point, primarily,

$$\bar{E},\ T,\ S,\ \bar{p},\ V, \text{ and } N.$$

Additionally, we have already seen some related quantities, such as the number of moles $\nu = N/N_A$, or the number density $n = N/V$. There will be more as we continue, and it will be useful to classify these parameters as either **extensive** or **intensive**.

The difference between these two types of parameters is simple: An intensive parameter is one which does not scale with the system, while an extensive parameter does. Suppose we take two systems that are in equilibrium with each other (they have the same pressures and temperatures) and combine them into one larger system. The intensive parameters will remain the same, while all of the extensive parameters will add together. We can see that of the quantities we have discussed up to now,

Intensive: $\bar{p}$ and T,
Extensive: $\bar{E}, S, V$, and N.

(Continued)

(Continued)

We can state this mathematically by scaling a system by some factor α: All extensive parameters will scale by that same factor, so that $S \to \alpha S$, $V \to \alpha V$, etc. At the same time, all intensive quantities will remain unchanged, so $T \to T$, $\overline{p} \to \overline{p}$, etc.

We can form intensive quantities quite easily from ratios of extensive quantities, which you have seen before this subject: The most common example is the mass density of a system. The mass m and volume V of a system are both extensive, but the mass density $\rho = m/V$ is intensive. Many intensive quantities constructed in such a manner are useful because their value will be a property of the system itself (as in, what makes up the system and how so) but not its size. Among the new variables introduced above, ν is extensive, and $n = N/V$ is intensive.

Exercise 6.6 Consider the **molar quantities** such as the molar entropy s, molar energy ε, and molar volume v. These are each the quantity per unit mole; how would you define them, and are they extensive or intensive?

Exercise 6.7 Consider the results for a monatomic ideal gas, Eqs. (6.27) and (6.31) and show that they are consistent with our identifications of which variables are extensive and intensive. (That is, if you scale the system by some factor α, the equations of state are consistent if you scale the extensive parameters by α and keep the intensive parameters fixed.)

6.1.4 The Entropy in the Ground state

While we have been discussing the entropy as a means to an end (a way to obtain the equations of state for a system), it has some interesting properties in its own right. The entropy is unique to a given macrostate of our system, as it is given by the (logarithm of the) number of microstates accessible to the system in that macrostate. The number of microstates, as we have seen before, is dependent on the energy of the system as well as external parameters like the volume. Imagine that we now fix the volume, so the only parameter we can change is the energy (that is, we can write $\Omega(E, V) \equiv \Omega(E)$).

As the energy is lowered, the system will (eventually) approach the quantum mechanical ground state.[8] For most systems, there is only one ground state, although we might have some **degeneracy** in the ground state energy, meaning there might be a small number of microstates that all have the same lowest energy, E_0. Let us assume that there are $\Omega(E_0) = \Omega_0$ such ground states (and of course, if there is no degeneracy, then $\Omega_0 = 1$). For most systems, generally the number of microstates increases as we increase the energy, so as E increase from E_0, then the number of states increases, or $\Omega(E) \geq \Omega_0$. Thus quite generally we can say about the entropy that

$$\text{as } E \to E_0, \quad S \to S_0 = k_B \ln \Omega_0 = \text{minimum}.$$

Exercise 6.8 Argue that S_0 might as well be treated as zero for our purposes (and this is what we will do from now on), considering how Ω depends upon energy for general macroscopic systems.

8 While we treat many systems classically, we know there's a limitation to this, and every system eventually would have to be treated quantum mechanically.

Assume you have a mole of molecules and compare the entropies at some larger energy to that of the ground state.

While we have seen in Chapter 5 that it is possible for the absolute temperature to be negative, we won't study those systems here, and thus we can say for any system of interest here that β (T) decreases (increases) as the energy increases

$$\frac{\partial \beta}{\partial E} < 0, \text{ or } \frac{\partial T}{\partial E} > 0.$$

This implies that as the energy of our system approaches the ground state energy, the temperature goes to absolute zero (from the positive side). This statement is important enough that it is our final law of thermodynamics.

The Third Law of Thermodynamics states that as a system approaches absolute zero, the entropy vanishes, or

$$\text{as } T \rightarrow 0, \quad S \rightarrow 0. \tag{6.32}$$

This is often written as

$$\text{as } T \rightarrow 0^+, \quad S \rightarrow S_1, \tag{6.33}$$

where S_1 is a constant that depends on the system (which we could evaluate explicitly if we had the microscopic details).

The notation 0^+ reminds us that we are approaching absolute zero from the positive side. The reason for this notation is that a system can never reach absolute zero, which is something you have possibly heard before. There are many ways this can be understood, but mostly it is from a practical perspective: Even if you could lower the energy of a system to reach absolute zero, there will always be quantum fluctuations that give rise to some thermal energy. As the temperature is a measure of the average thermal energy, T will always have a (very small) finite value.[9]

Summary of the thermodynamic laws and statistical relations
At this point we have pieced together all of the fundamental ideas we will need to continue our study of thermodynamics and statistical mechanics: four laws of thermodynamics and important equations that define our system statistically. This is a good time to summarize them all in one place. Through a careful study of the microscopic system, with a few assumptions about how systems should behave, we have been able to derive the laws of thermodynamics, which is often the starting point for many who study the subject. From a thermodynamic perspective these are stated as postulates from which you can do most of what we will do for the rest of this chapter. These laws govern the macroscopic behavior of a system are given as follows.

Zeroth law of thermodynamics: If two systems are in thermal equilibrium with a third system, they are in thermal equilibrium with each other. Or, if we have three systems A_1, A_2, and

(Continued)

9 From a theoretical perspective, remember β is a more valid parameter to consider as that relates to the change in the entropy with the energy. T becoming zero means that β has reached ∞, which is not possible (one can only approach it).

(Continued)

A_3, and we know that

$$T_1 = T_2 \quad \text{and} \quad T_1 = T_3,$$

then we must have

$$T_2 = T_3.$$

First law of thermodynamics: This is a statement of the conservation of energy. For a completely isolated system, then $\bar{E} =$ constant. However, if it is not isolated, then the change in the energy of the system is given by the heat added to the system minus the work done by the system. In infinitesimal form this is

$$d\bar{E} = đQ - đW,$$

and a finite energy change would be given by

$$\Delta\bar{E} = Q - W.$$

(When in Chapter 10 we allow the number of molecules in the system to change, then there will be an additional term on the right-hand side.)

Second law of thermodynamics: For any thermally isolated system undergoing a spontaneous process, the entropy can never decrease, or

$$\Delta S \geq 0.$$

If the system isn't isolated, then for a quasistatic process, we can calculate the infinitesimal entropy change in terms of the heat flow,

$$dS = \frac{đQ}{T}.$$

Third law of thermodynamics: As our system approaches absolute zero, then the entropy reaches a minimum value, or as we put it earlier,

$$\text{as } T \to 0^+, \quad S \to S_1,$$

where the notation indicates that we are approaching $T = 0$ from the positive side, and S_1 is a constant that could be known if we knew the microscopic details of the system. Remember, this form only is used because a system generally cannot reach absolute zero, but often we will just say that as $T \to 0$, then $S \to 0$.

The laws above are written in terms of general macroscopic quantities (such as $đW$), but for most of our applications, the work done by a system will arise from volume changes, so for a quasistatic process the first law is given by

$$d\bar{E} = TdS - \bar{p}dV.$$

The important thermodynamic variables that we will consider for now are

$$\bar{E},\ S,\ T,\ \bar{p},\ \text{and } V.$$

(The number of molecules N is also important, but for now it will not vary, so I omit it from this list.)

However, we don't want to forget about the basic statistical relations from which these laws come, as that allows us to access many more physical systems if need be (such as those with negative absolute temperature). These relations allow us to connect the microscopic (statistical) world to the macroscopic (thermodynamic) world. We first have the connection between probability and Shannon (Gibbs) entropy, which is just the missing information from Chapter 3,

$$S = -k_B \sum_r P_r \ln P_r,$$

or applying the fundamental postulate of statistical mechanics (for the microcanonical ensemble),

$$S = k_B \ln \Omega.$$

Because the probability of being in a given state is proportional to Ω, we have

$$P \propto \Omega \propto e^{S/k_B}.$$

If we know the number of states from a microscopic calculation (from Chapter 4), then essentially everything in thermodynamics can be determined from this relation. For a lot of classical thermodynamics, of course, it is easier to work with the macroscopic parameters, and that will be how we approach the rest of this chapter.

6.2 The First Derivatives

We have already seen how important changes in thermodynamic quantities are, as we generally ask how one quantity changes while we change another. For example, if we fix the temperature of an ideal gas, we know from the ideal gas equation that the pressure depends inversely on the volume like $\overline{p} \propto 1/V$. If we wanted to formalize this more, we would say we're calculating

$$d\overline{p} = \left(\frac{\partial \overline{p}}{\partial V}\right)_T dV.$$

That is, if we changed the volume of a system by an amount dV, what would the resulting change in the pressure, $d\overline{p}$, be?[10] From a theoretical perspective, if we know $\overline{p}(V, T)$, we simply evaluate this first derivative. Experimentally, we just fix the temperature while changing the volume, and then measure how the pressure changes. There are plenty of examples, however, where such changes would be difficult to measure. For example, perhaps we wish to know how the entropy of a system changes when we change the pressure at fixed volume, or

$$dS = \left(\frac{\partial S}{\partial \overline{p}}\right)_V d\overline{p}.$$

This can often be difficult to both calculate theoretically *and* to measure (how do we measure a change in the entropy?). Luckily, it turns out that this, and any other first derivatives can be written in terms of just three: the heat capacity, the coefficient of thermal expansion, and the isothermal compressibility. We will learn *how* to do this in Section 6.4, but first let's define these three quantities as well as some other new quantities that come from the Legendre transform in Section 6.3.2.

10 The treatment and manipulation of partial derivatives in thermodynamics is often confusing, as it's different than how we usually consider them in other areas of physics. In Appendix E, I work through some examples to connect these different treatments.

6.2.1 Heat Capacity

The first of our three easily measurable quantities is the **heat capacity at constant *y***. This tells us how much energy in the form of heat $đQ$ is required to change the temperature of a system by an amount dT while holding some parameter y constant, or

$$C_y \equiv \left(\frac{đQ}{dT}\right)_y . \tag{6.34}$$

Generally this is dependent on the temperature of the system and often the parameter held constant. This is not *technically* a derivative, given that an inexact differential ($đQ$) is in the numerator; however, for a quasistatic process, we will see that we can write this as a true derivative, so the notation in Eq. (6.34) is useful.

The two most common heat capacities we will work with are at constant volume (an **isochoric process**) and constant pressure (an **isobaric process**),

$$C_V = \left(\frac{đQ}{dT}\right)_V \text{ and } C_p = \left(\frac{đQ}{dT}\right)_p , \tag{6.35}$$

respectively.

Exercise 6.9 Argue that the heat capacity (regardless of what is held constant) is an extensive quantity.

As the heat capacity is an extensive quantity, it is useful to define an intensive form of it, known as the **specific heat at constant *y***. There are two common specific heats we will consider: the *molar specific heat*,

$$c_y^{(v)} \equiv \frac{C_y}{v}, \tag{6.36}$$

and the *specific heat per mass*

$$c_y^{(m)} \equiv \frac{C_y}{m}. \tag{6.37}$$

The second of these is common in chemistry and in introductory treatments of thermodynamics, and also just more useful to understand (as the mass of a substance seems more tangible, as we are more familiar with it); however, we will often use the molar specific heat as well. One sidenote, the notation given in these equations will generally not be used; I will usually drop the (m) or (v) superscript, and which specific heat we are considering will (hopefully) be clear from the context.[11]

A quick comment on the units. In the SI system, the heat capacity has units of J/K (or dimensions of energy per degree), while the specific heats have units of J/mol/K or J/(kg·K), respectively. Often you will also see the **calorie** (not to be confused with the unit used in the United States for dietary information, where 1 Calorie = 1 kilocalorie = 1000 calories) used as a unit of energy, where

$$1\,\text{cal} = 4.1840\,\text{J}, \tag{6.38}$$

11 It is exceptionally rare to use both specific heats in the same problem, which is why we can get away with using the same notation for both.

and as such you might see cal/K for heat capacity. When solving problems with numerical values for the heat capacity or specific heats, just be sure to note the units used to realize which (C, $c^{(v)}$, or $c^{(m)}$) is under consideration.

One other comment here is that usually instead of using kelvin, we will sometime see the degree Celsius used, such as J/°C. The reason for this is the heat capacities are only relevant when there are changes in the temperature (by definition), and remember the size of one degree Celsius is the same as the size of one kelvin. Since Celsius is more familiar (and more relevant for many everyday experiments), this other notation is common.

Exercise 6.10 The specific heat per mass of copper is 0.385 J/(g · K).

1. Convert this to a molar specific heat (the molar mass of copper is 63.546 g/mol).
2. Another specific heat we can define is the **specific heat per molecule**. Calculate this for copper.

As you can see, the specific heat per molecule is such a small number, it will be cumbersome for numerical calculations but there are reasons to consider it for some problems (e.g., Example 8.2).

In order to measure the specific heat of a substance, we need to either fix the volume (common for gases and liquids) or fix the pressure (common for solids). In the former case, if the volume is fixed, no work can be done so we have

$$đQ = d\bar{E}.$$

If we can measure the energy change of the system as we change the temperature, we can get the heat capacity (or specific heat) at constant volume.

In the second case, by fixing the pressure, work can be done so we have

$$đQ = d\bar{E} + \bar{p}dV.$$

Generally speaking, if the energy of a system changes by some amount, we can see that the heat added at constant volume will be less than that at constant pressure, given that some of the energy goes to work done by the system in the latter case. Thus we generally expect

$$c_p > c_V,$$

and we'll see this when we get to various examples.

The second law of thermodynamics allows us to write the heat capacity in terms of the entropy change, at least for quasistatic processes. In this case, we have $đQ = TdS$, so we can write

$$C_y = T\left(\frac{\partial S}{\partial T}\right)_y. \tag{6.39}$$

The molar specific heat can be written as

$$c_y = T\left(\frac{\partial s}{\partial T}\right)_y, \tag{6.40}$$

where you determined in Exercise 6.6 that the molar entropy is given by

$$s = \frac{S}{\nu}. \tag{6.41}$$

If you want to determine the specific heat per unit mass you could define a similar intensive quantity, the entropy per unit mass, to use in Eq. (6.40).

Using this along with the first law of thermodynamics, the heat capacity at constant volume can be written as

$$C_V = T\left(\frac{\partial S}{\partial T}\right)_V = \left(\frac{\partial \bar{E}}{\partial T}\right)_V, \tag{6.42}$$

because $d\bar{E} = TdS$ if V is constant ($dV = 0$). For systems under consideration here, the energy increases when the temperature does, $(\partial\bar{E}/\partial T)_V > 0$, and so the heat capacity at constant volume is positive. With our everyday understanding of heat, this makes sense: If you add energy as heat to an object, then the temperature should go up, leading to a positive heat capacity. We will also show in Chapter 10 that this is a general requirement for stable equilibrium.

Exercise 6.11 Use Eq. (6.42) to calculate C_V for a classical monatomic ideal gas. Also calculate c_V, the molar specific heat, for this gas. You should find them to both be constant: $C_V = \frac{3}{2}Nk_B$ and $c_V = \frac{3}{2}R$.

As you can see, if we know the microscopic structure of the system, we can easily calculate the heat capacities (or specific heats). Additionally, they are generally easy to measure, as we will discuss.

While generally the heat capacities are functions of the temperature, there are times where they are not (such as the case of the classical monatomic ideal gas). If they are independent of the temperature, the heat absorbed by the system as it goes from some initial T_i to some final T_f can be easily calculated as[12]

$$Q = \int_i^f mcdT = mc(T_f - T_i). \tag{6.43}$$

Here we use the specific heat per unit mass as this is often what is used in chemistry and introductory physics texts (and it's also easy for us to measure the mass of a substance but not always the number of moles!). Note also I am not distinguishing between c_V and c_p, because either would be valid, so it's easiest to just omit the subscript.

We can use this to determine the final temperature of two (or more) systems after coming into thermal contact with each other. Assume each system, A_1 and A_2, has initial temperatures T_1 and T_2, respectively, and after reaching equilibrium they will both be at temperature T_f. Then $Q_1 = m_1c_1(T_f - T_1)$ and $Q_2 = m_2c_2(T_f - T_2)$, and because

$$Q_1 + Q_2 = 0, \tag{6.44}$$

if we knew the specific heats of systems A_1 and A_2, then we could use this to determine the final temperature the systems will have upon reaching equilibrium. Alternatively, if we knew the specific heat of one system and not the other, then we could *measure* the final temperature of the combined system to determine the specific heat of the other system, which is a common introductory experiment. As mentioned above, we didn't include a subscript on the specific heat, although c_p is usually more easily measured experimentally (it is often easier to hold the pressure constant). We will derive an expression to relate the two (see Eq. (6.92) later in this chapter), so in the end it doesn't really matter which we are considering, so long as it's clear.

12 This is the most common example seen in chemistry and introductory physics classes, but is only valid in a limited number of cases.

What about the change in the entropy? For a quasistatic process, this can be determined from the heat capacity, by integrating

$$\Delta S = \int_i^f \frac{đQ}{T} = \int_i^f \frac{C_y(T)dT}{T}. \tag{6.45}$$

For the special case where the heat capacity is constant as a function of the temperature, then[13]

$$\Delta S = S_f - S_i = C_y \int_i^f \frac{dT}{T} = C_y \ln\left(\frac{T_f}{T_i}\right). \tag{6.46}$$

Because entropy is a state function, we can use this expression even if we know that the process to bring our system from the initial state to the final state is not quasistatic. Also we see that if $T_f < T_i$, then $\Delta S < 0$, which is possible if the system isn't isolated. If we look at the total entropy change of the combined systems A_1 and A_2 where Eq. (6.44) is true, we would see that

$$\Delta S = \Delta S_1 + \Delta S_2 \geq 0,$$

and this will be checked in several problems.

Exercise 6.12 If we look at Eq. (6.43), it appears as though the heat absorbed is also a state function—it only depends upon the initial and final temperatures! Explain why this is not the case, that Q *does* in fact depend on the process used to go from $T_i \to T_f$. That is, where is the dependence on the process used?

If we have more than two systems in contact, we can generalize Eq. (6.44) to

$$Q_1 + Q_2 + Q_3 + \cdots = 0,$$

and additionally we can determine the change in the entropies of each system as well as the total change in the entropy.

Example 6.2 A common introductory physics lab is to determine the specific heat at constant pressure of a metal by heating some metal shots (small balls no more than a few millimeters in diameter) to around 100°C and drop them in room temperature water. The temperature of the water–metal mixture is then measured until it reaches an equilibrium point.

Such a system has three components: the metal shots (suppose we have aluminum shots), the water, and the dewar (or container) used to hold the water. All three must be included in our analysis of this system; however, the specific heat of the dewar is often not known. In some cases, it is common for the manufacturer to quote a **water equivalent** of the dewar, as a mass M_d, instead of its specific heat. The water equivalent is the mass of water that has the same heat capacity as the dewar, or

$$M_d c_w = m_{\text{dewar}} c_{\text{dewar}},$$

where c_w is the specific heat of water. We just have to add the water equivalent to the mass of the water, $m_w \to m_w + M_d$.

Suppose we have 100 g of water in a dewar with a water equivalent of 10 g just around 20°C, and we place 200 g of aluminum that is at 99°C into the water. When the system reaches equilibrium, we find the combined system to be at a temperature of 42.0°C. What is the specific heat of aluminum?

13 Be careful! Often when calculating Q in these sorts of problems we can get away with using Celsius for temperature, but when calculating the entropy change you *must* use kelvin. It is easy to see that this equation can be a problem at 0°C if we use the wrong units.

We assume the heat capacities are all constant in this temperature range, and use Eqs. (6.43) and (6.44). Some simple algebra shows that the specific heat of aluminum c_a is given by

$$c_a = \frac{c_w(m_w + M_d)(T_f - T_{iw})}{m_a(T_{ia} - T_f)}, \tag{6.47}$$

where $c_w = 4.186$ J/(g°C), T_{iw} and T_{ia} are the initial temperatures of the water and aluminum, respectively, and T_f is the final equilibrium temperature of the system. Putting in the numbers above, we get $c_a = 0.889$ J/(g°C).

Exercise 6.13 Fill in the steps in Example 6.2 to obtain Eq. (6.47).

Exercise 6.14 For the problem in Example 6.2, show that the entropy change in the aluminum is negative while that of the water + dewar is positive, and the total entropy change is positive. You should find

$$\Delta S_{w+d} = 33.3 \text{ J/K}, \quad \Delta S_a = -29.5 \text{ J/K}, \quad \text{and} \quad \Delta S_{\text{total}} = 3.78 \text{ J/K}.$$

On a related note, let's briefly consider the heat absorbed during a phase transition (such as ice to water and liquid water to vapor to be studied in Chapter 10). During a phase transition, there is no temperature change, so the heat capacity is ill-defined ($đQ/dT$ doesn't make sense if $dT = 0$). As you heat up ice, for example, once it reaches 0°C, you must still add heat to it to convert it to water, but it remains at this temperature until the transition is complete. As such, the heat absorbed is known as the **latent heat of transformation**. We will use L or ℓ to denote this quantity,

$$Q_{\text{phase transition}} = L \Rightarrow Q = m\ell^{(m)} = \nu\ell^{(v)}. \tag{6.48}$$

L is extensive, as it is just the heat absorbed during the transformation, while the two versions of ℓ are intensive, and are the latent heat *per unit mass* or *per mole* (respectively). The latent heat can be negative (if you are removing energy from the system for a liquid to solid transition) or positive (say for the solid to liquid transition). Some are related to each other: For example, the **latent heat of melting** (for the solid to liquid transition) L_m is related to the **latent heat of freezing** (for the liquid to solid transition): $L_f = -L_m$.

If, as two systems approach equilibrium, one of the systems undergoes a phase change, we would have to include Eq. (6.48) when determining how much heat transfers between the two systems to reach equilibrium. Thus, while L is not a derivative as we are discussing in this section, it is an important quantity that is connected to the heat capacity when considering systems in contact reaching equilibrium, hence the reason to include it here.

During a phase transition, the system can be treated as a heat reservoir—it's a *perfect* reservoir because the temperature remains precisely constant. Thus,

$$\Delta S = \frac{L}{T} \tag{6.49}$$

for a phase transition.

6.2.2 Coefficient of Thermal Expansion

Another easily measurable first derivative is called the **coefficient of thermal expansion**, which is defined as

$$\alpha \equiv \frac{1}{V}\left(\frac{\partial V}{\partial T}\right)_p. \tag{6.50}$$

The derivative itself gives us how the volume of a system changes if we change the temperature while holding the pressure constant. Including the prefactor of $1/V$ makes this an intensive quantity, so we can consider α the *fractional* change in the volume when we change the temperature. We can write this in terms of the **molar volume** (again from Exercise 6.6), $v = V/\nu$: $\alpha = (1/v)(\partial v/\partial T)_p$.

This is another example of an isobaric process, as it is performed at constant mean pressure. This expansion is also something that is familiar to us: On a really hot day, sometimes it is hard to open a door, or to take a ring off your finger, because of the expansion in the volume with the increase in temperature. Given this everyday (or every-hot-day) experience, we generally expect that α is positive for most systems.

Exercise 6.15 Calculate α for an ideal gas and show that it is $1/T$.

6.2.3 Isothermal Compressibility

The **isothermal compressibility** is defined both in terms of the volume V and the molar volume v as

$$\kappa \equiv -\frac{1}{V}\left(\frac{\partial V}{\partial p}\right)_T = -\frac{1}{v}\left(\frac{\partial v}{\partial p}\right)_T, \tag{6.51}$$

and is the fractional *decrease* in the volume as we increase the pressure at constant temperature, or during an **isothermal process**. As we compress a system (and thus *increase* the mean pressure on it), we expect the volume to decrease; the negative sign ensures that $\kappa > 0$.

Exercise 6.16 Show that $\kappa = 1/\overline{p}$ for an ideal gas.

Exercise 6.17 Consider these derivatives: C_V (or C_p), α, and κ. Using your everyday intuition, argue why it is clear that these are easy to measure (at least, in principle).

Because these derivatives—C_V (or C_p), α, and κ—are easy to measure and any other first derivative we can come up with can be written in terms of them, they are extremely useful. (You might be thinking that technically C_V and C_p are different derivatives; however, they are the related to each other, so we consider this group be a set of three derivatives.[14])

6.3 The Legendre Transform and Thermodynamic Potentials

In addition to the three first derivatives defined in Section 6.2, remember that we need one more tool before we can get to the procedure that allows us to reduce derivatives in terms of them. This tool is a rarely discussed transform, and to understand that we need to first introduce the concept of "naturally independent variables."

14 I won't choose one heat capacity over the other though: C_V is usually easier to calculate while C_p is easier to measure, but we will just write our expressions in terms of whichever makes them simpler.

6.3.1 Naturally Independent Variables

As we have seen, for fixed N, there are five important variables: $\bar{E}$, S, T, $\overline{p}$, and V, and they aren't independent of each other. Starting from the number of states accessible to the system, we were able to explicitly determine the entropy as a function of energy and volume, $S(\bar{E}, V)$. From these, we could then determine the other two variables listed with

$$\frac{1}{T} = \left(\frac{\partial S}{\partial \bar{E}}\right)_V, \text{ and } \frac{\overline{p}}{T} = \left(\frac{\partial S}{\partial V}\right)_E$$

in terms of temperature and volume. Using these equations of state, we could easily eliminate $\bar{E}$ or V in S to write the entropy as a function of other variables, such as $S(V, T)$, or $S(\bar{E}, T)$. However, because the energy and volume appear in the entropy from our initial calculation, we say that S is **naturally a function of $\bar{E}$ and V**, or $\bar{E}$ and V are the **natural independent variables** S depends upon.

The equations of state are two constraints that imply that of the five variables we have, we can always consider one to be a function of only two others. That is, while we could say $S(\bar{E}, V, T, \overline{p})$, the equations of state give us two equations with these four unknowns and we can always reduce S to be a function of only two variables. And at any moment, we can choose whichever variables we want to use as the independent variables. Choosing $\bar{E}$ and V, we can then write for a change in S,

$$dS = \left(\frac{\partial S}{\partial \bar{E}}\right)_V d\bar{E} + \left(\frac{\partial S}{\partial V}\right)_E dV, \tag{6.52}$$

which is merely a mathematical statement true for any function of two variables.

However, $\bar{E}$ and V are special for the entropy, because the partial derivatives of S with respect to each of these are known, and we can write Eq. (6.52) as

$$dS = \frac{1}{T}d\bar{E} + \frac{\overline{p}}{T}dV. \tag{6.53}$$

We can use this to determine the natural variables for the mean energy by a simple rewriting to give us the first law,

$$d\bar{E} = TdS - \overline{p}dV.$$

Consider the function $\bar{E}(S, V)$: We can write $d\bar{E}$, from a purely mathematical point of view, as

$$d\bar{E} = \left(\frac{\partial \bar{E}}{\partial S}\right)_V dS + \left(\frac{\partial \bar{E}}{\partial V}\right)_S dV. \tag{6.54}$$

Comparing these two equations we have the equalities

$$\left(\frac{\partial \bar{E}}{\partial S}\right)_V = T, \text{ and } \left(\frac{\partial \bar{E}}{\partial V}\right)_S = -\overline{p}. \tag{6.55}$$

That is, the derivative of $\bar{E}$ with respect to S holding V constant in known explicitly (and similarly for the other derivative shown).[15] These relations allow us to state that $\bar{E}$ is naturally a function of S and V, just as S was naturally a function of $\bar{E}$ and V.

Returning to our discussion of exact differentials (like $d\bar{E}$), we know that the mixed second derivatives are equal,

$$\left(\frac{\partial}{\partial V}\left(\frac{\partial E}{\partial S}\right)_V\right)_S = \left(\frac{\partial}{\partial S}\left(\frac{\partial E}{\partial V}\right)_S\right)_V,$$

15 In this form, we can see that the extensive and intensive nature of the variables are clear; the intensive variables $\overline{p}$ and T are derived from ratios of the extensive variables.

which allows us to say

$$\left(\frac{\partial T}{\partial V}\right)_S = -\left(\frac{\partial \bar{p}}{\partial S}\right)_V. \tag{6.56}$$

This is the first of four **Maxwell relations**,[16] a general term for the relationship between derivatives that arise from the equality of the mixed second derivatives. This particular relation allows us to relate the change in temperature with respect to volume when holding the entropy fixed (an **isentropic process**) to the change in mean pressure with entropy at constant volume (which recall is an isochoric process). There are three other Maxwell relations that we can derive, but to more easily obtain them, we will use the technique discussed in Section 6.3.2.

6.3.2 Legendre Transform

Transforms are useful in mathematics and physics as they allow us to convert a function of one variable into a different function of a different variable while retaining all of the same information. One form tends to be more useful than the other in certain circumstances. Probably the most famous and often used transform is the Fourier transform,[17] which allows us to write a function of time (for example) as a different function of frequency. Viewing this function in terms of its dependence upon the frequency instead of time makes it useful for signal processing across a wide variety of fields. This is not the what we wish to discuss here; we will just move on to the transform of interest.

The **Legendre transform**[18] is actually an important mathematical tool in physics that is used often yet rarely explained. Just like the Fourier transform, we begin with a function of a variable such as $Y(X)$ and we would like to find another function of a different variable, $\psi(P)$, such that Y and ψ contain the same information but ψ can be more useful in cases when Y may not be. The question is what should we choose for P to make this useful for our purposes here? Considering we will be applying this to thermodynamics, we already know that the derivative of Y with respect to X is a useful quantity (consider Y to be $\bar{E}$ and X to be S; then this derivative is just T, so it has physical meaning). While we will go through the necessary steps to understand and use the Legendre transform, I would encourage you to read Ref. [2], which gives some great insight into this tool.

Considering Y to be a function of X, $Y(X)$, we define the slope of this curve to be

$$P = \left(\frac{\partial Y}{\partial X}\right). \tag{6.57}$$

As this is only a function of a single variable, we don't need the partial derivative (here $P = dY/dX$), but we'll use this notation with the idea that there may be other variables in our function that we are omitting, and thus this will make it easier to generalize to functions of multiple variables. With this notation we can write

$$dY = \left(\frac{\partial Y}{\partial X}\right) dX = PdX. \tag{6.58}$$

The question on our mind is whether or not we could consider our function Y a function of the slope. That is, could we say $Y = Y(P)$, such that Y is a function of P and still contain all of the same information?

This is clearly not possible, as we can see in Figure 6.4: Any function can be shifted by a constant and still have the same derivative (we show one point specifically with the tangent line to the graph)

16 James Clerk Maxwell (1831–1879).
17 Jean-Baptiste Fourier (1768–1830).
18 Adrien-Marie Legendre (1752–1833).

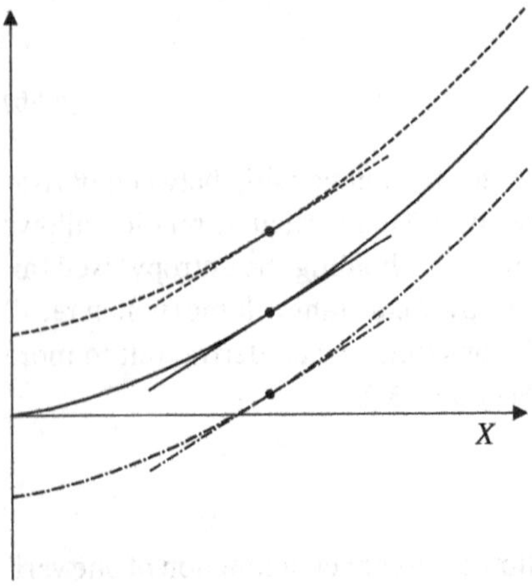

Figure 6.4 Three functions that all have the same slope (derivative) but are distinct.

at every given point X. We need a new function related to $Y(X)$ that is a function of P, and that is what we define as $\psi = \psi(P)$.

The method proposed by Legendre is to consider the tangent line that goes through the point (X, Y) and has the slope P as in Figure 6.4. Then extend that line to the Y-axis, is in Figure 6.5, and define the intercept (denoted by stars in the figure) as $\psi = Y(0)$. P in this case is just the slope of this straight line, or

$$P = \frac{Y - \psi}{X - 0}. \tag{6.59}$$

We get the Legendre transform by solving for ψ,

$$\psi = Y - PX. \tag{6.60}$$

Since we know that Y is a function of X and its differential, dY, we can calculate the differential of ψ using the product rule,

$$d\psi = dY - dPX - PdX.$$

Because $dY = PdX$, this simplifies to

$$d\psi = -XdP, \tag{6.61}$$

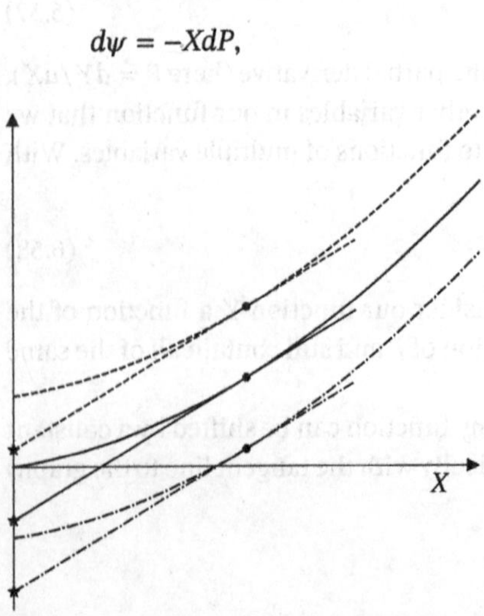

Figure 6.5 The same as Figure 6.4 but with the tangent lines extended to the Y-axis (marked by the stars).

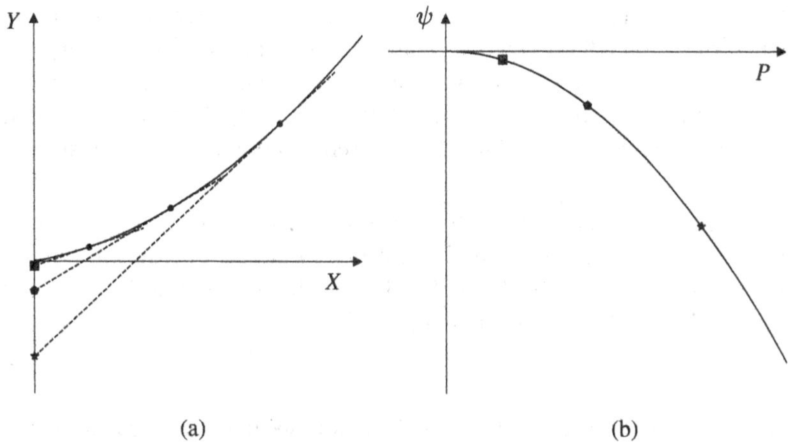

Figure 6.6 (a) One of the curves in Figures 6.4 and 6.5, with different values of X marked with circles. For each point, the tangent lines are continued to the Y-axis (marked with different shapes). (b) The Legendre transform for this function, $\psi(P)$, with the corresponding points P marked (with matching shapes) for each X.

so we can say

$$\left(\frac{\partial \psi}{\partial P}\right) = -X. \tag{6.62}$$

Equation (6.61) tells us that ψ is naturally a function of the slope P, and this is confirmed by Eq. (6.62): The derivative of ψ with respect to P is nothing more than $-X$. To visualize how this can give a one-to-one correspondence between the functions Y and ψ, I show three values of X marked with circles on the left plot of Figure 6.6, with dashed tangent lines drawn to the Y-axis, marked by different shapes. The corresponding plot of the Legendre transform ψ vs. P is shown on the right, with those values of ψ which correspond to the appropriate Y-intercept (with corresponding shapes marked). We can see that for each X, there is a single point P on the right plot. It's not as obvious that the figure on the right "contains all of the same information" as that on the left, but we will see how this comes about when we consider some physical examples.

One last comment before applying this to thermodynamics: Everything we discussed above readily generalizes for a function of multiple variables. If we have Y as a function of several variable, X_1, X_2, etc., then we can define

$$P_i = \left(\frac{\partial Y}{\partial X_i}\right)_{X_{j\neq i}},$$

where we just keep all of the other variables constant. We have many possible Legendre transforms in this case, such as

$$\psi_i = Y - P_i X_i, \tag{6.63}$$

where ψ_i is naturally a function of P_i instead of X_i, but still naturally a function of the X_j's for $j \neq i$. We also could "transform away" several variables at a time; there is no reason to stick to just one variable in Eq. (6.63). This will be useful when applying this to thermodynamics, as we already know that our thermodynamic functions depend on more than one variable.

Exercise 6.18 Figures 6.4 and 6.5 show the function $Y = X^2 + X + C$ for different C where C is a constant. Given this, determine P as well as the Legendre transform $\psi(P)$, and show that ψ will be unique for every choice of the constant C.

Computer Exercise 6.1 You can visit the Legendre Transforms section of the companion site to use the Jupyter code that created the figures here. Specifically, the code uses the function Y in Exercise 6.18, which allows you to check your result. You can see that the Legendre transforms are indeed unique when you shift the function by a constant. Additionally, using the method you used in that exercise, you can modify the functions used in the code to check other Legendre transforms.

Exercise 6.19 Perhaps the most famous Legendre transform is the Hamiltonian, which comes from the Legendre transform of the Lagrangian in classical mechanics. In one dimension, the Lagrangian, naturally a function of x and $\dot{x} = dx/dt$, is defined as the difference of the kinetic energy and the potential energy (which we assume is only a function of x),

$$L(x, \dot{x}) = K(\dot{x}) - U(x).$$

(a) We know the kinetic energy is $K = \frac{1}{2}m\dot{x}^2$. Evaluate $(\partial L/\partial \dot{x})_x$, and see that it corresponds to the momentum, p.
(b) Use this to determine the Hamiltonian, which is naturally a function of p and x.
Hint: The Hamiltonian is identified as the total energy of the system, so to make this work out, a slightly different Legendre transform is usually used, namely, $H = \dot{x}p - L$.

6.3.3 Thermodynamic Potentials

Now let's apply the Legendre transform to thermodynamics, specifically to the internal energy, so that we can obtain other "energies," known as **thermodynamic potentials**, that each can give us a unique insight to the system. For now we will just define them as Legendre transforms of the energy in a mathematical sense, and then see how they each have special roles to play in various applications.

6.3.3.1 $\bar{E}$: Internal Energy

This first thermodynamic potential is just the familiar quantity that we are starting with, the energy or as we often refer to it for thermodynamics, the **internal energy**.[19] I only include it here to set the stage for the later Legendre transforms. As we have seen, it is naturally a function of the entropy and volume,

$$\bar{E} = \bar{E}(S, V),$$

and with the first law,

$$d\bar{E} = TdS - \bar{p}dV, \tag{6.64}$$

we have the Maxwell relation from above, Eq. (6.56) along with the relationships between $\bar{p}$ and T and the derivatives of $\bar{E}$.

We don't really need to discuss the meaning of this quantity; however, let's remind ourselves what it truly means. The internal energy is the energy required to set up an isolated system—that is, to create the system from nothing. This will be more clear after we discuss the first Legendre transform of the energy.

19 I have often just referred to this as the energy or sometimes the thermal energy. In this chapter I will more often use this term now to distinguish it from the other energies I will define.

6.3.3.2 *F*: Helmholtz Free Energy

The first Legendre transform of the internal energy we consider is the **Helmholtz free energy**, F, which arises by eliminating S as one of the two variables.[20] To do so, we must evaluate $(\partial\bar{E}/\partial S)_V$, which we know is the temperature T. The Legendre transform is

$$F \equiv \bar{E} - \left(\frac{\partial \bar{E}}{\partial S}\right)_V S = \bar{E} - TS. \tag{6.65}$$

Differentiating this, we get $dF = d\bar{E} - SdT - TdS$, or with the first law ($d\bar{E} = TdS - \bar{p}dV$),

$$dF = -SdT - \bar{p}dV. \tag{6.66}$$

This form makes it clear that $F = F(T, V)$ as we expected, and the first derivatives of F can be read directly from this equation,

$$\left(\frac{\partial F}{\partial T}\right)_V = -S, \text{ and } \left(\frac{\partial F}{\partial V}\right)_T = -\bar{p}. \tag{6.67}$$

Equating the mixed second derivatives,

$$\left(\frac{\partial}{\partial V}\left(\frac{\partial F}{\partial T}\right)_V\right)_T = \left(\frac{\partial}{\partial T}\left(\frac{\partial F}{\partial V}\right)_T\right)_V,$$

we get our second Maxwell relation,

$$\left(\frac{\partial \bar{p}}{\partial T}\right)_V = \left(\frac{\partial S}{\partial V}\right)_T. \tag{6.68}$$

Even with everything above merely being a mathematical discussion, we can see that we can obtain some physics. For a given system, this Maxwell relation allows us to say that the way the mean pressure (easy to measure) changes when we change the temperature at constant volume is the same as how the entropy (difficult to measure) changes with the volume isothermally. Additionally, later we will see that often equations of state $\bar{p}(V, T)$ are determined empirically, and with such a determination, this second Maxwell relation can be used to determine how the entropy depends on the volume.

Exercise 6.20 Why do we know the mixed second derivatives must be equal? That is, argue that F, like $\bar{E}$, is a state function so that dF is an exact differential. Go one step further, and argue this must be true of *any* Legendre transform of the internal energy.

While we have introduced this as a mathematical tool, F has physical meaning that is useful in thermodynamic applications. For example, for an isothermal process (T = constant), then Eq. (6.66) implies that the Helmholtz free energy is nothing more than the work done during the process, and thus it is something we can measure. The internal energy is only the work done at constant *entropy*, not constant *temperature*. F can also be considered the energy that is required to set up a system while in contact with a heat reservoir at temperature T. Some of the energy that is required to create this system could be obtained from a transfer of heat, TS, from the other system, which gives rise to $F = \bar{E} - TS$.

6.3.3.3 *H*: Enthalpy

The next Legendre transform we will discuss is the **enthalpy**, H, a well-known quantity in chemistry. In this case, we will switch the dependence on V for $\bar{p} = -(\partial\bar{E}/\partial V)_S$, by defining

20 Hermann von Helmholtz (1821–1894).

$$H \equiv \bar{E} - \left[-\left(\frac{\partial \bar{E}}{\partial V}\right)_S\right] V = \bar{E} + \bar{p}V, \tag{6.69}$$

so that the natural variables for the enthalpy are entropy and pressure, or $H = H(S, \bar{p})$. We can see this explicitly by calculating $dH = d\bar{E} + \bar{p}dV + Vd\bar{p}$, or

$$dH = TdS + Vd\bar{p}. \tag{6.70}$$

Again equating the mixed second derivatives of

$$\left(\frac{\partial H}{\partial S}\right)_p = T, \quad \text{and,} \quad \left(\frac{\partial H}{\partial \bar{p}}\right)_S = V, \tag{6.71}$$

we get another Maxwell relation,

$$\left(\frac{\partial T}{\partial \bar{p}}\right)_S = \left(\frac{\partial V}{\partial S}\right)_p. \tag{6.72}$$

Exercise 6.21 Derive Eq. (6.72) by calculating the mixed partial derivatives of Eq. (6.71).

The enthalpy is often also known as the **heat of reaction**, as for constant pressure, it is nothing more than the heat absorbed (or released) by a system. Many chemical reactions (especially in undergraduate courses) occur at constant pressure; the enthalpy is the heat absorbed by the system during such reactions. To construct this system, we would need to provide an amount of work $\bar{p}V$, with $\bar{p}$ the (constant) pressure of the surroundings, so that $H = \bar{E} + \bar{p}V$ is the heat absorbed by the system at constant pressure. $\bar{E}$, on the other hand, is the heat absorbed by the system at constant volume.

6.3.3.4 G: Gibbs Free Energy

Finally, we could eliminate both S and V for T and $\bar{p}$, respectively, to give us G, the **Gibbs free energy**,

$$G \equiv \bar{E} - \left[-\left(\frac{\partial \bar{E}}{\partial V}\right)_S\right] V - \left(\frac{\partial \bar{E}}{\partial S}\right)_V S = \bar{E} + \bar{p}V - TS, \tag{6.73}$$

giving us $G = G(T, \bar{p})$. Following the same steps as above we get

$$dG = -SdT + Vd\bar{p}, \tag{6.74}$$

so that

$$\left(\frac{\partial G}{\partial T}\right)_p = -S, \quad \left(\frac{\partial G}{\partial \bar{p}}\right)_T = V. \tag{6.75}$$

From these, we get our fourth Maxwell relation,

$$\left(\frac{\partial S}{\partial \bar{p}}\right)_T = -\left(\frac{\partial V}{\partial T}\right)_p. \tag{6.76}$$

The Gibbs free energy will be useful when we study phase transitions in Chapter 10, which often occur at constant pressure and temperature. In this case, dT and $d\bar{p}$ are both zero and Eq. (6.74) implies $dG = 0$, or G is constant. We will find that the equation of state, $\bar{p}(T, V)$, itself cannot be used by itself to understand a system during a phase transition; however, the constancy of G will provide important insight into these processes.

6.3.3.5 Maxwell Relations

Let's summarize the different Maxwell relations from above,

$$-\left(\frac{\partial T}{\partial V}\right)_S = \left(\frac{\partial \overline{p}}{\partial S}\right)_V \qquad \text{MR 1,}$$

$$\left(\frac{\partial \overline{p}}{\partial T}\right)_V = \left(\frac{\partial S}{\partial V}\right)_T \qquad \text{MR 2,}$$

$$\left(\frac{\partial T}{\partial \overline{p}}\right)_S = \left(\frac{\partial V}{\partial S}\right)_p \qquad \text{MR 3,}$$

$$-\left(\frac{\partial S}{\partial \overline{p}}\right)_T = \left(\frac{\partial V}{\partial T}\right)_p \qquad \text{MR 4,}$$

where we have labeled them for easier reference later. These relations are not each new pieces of information, but rather they enumerate the various relationships that the variables

$$S,\ V,\ T,\ \text{and}\ \overline{p}$$

have with each other. Given $S(\bar{E}, V)$ and our equations of state $\bar{E}(T, V)$ and $\overline{p}(V, T)$, we could determine these various relationships for any system. However, the Maxwell relations allow for a deeper and more general way to relate these variables. We will see that with these relations, we will be able to determine many more properties of a system even if we do not have the equations of state. More importantly, these relations allow us to use equations of state that are determined empirically to learn more about our system.

Exercise 6.22 Show that you can define the Gibbs free energy as a Legendre transform of the Helmholtz free energy or the enthalpy.

Thermodynamic Square

The various differentials above $d\bar{E}$, dF, dG, and dH along with the Maxwell relations are a lot to remember. A useful mnemonic to help remember these relations is the **thermodynamic square**:

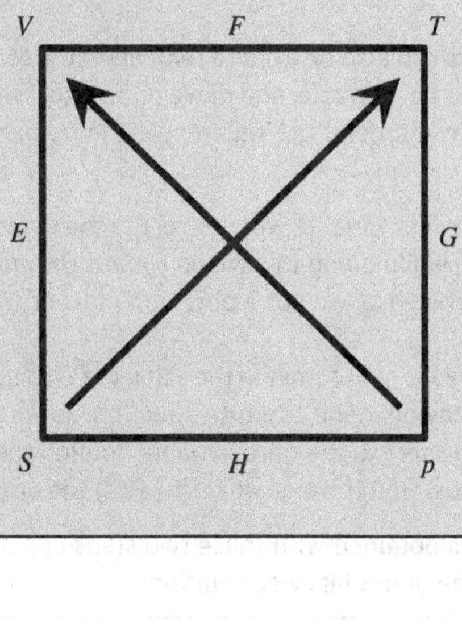

(Continued)

(Continued)

There are two rules to drawing this correctly to make it useful:

1. To get the variables in the correct order, starting from the upper left and going counter-clockwise, we remember the phrase:

   Virtually **E**very **S**tudent **H**ates **P**rofessors **G**iving **T**hem **F**'s.

   and fill in the first letters of this sentence on the corners and sides as shown.
2. Then we draw the two arrows as shown, from $\overline{p}$ to V and from S to T. I tend to remember Wonder Woman holding her arms crossed, so that her hands in this case point in the direction of the arrows.

When drawn properly there is an energy (potential) in the middle of each side, and its two naturally independent variables flanking it (as we can see is true here).

From this square, you can write the differential relations (with the proper signs) with the following technique. As stated above, the energy is flanked by its naturally independent variables, so you take the differential of the potential, and then the differentials of the variables on either side, so you have something like (for F),

$$dF = (\quad)d\overline{p} + (\quad)dT.$$

What goes inside the empty parentheses next to each differential are those that are connected by the arrows to the variables in the differentials. $\overline{p}$ is connected to V and T to S, so then we write,

$$dF = Vd\overline{p} + SdT.$$

Finally though we need the correct signs. If we have to go *against* the arrow when going from the variable in the differential to the other variable (as is the case when going from $T \rightarrow S$ but not for the case from $\overline{p} \rightarrow V$), then we get a minus sign, and thus in this example,

$$dF = Vd\overline{p} - SdT.$$

which agrees with Eq. (6.66).

The thermodynamic square can also be used to remember the Maxwell relations. To get any of these, we start from any corner variable, and move either clockwise or counter-clockwise to the next corner, and then travel around the square, picking up variables from each corner as follows.

1. First take the partial of the first variable with respect to the next variable, while holding the third variable constant. If, while doing this we go *against the arrow*, we put a minus sign in front of this derivative; otherwise, we put a plus sign in front. This will give us one partial derivative.
2. Then, we continue to the remaining corner in the same direction, and follow the same rule as in the previous step, but moving in the opposite direction. So we get the partial derivative of the fourth variable with respect to the third variable holding the second variable constant, again putting a minus (plus) sign if we go against (with) the arrow.

These two partial derivatives obtained with these two steps are equal to each other and the resulting equation will be one of our Maxwell relations.

This description is complicated to see in words, so let's consider an example. Let's start at the lower left (S) and go counter-clockwise, from $S \to \overline{p} \to T$. This gives, using step 1,

$$+\left(\frac{\partial S}{\partial \overline{p}}\right)_T,$$

where the plus sign is there because we are traveling with the arrow. (I know we don't *need* the plus sign, but I find it helpful to remember that I did correctly account for the proper sign.) Then, the fourth variable (the next if we continue counter-clockwise) is V, and our return journey (moving clockwise now, from $V \to T \to \overline{p}$) gives

$$-\left(\frac{\partial V}{\partial T}\right)_p.$$

Setting this equal to the previous derivative obtained, we have just written down Eq. (6.76), or MR 4.

Exercise 6.23 Calculate, using this mnemonic, the other three Maxwell relations.

Note there are eight possible equations we can obtain, given that we can start from any of the four corners, and from each corner start out going clockwise or counter-clockwise. It is easy to see that this will still only lead to four unique Maxwell relations, which you can (and should) easily show. With the example above, you could get the same Maxwell relation if you started with V and moved clockwise initially instead of counter-clockwise.

6.3.4 Fundamental Relations and the Equations of State

We have spent a lot of time deriving many different equations and relations, so before moving to the next section, let's pause to review them and their relevance to thermodynamic problems.

For our problems, without defining it as such, we have started from a **fundamental relation**. This is an equality which allows us to obtain *all* other relevant information about a theory. For the microcanonical ensemble, this is given by the entropy,

$$S = S(\bar{E}, V, \ldots),$$

where we include the … to allow for other variables (including N, the number of molecules in our system, as well as other external variables), and we have kept the naturally independent variables explicit. The fundamental relation can be derived from first principles (in this case by counting the number of accessible microstates in the energy range $\bar{E}$ to $\bar{E} + \delta E$ for some volume V and other external parameters).

From the fundamental relation, we can derive the **equations of state**. They can be determined from the fundamental equation by using (assuming V is our only external parameter)

$$\frac{1}{T} = \left(\frac{\partial S}{\partial \bar{E}}\right)_V \quad \text{and} \quad \frac{\overline{p}}{T} = \left(\frac{\partial S}{\partial V}\right)_E.$$

For example, for 1 mol[21] of a monatomic classical ideal gas, the fundamental relation is

$$S = R \ln V + \frac{3}{2} R \ln E + A, \tag{6.77}$$

21 Remember that $N = \nu N_A$ and $N_A k_B = R$, so setting $\nu = 1$ gives us this equation.

where A is just a constant (independent of any thermodynamic parameters). As we saw before, this leads to two equations of state,

$$\bar{E} = \frac{3}{2}RT, \qquad \bar{p}V = RT. \tag{6.78}$$

Exercise 6.24 The equation of state for $\bar{E}$ is not a function of S and V explicitly, but these are the natural variables for the energy. Use the expression for the entropy in Eq. (6.77) to solve for $\bar{E}$ to get it in terms of S and V.

Exercise 6.25 Determine the other thermodynamic potentials for the monatomic classical ideal gas. Rewrite each in terms of their naturally independent variables.

Any thermodynamic potential that is written in terms of its natural variables can be considered a fundamental relation. For example, because $T = (\partial\bar{E}/\partial S)_V$ and $\bar{p} = -(\partial\bar{E}/\partial V)_S$, then we can say that $\bar{E}(S, V)$ is a fundamental relation.

Exercise 6.26 Verify the Maxwell relation

$$\left(\frac{\partial\bar{p}}{\partial T}\right)_V = \left(\frac{\partial S}{\partial V}\right)_T$$

explicitly for 1 mol of a classical monatomic ideal gas.

Hint: For the ideal gas, remember that if the temperature is constant, so is the internal energy.

While we consider the potentials to be "naturally" dependent on certain variables, any thermodynamic function (not just the potentials) can be identified as a function of any other two variables. This is important to note because we might want to determine, say, how $\bar{E}$ depends upon $\bar{p}$ and V because those are the variables we might be able to control experimentally, even though they are not the naturally independent variables for $\bar{E}$. In this case, we could very easily say that $\bar{E} = \bar{E}(\bar{p}, V)$ and write the purely mathematical relationship,

$$d\bar{E} = \left(\frac{\partial\bar{E}}{\partial\bar{p}}\right)_V d\bar{p} + \left(\frac{\partial\bar{E}}{\partial V}\right)_p dV.$$

These derivatives are not immediately known, however, as I have been saying for most of this chapter, there is a simple recipe which we will study in Section 6.4 that will allow us to easily write these derivatives in terms of the heat capacity, the coefficient of thermal expansion, and the isothermal compressibility.

6.4 Derivative Crushing

As I have mentioned several times, we are usually interested in *changes* in properties of systems when studying thermodynamics. As long as the volume is the only external parameter, this means that we are generally interested (as the partial derivatives defined in Section 6.2) in seeing how one parameter changes when changing another while holding a third parameter fixed. Given the constraints of the equations of state, any of the observables we have considered thus far—$\bar{E}$, F, G, H, S, V, $\bar{p}$, and T—can be written as a function of any two other variables. Consider a case where we are interested in how the enthalpy of a system changes when we change the volume at fixed entropy (as in a quasistatic adiabatic process, or an isentropic process), then we are interested in

$$\left(\frac{\partial H}{\partial V}\right)_S.$$

H is not naturally dependent on these variables, but in some cases, such as in the case of the monatomic classical ideal gas, we could write H explicitly in terms of these variables (see Exercise 6.25) given the equations of state.[22]

In general, however, this is not always possible, but we can write $H = H(V, S)$ and look at the infinitesimal change in the enthalpy, or

$$dH = \left(\frac{\partial H}{\partial V}\right)_S dV + \left(\frac{\partial H}{\partial S}\right)_V dS. \tag{6.79}$$

In this example, $dS = 0$, and from this we can determine the finite enthalpy change by evaluating

$$\Delta H = \int_i^f \left(\frac{\partial H}{\partial V}\right)_S dV. \tag{6.80}$$

In general, this derivative is not easy to determine, but we can use a process known as **derivative crushing**[23] to rewrite it in terms of the heat capacity at constant pressure C_p, the coefficient of thermal expansion α defined in Eq. (6.50), and the isothermal compressibility κ defined in Eq. (6.51). As stated multiple times, these quantities are in general easy to measure, and thus are usually tabulated for many systems.[24] Many other textbooks use similar methods for reducing derivatives; however, almost none (except for Ref. [3]) use this methodical approach, which is useful for ensuring we don't go around in circles while trying to simplify such derivatives.

Before getting to the recipe itself, it will be useful to list three properties among partial derivatives from multivariable calculus, assuming we have three variables X, Y, and Z that are related by some function f such that $f(X, Y, Z) = 0$:

$$\left(\frac{\partial X}{\partial Y}\right)_Z = \frac{1}{\left(\frac{\partial Y}{\partial X}\right)_Z}, \tag{6.81}$$

$$\left(\frac{\partial X}{\partial Y}\right)_Z = \frac{\left(\frac{\partial X}{\partial W}\right)_Z}{\left(\frac{\partial Y}{\partial W}\right)_Z}, \tag{6.82}$$

$$\left(\frac{\partial X}{\partial Y}\right)_Z = -\frac{\left(\frac{\partial Z}{\partial Y}\right)_X}{\left(\frac{\partial Z}{\partial X}\right)_Y}. \tag{6.83}$$

The second of these also includes a fourth variable W, upon which the other variables could also depend. We will use these identities while crushing derivatives to move quantities around. I will often use the phrase: "put a quantity in the numerator," which refers to a quantity in the derivative in any of these three positions. For example, Eqs. (6.81) and (6.82) both put Y into the numerator (even though the derivative with Y is in the denominator in both cases). We will use Eq. (6.82) when we are asked to "put a ∂W under the ∂X," which requires us

22 With the variables we have considered until now, there are 344 possible first derivatives to consider!

23 This term comes to me from Claude Bernard, and he learned it from Paul Bamberg at Harvard. In section 7.3 of Ref. [3], this procedure is described, but this term is not used. I find this term motivating for some reason, so we'll stick with it.

24 I choose C_p because that tends to be easier to measure experimentally, although as we will derive a relationship among C_p and C_V, leaving our results in terms of C_V will be fine.

to also add it under the ∂Y. Finally, Eq. (6.83) puts Z (the quantity that originally is held constant) into the numerator (which seems odd at first given this expression). Putting variables into the numerator and putting differentials under others are not rigorous statements (and many mathematicians might hate this terminology), but they will be useful for this process.

Exercise 6.27 The derivative relations in Eqs. (6.81) and (6.82) are easy to understand by inspection; however, Eq. (6.83) is less obvious. Treat X as a function of Y and Z, and Y as a function of X and Z, and write down dX and dY. Substitute your expression for dY into your expression for dX and set $dX = 0$ to derive this third relation. Section 5.4 of Ref. [4] (or any textbook on mathematical physics) will be helpful here.

Whenever asked to crush a derivative, follow these steps *in this precise order*. We will work through several examples after discussing them.

1. If the derivative contains any potentials ($\bar{E}$, F, G, or H), bring them one by one to the numerator and eliminate them using the differential relations we have in Eqs. (6.64), (6.65), (6.70), or (6.73).
2. If the chemical potential[25] μ is in our derivative (relevant when we allow the number of molecules to change), bring it to the numerator and eliminate it with the Gibbs–Duhem relation in Eq. (10.63), which we will derive in Section 10.5.2. As we will not need this now, we will wait until later to discuss it.
3. If the derivative contains the entropy S, first bring it to the numerator and then do one of the following:
   a. if one of the four Maxwell relations now eliminates it, invoke it, or
   b. if not, put a ∂T under the ∂S with Eq. (6.82), and the entropy will be eliminated in favor of one of the heat capacities,
   $$C_V = T\left(\frac{\partial S}{\partial T}\right)_V \quad \text{or} \quad C_p = T\left(\frac{\partial S}{\partial T}\right)_p.$$
4. Any remaining derivatives will contain the volume V, so bring it to the numerator, and these derivatives can be written in terms of either
   $$\alpha = \frac{1}{V}\left(\frac{\partial V}{\partial T}\right)_p \quad \text{or} \quad \kappa = \frac{1}{V}\left(\frac{\partial V}{\partial p}\right)_T.$$
5. (optional) We can eliminate C_V for C_p via
   $$C_V = C_p - \frac{TV\alpha^2}{\kappa},$$
   which is an expression that we will derive in Example 6.3 as a useful application of derivative crushing.

A good way to remember this recipe is that we will always want to eliminate each variable *alphabetically*: $\bar{E}$, F, G, H, μ, S, and then V.[26] Eliminating a variable requires us to move it to the numerator (or put a ∂T under it) via one of the derivative rules given in Eqs. (6.81)–(6.83), then use

25 I won't define this until Chapter 10, so as you can imagine we will be skipping this step in all of our problems for now.

26 Again we aren't considering μ yet, but as this is just a Greek "m," our alphabetical order is still valid. This is also another reason I prefer to use E for the internal energy instead of U, incidentally.

the thermodynamic square as needed (for the relationships of the differentials of the potentials or the Maxwell relations). If the thermodynamic square is not useful, then the derivative will be one of C_p, C_V, α, or κ.

Example 6.3 Our first application of this approach will be to derive the relationship used in step 5. This actually involves several different stages that are not at all obvious at first, but with practice, these methods become second nature.

We start first by considering the entropy to be a function of temperature and pressure, $S = S(T, \overline{p})$, and thus dS can be written as

$$dS = \left(\frac{\partial S}{\partial T}\right)_p dT + \left(\frac{\partial S}{\partial \overline{p}}\right)_T d\overline{p}. \tag{6.84}$$

To crush the derivatives in this equation, first we see that there are no potentials (and of course no chemical potential) so we can skip steps 1 and 2. The third step is to eliminate the entropy, which is already in the numerator in both cases. For the first derivative, there is no Maxwell relation to eliminate it, but it *already* has a ∂T under it; it is merely the heat capacity at constant pressure divided by the temperature, so this becomes

$$dS = \frac{C_p}{T} dT + \left(\frac{\partial S}{\partial \overline{p}}\right)_T d\overline{p}. \tag{6.85}$$

Looking at the thermodynamic square, we find the entropy in the second derivative can be eliminated with a Maxwell relation (MR 4), to get

$$dS = \frac{C_p}{T} dT - \left(\frac{\partial V}{\partial T}\right)_p d\overline{p}. \tag{6.86}$$

Moving to step 4, we eliminate the volume (already in the numerator) using the coefficient of thermal expansion, where $(\partial V/\partial T)_p = V\alpha$. At this point, dS becomes

$$dS = \frac{C_p}{T} dT - V\alpha d\overline{p}. \tag{6.87}$$

The first term here has C_p in it, which is part of the relationship we are considering. Given that we also want C_V, then it makes sense to consider $\overline{p} = \overline{p}(T, V)$—I choose T as a variable because C_V is a derivative with respect to T, and V because this is what is held constant in C_V. We can write

$$d\overline{p} = \left(\frac{\partial \overline{p}}{\partial T}\right)_V dT + \left(\frac{\partial \overline{p}}{\partial V}\right)_T dV, \tag{6.88}$$

and again we can crush these derivatives. They are both easier than before because we can start with step 4, and the derivative in the second term is easier, so we'll start with that one. We move the volume to the numerator with Eq. (6.81) and then we have the isothermal compressibility (times the volume),

$$\left(\frac{\partial \overline{p}}{\partial V}\right)_T = \frac{1}{\left(\frac{\partial V}{\partial \overline{p}}\right)_T} = -\frac{1}{V\kappa}.$$

The derivative in the first term requires the rule in Eq. (6.83) to move the volume to the numerator, and this immediately results in the ratio of α and κ,

$$\left(\frac{\partial \bar{p}}{\partial T}\right)_V = -\frac{\left(\frac{\partial V}{\partial T}\right)_p}{\left(\frac{\partial V}{\partial \bar{p}}\right)_T} = -\frac{\frac{1}{V}\left(\frac{\partial V}{\partial T}\right)_p}{\frac{1}{V}\left(\frac{\partial V}{\partial \bar{p}}\right)_T} = \frac{\alpha}{\kappa}. \tag{6.89}$$

Putting these derivatives into our expression for $d\bar{p}$ and then substituting $d\bar{p}$ into our expression for dS, we get

$$dS = \left(\frac{C_p}{T} - V\frac{\alpha^2}{\kappa}\right) dT + \frac{\alpha}{\kappa} dV, \tag{6.90}$$

from which we can read off

$$\left(\frac{\partial S}{\partial T}\right)_V = \frac{C_p}{T} - V\frac{\alpha^2}{\kappa}.$$

For a quasistatic process,

$$C_V = T\left(\frac{\partial S}{\partial T}\right)_V,$$

so we immediately find

$$C_V = C_p - \frac{VT\alpha^2}{\kappa}, \tag{6.91}$$

which is the relation we were looking for. We can also write this in terms of the molar volume v and the molar specific heats,

$$c_V = c_p - \frac{vT\alpha^2}{\kappa}. \tag{6.92}$$

Example 6.4 As another example, let us return to Eq. (6.80), where we considered the derivative

$$\left(\frac{\partial H}{\partial V}\right)_S.$$

We have from step 1 and Eq. (6.70),

$$\left(\frac{\partial H}{\partial V}\right)_S = V\left(\frac{\partial \bar{p}}{\partial V}\right)_S + T\left(\frac{\partial S}{\partial V}\right)_S.$$

The second term vanishes because S is constant, so let's look at the derivative in the first term. We move S to the numerator with Eq. (6.83),

$$\left(\frac{\partial \bar{p}}{\partial V}\right)_S = -\frac{\left(\frac{\partial S}{\partial V}\right)_p}{\left(\frac{\partial S}{\partial \bar{p}}\right)_V}.$$

Neither of these can be simplified with a Maxwell relation, so we put a ∂T under both derivatives (be careful with problems like this as it's easy to make a mistake!),

$$\left(\frac{\partial \bar{p}}{\partial V}\right)_S = -\frac{\left(\frac{\partial S}{\partial T}\right)_p \left(\frac{\partial \bar{p}}{\partial T}\right)_V}{\left(\frac{\partial V}{\partial T}\right)_p \left(\frac{\partial S}{\partial T}\right)_V},$$

and we can see both C_p and C_V here to write

$$\left(\frac{\partial \bar{p}}{\partial V}\right)_S = -\frac{C_p/T}{\left(\frac{\partial V}{\partial T}\right)_p}\frac{\left(\frac{\partial \bar{p}}{\partial T}\right)_V}{C_V/T}.$$

For step 4, we realize the derivative in the denominator of the first factor is simply $V\alpha$, while that in the numerator of the second factor we have already found in Eq. (6.89), so

$$\left(\frac{\partial \bar{p}}{\partial V}\right)_S = -\frac{C_p}{C_V V\alpha}\frac{\alpha}{\kappa} = -\frac{C_p}{C_V V\kappa},$$

and thus

$$\left(\frac{\partial H}{\partial V}\right)_S = -\frac{C_p}{C_V\kappa}.$$

Thus if we know these different parameters (either by calculating them from the equations of state or from empirical sources), we now can determine how the enthalpy changes with the volume for an isentropic process.

Note we don't always need (and rarely do) every step of the procedure. However, remembering to follow this recipe every time will allow for relatively quick solutions to many problems.

Example 6.5 Let's look at a thermodynamic problem that is more realistic and will require us to use this procedure. Imagine we have a gas of N molecules that we quasistatically expand from V_1 to V_2. It is observed that the temperature increases linearly with the volume from T_1 to T_2. What is the heat absorbed by the system during this process?

Since this is a quasistatic process, we can integrate $đQ = TdS$ to get the heat absorbed, which means we first need to determine dS. As we are controlling the volume while measuring the temperature, we treat S as a function of T and V, or $S = S(T, V)$, so that

$$dS = \left(\frac{\partial S}{\partial T}\right)_V dT + \left(\frac{\partial S}{\partial V}\right)_T dV.$$

Now we have derivatives to crush and we can start with step 3 (with the entropy already in the numerator for both derivatives!). The first derivative is just C_V/T, while the second can be rewritten with a Maxwell relation (MR 2),

$$\left(\frac{\partial S}{\partial V}\right)_T = \left(\frac{\partial \bar{p}}{\partial T}\right)_V = \frac{\alpha}{\kappa}.$$

In the second equality, we used Eq. (6.89) to skip our derivative crushing steps. This means that

$$dS = \frac{C_V}{T}dT + \frac{\alpha}{\kappa}dV. \tag{6.93}$$

We know how the temperature depends on the volume for this problem,

$$T = A + BV,$$

where the constants

$$A = \frac{T_1V_2 - T_2V_1}{V_2 - V_1}, \qquad B = \frac{T_2 - T_1}{V_2 - V_1}$$

are easily determined from the initial and final states. Thus, we can write

$$dT = BdV,$$

so

$$dS = \left(B\frac{C_V}{T} + \frac{\alpha}{\kappa}\right) dV.$$

Integrating to find the heat absorbed is straightforward,

$$Q = \int_{V_1}^{V_2} \left[BC_V(V) + \frac{\alpha}{\kappa}T(V)\right] dV, \tag{6.94}$$

where I have made the volume dependence of C_V and T explicit. Without more information about the system, we cannot simplify this further, but let's suppose that we have determined that, for the range of volumes here, C_V, α, and κ are independent of the volume. In this case we can easily evaluate this to get

$$Q = \left(BC_V + \frac{\alpha}{\kappa}A\right)(V_2 - V_1) + \frac{\alpha}{2\kappa}B\left(V_2^2 - V_1^2\right). \tag{6.95}$$

Exercise 6.28 The **adiabatic bulk modulus** is defined as

$$\beta_S = -V\left(\frac{\partial \bar{p}}{\partial V}\right)_S, \tag{6.96}$$

and it describes the resistance a system has when it is compressed while thermally isolated. Using the recipe above, crush this derivative.

6.5 More About the Classical Ideal Gas

Let's apply our new techniques to the classical ideal gas. In this case we can calculate pretty much any derivative we want easily as we have the equations of state. However, it is instructive to look at this system from a more mathematical perspective to see how everything fits together.

Recall that for an infinitesimal process, we generally have

$$đQ = d\bar{E} + \bar{p}dV.$$

Let's use this to study the molar specific heats and how they relate to the energy. For example, with this expression we can write the molar specific heat at constant volume as

$$c_V = \frac{1}{\nu}\left(\frac{\partial \bar{E}}{\partial T}\right)_V.$$

We saw in Exercise 5.7 that $\bar{E}$ doesn't depend on the volume for an ideal gas, so this partial derivative can be written as a total derivative to obtain

$$d\bar{E} = \nu c_V dT. \tag{6.97}$$

We can also get this relationship by our crushing method above as follows (which will be useful for non-ideal gases). If we treat $\bar{E}$ as a function of V and T, then

$$d\bar{E} = \left(\frac{\partial \bar{E}}{\partial T}\right)_V dT + \left(\frac{\partial \bar{E}}{\partial V}\right)_T dV. \tag{6.98}$$

Crushing the derivative in the second term, we find

$$\begin{aligned}\left(\frac{\partial \bar{E}}{\partial V}\right)_T &= T\left(\frac{\partial S}{\partial V}\right)_T - \bar{p}\left(\frac{\partial V}{\partial V}\right)_T \\ &= T\left(\frac{\partial \bar{p}}{\partial T}\right)_V - \bar{p}. \end{aligned} \tag{6.99}$$

This is quite general, and we could crush this further, but for the ideal gas,

$$\bar{p} = \frac{\nu RT}{V} \Rightarrow \left(\frac{\partial \bar{p}}{\partial T}\right)_V = \frac{\nu R}{V},$$

so we see that

$$\left(\frac{\partial \bar{E}}{\partial V}\right)_T = T\left(\frac{\partial \bar{p}}{\partial T}\right)_V - \bar{p} = 0.$$

This is true for *any* ideal gas, so we can say $\bar{E}(T,V) = \bar{E}(T)$ quite generally, with only knowing how $\bar{p}$ depends on V and T! We saw this from our microscopic derivation of the number of states, and it's nice to see that this is confirmed from a general thermodynamic perspective as well.

Looking at the molar specific heat at constant pressure, we have

$$c_p = \frac{1}{\nu}\left(\frac{\partial \bar{E}}{\partial T}\right)_p + \frac{\bar{p}}{\nu}\left(\frac{\partial V}{\partial T}\right)_p.$$

Instead of crushing this, let's simplify it some for the ideal gas. From the ideal gas equation, we can write

$$\left(\frac{\partial V}{\partial T}\right)_p = \frac{\nu R}{p},$$

so

$$\frac{\bar{p}}{\nu}\left(\frac{\partial V}{\partial T}\right)_p = R.$$

Also, given that the internal energy is only a function of the temperature, holding V fixed is equivalent to holding $\bar{p}$ fixed, so we can say

$$c_p = \frac{1}{\nu}\left(\frac{\partial \bar{E}}{\partial T}\right)_V + R,$$

or

$$c_p = c_V + R. \tag{6.100}$$

This is a general relationship between the molar specific heats for any ideal gas. We can calculate them directly for a monatomic ideal gas by using $\bar{E} = \frac{3}{2}\nu RT$, so

$$c_V = \frac{1}{\nu}\left(\frac{3}{2}\nu R\right) = \frac{3}{2}R,$$

which implies $c_p = \frac{5}{2}R$. A useful parameter for gases is the **specific heat ratio,** γ,

$$\gamma \equiv \frac{c_p}{c_V}. \tag{6.101}$$

For an ideal gas, $\gamma = 1 + R/c_V$, and specifically for a monatomic ideal gas,

$$\gamma = \frac{5}{3}. \tag{6.102}$$

Exercise 6.29 For a diatomic ideal gas, we will argue later that

$$\bar{E} = \frac{5}{2}\nu RT. \tag{6.103}$$

Use this to show

$$c_V = \frac{5}{2}R \text{ and } c_p = \frac{7}{2}R, \tag{6.104}$$

and $\gamma = 7/5 = 1.4$.

In Table 6.1 I show the specific heats at 25°C for various monatomic and diatomic gases. The second and third columns (taken from Table A-1 of Ref. [5]) are the specific heats per mass (at constant pressure and volume, respectively), the fourth and fifth columns are the molar specific heats (at constant pressure and volume, respectively) calculated using the molar masses which you can look up in the periodic table in Appendix B, and the final column is γ, the specific heat ratio (calculated directly from the fourth and fifth columns). There are several interesting things to note in the table. For one, the specific heats per mass vary wildly, but the molar specific heats are more universal, hence the reason we often use the latter in calculations. Additionally, we see that results for the noble gases agree exceptionally well with the ideal gas results, which we understand because the noble gases are very non-reactive, so the ideal gas approximation is quite good. The diatomic gases in this table do not agree as well with our approximation, but even the worst case (chlorine) is only off by about 6%. While not a perfect approximation, it is clear the ideal gas assumption gives a good qualitative picture for a variety of gases.

Let's now look at the other important first derivatives, which you should have calculated in Exercises 6.15 and 6.16:

$$\alpha = \frac{1}{V}\left(\frac{\partial V}{\partial T}\right)_p = \frac{\nu R}{\bar{p}V} = \frac{1}{T}, \tag{6.105}$$

Table 6.1 Specific heats for various monatomic and diatomic gases.

Gas	$c_p^{(m)}$ (J/g/K)	$c_V^{(m)}$ (J/g/K)	$c_p^{(v)}$ (J/mol/K)	$c_V^{(v)}$ (J/mol/K)	γ
Argon, Ar	0.5203	0.3122	20.78	12.47	1.667
Helium, He	5.193	3.116	20.79	12.47	1.667
Krypton, Kr	0.2480	0.1488	20.78	12.47	1.667
Neon, Ne	1.030	0.6180	20.79	12.47	1.667
Xenon, Xe	0.1583	0.09499	20.78	12.47	1.666
Carbon monoxide, CO	1.039	0.7417	29.10	20.78	1.401
Chlorine, Cl_2	0.4781	0.3608	33.90	25.58	1.325
Fluorine, F_2	0.8237	0.6050	31.30	22.99	1.362
Hydrogen, H_2	14.30	10.18	28.83	20.52	1.405
Nitrogen, N_2	1.040	0.7429	29.13	20.81	1.4
Oxygen, O_2	0.9180	0.6582	29.38	21.06	1.395

Columns two and three are taken from Table A-1 of Ref. [5] Y.A. Çengel and A.J. Ghajar 2019/McGraw-Hill Education, while the fourth and fifth columns were calculated from those using the molar mass (which you can determine from the periodic table in Appendix B).

and

$$\kappa = -\frac{1}{V}\left(\frac{\partial V}{\partial \bar{p}}\right)_T = \frac{1}{\bar{p}}. \tag{6.106}$$

A quick calculation shows that the relationship between specific heats in Eq. (6.92) can be easily verified for an ideal gas. Using Eqs. (6.105) and (6.106), we have

$$\frac{Tv\alpha^2}{\kappa} = \frac{Tv\bar{p}}{T^2} = R,$$

where in the second equality we used the ideal gas equation. Equation (6.100) clearly holds for the ideal gas.

6.6 First Derivatives Near Absolute Zero

As absolute zero is a special temperature, let's examine what we expect the heat capacities, α, and κ to behave like as we approach this point. From the third law, we know that the entropy goes to a well-defined constant S_1 as T approaches zero from the positive side. The entropy, for a quasistatic process, can be calculated from the heat capacity with

$$\Delta S = S(T) - S_1 = \int_0^T \frac{C_y(T')}{T'} dT',$$

where we will use whichever heat capacity is relevant for a given situation. Because the entropy is well-defined at absolute zero (we can calculate it by counting the number of microstates), it must be true that the heat capacities not only go to zero as $T \to 0$, but they must vanish at least linearly with the temperature. Otherwise the integral to calculate ΔS will be ill-defined. Thus we say for $y = V$ or $\bar{p}$,

$$C_y(T) \propto T^a \text{ near } T = 0^+, \quad \text{for } a \geq 1. \tag{6.107}$$

Additionally, because we know the limiting value of the entropy is independent of all the parameters of the system, it is also independent of any pressure or volume variations, so we expect

$$\left(\frac{\partial S}{\partial \bar{p}}\right)_T \to 0 \text{ as } T \to 0^+. \tag{6.108}$$

Crushing this derivative we find

$$\left(\frac{\partial S}{\partial \bar{p}}\right)_T = -V\alpha, \tag{6.109}$$

which implies that the coefficient of thermal expansion should vanish at absolute zero, or

$$\alpha \to 0 \text{ as } T \to 0^+. \tag{6.110}$$

Exercise 6.30 Verify Eq. (6.109).

The isothermal compressibility κ is different, as it is a purely mechanical property. Even in the ground state, a system has a well-defined compressibility, which makes sense if we consider a solid—while it may be very small for such a system, κ could generally remain finite in the low-temperature limit. As such, we can safely assume that κ *does not vanish* at absolute zero.

Using the relationship between heat capacities and α and κ,

$$C_p - C_V = \frac{TV\alpha^2}{\kappa},$$

and the fact that it is reasonable to assume that the derivatives C_V, C_p, and α vanish at roughly the same rate, we expect

$$\frac{C_p - C_V}{C_V} \to 0 \text{ as } T \to 0^+. \tag{6.111}$$

That is $C_V \to C_p$ for low temperatures.

These relationships break down for the classical ideal gas. The heat capacities are constant, so they cannot vanish at zero temperature—as such the entropy is ill-defined at low T! Additionally, $\alpha = 1/T \to \infty$ as $T \to 0$ instead of vanishing (quite the opposite of what should happen). This should not surprise us, as the ideal approximation is quite simplistic. We will see time and again that many classical approximations will break down at low temperatures, where quantum effects will become much more important.

6.7 Empirical Determination of the Entropy and Internal Energy

Our focus on thermodynamics began with the microscopic picture, primarily because of the power of the fundamental point of view. By counting the number of accessible states, we could determine the entropy in terms of the energy and volume (the fundamental relation), from which the equations of state could be found. However, it is not always simple to obtain the fundamental relation from first principles, so while this approach is ideal, we often need to find other approaches to solving problems. Instead, often the equation of state, $\overline{p}(T, V)$, can be found empirically for a given system (such was the origin of the ideal gas equation). Additionally, as discussed in Section 6.2, the heat capacities are generally easy to determine empirically as well.

We will show in this section that if we know two things:

1. the heat capacity at constant volume as a function of T for *any one value* of $V = V_0$, and
2. the equation of state $\overline{p}(T, V)$,

then we can determine the entropy and the internal energy both up to an additive constant. As we are usually only interested in changes in the entropy or energy, this is sufficient to determine most relevant physical quantities.

To verify this, we start by considering the entropy as a function of temperature and volume, so we have $S = S(T, V)$ and

$$dS = \left(\frac{\partial S}{\partial T}\right)_V dT + \left(\frac{\partial S}{\partial V}\right)_T dV.$$

Let us partially crush this to write it as

$$dS = \frac{C_V}{T} dT + \left(\frac{\partial \overline{p}}{\partial T}\right)_V dV. \tag{6.112}$$

If we know the equation of state as a function of temperature and volume, then we can evaluate the derivative in the second term. C_V is in general a function of T and V; however as we will now see, we can obtain its volume dependence from $\overline{p}(T, V)$. We start with $C_V = T(\partial S/\partial T)_V$ and differentiate this with respect to the volume at fixed temperature,

$$\begin{aligned}\left(\frac{\partial C_V}{\partial V}\right)_T &= \left(\frac{\partial}{\partial V}\left[T\left(\frac{\partial S}{\partial T}\right)_V\right]\right)_T \\ &= T\left(\frac{\partial}{\partial T}\left(\frac{\partial S}{\partial V}\right)_T\right)_V \\ &= T\left(\frac{\partial^2 \overline{p}}{\partial T^2}\right)_V,\end{aligned} \tag{6.113}$$

where in the last line we used a Maxwell relation (MR 2). From the second derivative of the equation of state, we can thus determine how the heat capacity depends upon the volume.

Exercise 6.31 Fill in the steps to obtain Eq. (6.112).

Imagine along with the equation of state $\overline{p}(T, V)$, we have been able to determine $C_V(T, V_0)$, the heat capacity as a function of temperature for a single volume $V = V_0$. We can integrate Eq. (6.113) to obtain

$$\begin{aligned}C(T, V) &= C(T, V_0) + \int_{V_0}^{V} \left(\frac{\partial C_V(T, V')}{\partial V'}\right)_T dV' \\ &= C(T, V_0) + \int_{V_0}^{V} T\left(\frac{\partial^2 \overline{p}}{\partial T^2}\right)_V dV'.\end{aligned} \tag{6.114}$$

From this, we now know $C_V(T, V)$, and then along with $\left(\partial\overline{p}/\partial T\right)_V$, we can integrate Eq. (6.112) to obtain

$$\int_{T_0,V_0}^{T,V} dS = S(T, V) - S(T_0, V_0) = \int_{T_0}^{T} \frac{C_V(T', V)}{T'} dT' + \int_{V_0}^{V} \left(\frac{\partial p}{\partial T}\right)_V dV'. \tag{6.115}$$

While we can use any path to perform this integral (S is a state function after all), this can be understood as performing two processes in a particular order. First determine how the entropy changes for a system at constant temperature as its volume changes from $V_0 \to V$. Then fix the volume to V and change the temperature of the system from $T_0 \to T$. We can write this as

$$S(T, V) - S\left(T_0, V_0\right) = \left[S(T, V) - S\left(T_0, V\right)\right] + \left[S\left(T_0, V\right) - S\left(T_0, V_0\right)\right].$$

Now we have the entropy, but we only have it as a function of temperature and volume. To properly obtain the fundamental relation, or the entropy as a function of the internal energy and volume, $S(\bar{E}, V)$, we need to know how the energy depends on T and V. We can determine the internal energy with the same information we used to get the entropy, and in fact we have already done the work needed for this. Using the results in Eqs. (6.98) and (6.99), we can write

$$d\bar{E} = C_V dT + \left[T\left(\frac{\partial \overline{p}}{\partial T}\right)_V - \overline{p}\right] dV.$$

Which allows us to identify

$$\left(\frac{\partial \bar{E}}{\partial T}\right)_V = C_V, \tag{6.116}$$

and

$$\left(\frac{\partial \bar{E}}{\partial V}\right)_T = T\left(\frac{\partial \overline{p}}{\partial T}\right)_V - \overline{p}. \tag{6.117}$$

As we did for the entropy, we can integrate these equations to determine $\bar{E}(T, V) - \bar{E}(T_0, V_0)$ as long as we know $C_V(T, V_0)$ for one volume and the equation of state. This will make more sense with an example.

Example 6.6 We can move beyond the simple approximation of the ideal gas by considering the **Van der Waals gas**,[27] which is an empirical extension of the ideal gas equation of state. We can use this equation of state to come up with an expression for the entropy and internal energy of such a gas.

Recall one of the forms of the ideal gas equation is in terms of the molar volume $v = V/\nu$,

$$\bar{p}v = RT, \tag{6.118}$$

and we already know this is only valid in very simple cases. While it can lead to some good qualitative results in certain cases, it is very limiting. Not only is it not valid as we approach absolute zero which we saw in Section 6.6, but we will see in Chapter 10 that it does not allow phase transitions. It will be useful to have a more realistic model to work with. While this can be derived from statistical methods with suitable approximations (it will be easier when we consider the canonical ensemble, see Section 8.6), for now we'll use empirical arguments to modify it accordingly.

The first modification comes from realizing that the molecules do not have access to the entire volume in which it is contained. The ideal gas approximation assumes the molecules are point-like, but in actuality they take up some volume of their own. A given molecule would have access not to the entire volume V but some volume $V - B$, where B is a constant that depends on the gas, related to the size of each molecule. To account for this, in terms of the molar volume, we make the replacement

$$v \to v - b, \tag{6.119}$$

where $b = B/\nu$.

The second modification takes into account the molecular interactions which were completely neglected for the ideal gas. The molecules can be approximated as electric dipoles, and thus the interactions (to a first approximation) would be attractive dipole–dipole forces that would increase the mean pressure of the system. The potential energy leading to this attractive force has a $\sim 1/r^6 \sim 1/V^2$ dependence. Thus we should add to the pressure (keeping this in terms of the molar volume),

$$\bar{p} \to \bar{p} + \frac{a}{v^2}, \tag{6.120}$$

with a a constant dependent upon the system of interest related to the strength of these interactions. This leads to the **Van der Waals equation**,

$$\left(\bar{p} + \frac{a}{v^2}\right)(v - b) = RT, \tag{6.121}$$

with some examples for measured values of a and b for some gases shown in Table 6.2, taken from Ref. [6].

As we didn't obtain this equation of state from the entropy, we will use our approach from above to obtain the entropy (and internal energy) from the Van der Waals equation. We will use molar quantities for this calculation, but the results from above will still be valid. Solving Eq. (6.121) for $\bar{p}$ we have

$$\bar{p} = \frac{RT}{v - b} - \frac{a}{v^2},$$

so

$$\left(\frac{\partial \bar{p}}{\partial T}\right)_v = \frac{R}{v - b}, \tag{6.122}$$

---

27 Johannes van der Waals (1837–1923).

Table 6.2 Coefficients for the Van der Waals equation for a select set of gases, from Ref. [6] J.R. Rumble 2022/ Taylor & Francis.

Gas	a (m^6 Pa/mol^2)	b (m^3/mol)
Carbon dioxide, CO_2	0.3658	4.29×10^{-5}
Chlorine, Cl_2	0.6343	5.42×10^{-5}
Ethane, C_2H_4	0.5580	6.51×10^{-5}
Helium, He	0.00346	2.38×10^{-5}
Hydrogen, H_2	0.02452	2.65×10^{-5}
Nitrogen, N_2	0.1370	3.87×10^{-5}
Oxygen, O_2	0.1382	3.19×10^{-5}
Water, H_2O	0.5537	3.05×10^{-5}

and

$$\left(\frac{\partial c_V}{\partial v}\right)_T = \left(\frac{\partial^2 \overline{p}}{\partial v^2}\right)_T = 0. \tag{6.123}$$

From this we see that the specific heat is independent of the volume,

$$c_V(T, v) = c_V(T).$$

This allows us to write the differential of the molar entropy as

$$ds = \frac{c_V(T)}{T} dT + \frac{R}{v - b} dv, \tag{6.124}$$

and if we knew the temperature dependence of the specific heat, we could easily integrate this. For now, let's assume we are working in a range of temperatures where the specific heat is also independent of temperature. If this is true, then we have

$$\begin{aligned} s(T, v) - s(T_0, v_0) &= \int_{T_0}^{T} \frac{c_V}{T'} dT' + \int_{v_0}^{v} \frac{R}{v' - b} dv' \\ &= c_V \ln\left(\frac{T}{T_0}\right) + R \ln\left(\frac{v - b}{v_0 - b}\right), \end{aligned}$$

or

$$s(T, v) = c_V \ln T + R \ln(v - b) + \text{constant}. \tag{6.125}$$

Next we determine the molar energy, $\overline{\varepsilon} = \bar{E}/v$. From Section 6.5 we have

$$\left(\frac{\partial \overline{\varepsilon}}{\partial T}\right)_v = c_V(T),$$

and

$$\left(\frac{\partial \overline{\varepsilon}}{\partial v}\right)_T = \frac{RT}{v - b} - \overline{p} = \frac{a}{v^2}.$$

Again we can integrate these to get the molar energy if we knew the temperature dependence of the specific heat. Still considering the case where c_V is constant, then we get

$$\bar{\varepsilon}(T, v) = -\frac{a}{v} + c_V T + \text{constant}. \tag{6.126}$$

In the case of the Van der Waals gas, we have the same result for the temperature dependence of the gas as we did for the ideal gas, but now there is a volume dependence as well. Both Eqs. (6.125) and (6.126) reduce to the ideal gas results when we take the limits $a \to 0$ and $b \to 0$.

Exercise 6.32 Perform the necessary integrals to obtain Eq. (6.126).

Exercise 6.33 Set the constants in Eqs. (6.125) and (6.126) to zero, and eliminate T to get $s(\bar{\varepsilon}, V)$, so that we have the fundamental relation in terms of the appropriate variables. Show that you get the result for the ideal gas in the appropriate limit.

Exercise 6.34 Calculate α and κ for a Van der Waals gas.

6.8 Summary

- Heat reservoirs are systems that can be used as an infinite source or sink of energy (in the form of heat). More precisely, they are systems whose temperatures remain (almost exactly) constant no matter how much heat is gained or lost during an interaction.
- Just as we can obtain the temperature from the fundamental relation $S(\bar{E}, V)$, we can also obtain the mean pressure of a system. The resulting relationships are called equations of state.
- Various thermodynamic potentials can be defined as Legendre transforms of the internal energy, replacing S or V (or both) for T or $\bar{p}$ (respectively) as the natural independent parameters.
- Many quantities of interest in thermodynamics take the form of first derivatives, and all first derivatives (so long as volume is the only external parameter) can be written in terms of the heat capacity, the coefficient of thermal expansion, and the isothermal compressibility (all of which tend to be simple to measure experimentally). This process, derivative crushing, can be done in a systematic way.
- These derivatives have known behaviors in the zero-temperature limit, which is a useful test to see if thermodynamic models are valid at low temperatures.
- In many realistic cases, it is difficult to obtain the equations of state from the microscopic picture. In such cases, we can instead use an empirically determined equation of state—specifically $\bar{p}(V, T)$—as well as the heat capacity at constant volume as a function of T for one volume, and determine the internal energy and entropy.

Problems

6.1 Return to Problem 4.12, where 1 mol of monatomic ideal gas expands quasistatically from $p_i = 3.2$ MPa and $V_i = 10^{-3}\,\text{m}^3$ to $p_f = 0.1$ MPa and $V_f = 8 \times 10^{-3}\,\text{m}^3$. **By integrating** $dS = đQ/T$ for each of the four paths (a)–(d) shown, find $S_f - S_i$ and show that it is the same for all. You may use $\bar{p}V = RT$, $C_V = (3/2)R$, and $C_p = (5/2)R$. Path (d) is defined to be the adiabatic path, where $\bar{p} = KV^{-5/3}$ for constant K.

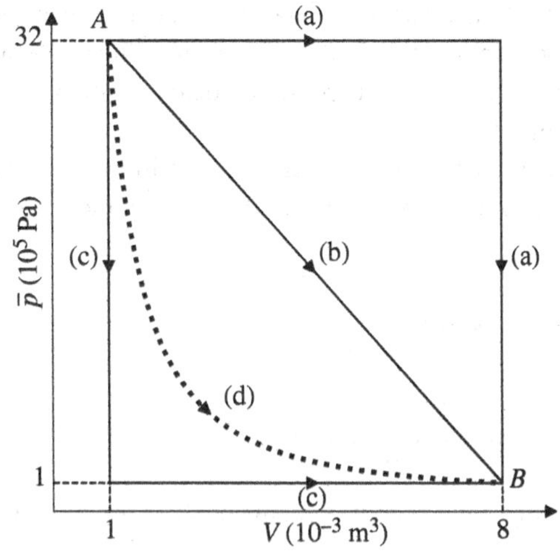

6.2 A rubber band is made out of long molecular chains ("polymers") which are folded back over themselves in an arbitrary way, as indicated by the figures shown below. Both of these examples have 16 links in the chain and a net length of six links, but they are folded up differently. The potential energy of the rubber band is due to the (weak) attraction between links which are lying next to each other because of the folding. To a first approximation, the potential energy in a particular state does not depend on the particular arrangement of folds but only on the net length of the chain and the total number of links. Let ℓ be the length of the rubber band; $U(\ell)$, its potential energy; E, its mean total energy; and t, its mean tension.

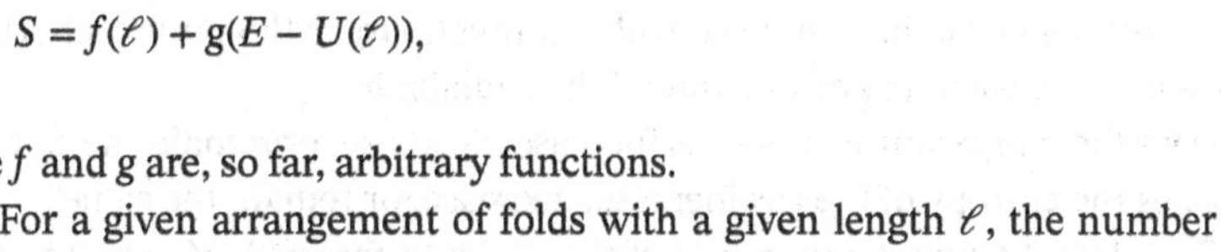

(a) Argue that the entropy S has the form

$$S = f(\ell) + g(E - U(\ell)),$$

where f and g are, so far, arbitrary functions.
Hint: For a given arrangement of folds with a given length ℓ, the number of ways of arranging the system in phase space just depends on the total kinetic energy of the system.

(b) Show that the quantity $E - U(\ell)$ is a function of the temperature T only.

(c) Argue that S *decreases* when the rubber band is stretched isothermally.

(d) Show that the rubber band must get warmer when it is stretched adiabatically.
Hint: You may assume that the heat capacity at fixed length C_ℓ is "normal," i.e., that it is positive.

(e) Show that the tension t increases when the temperature T is increased at fixed ℓ.
Hint: Use a Maxwell relation with the replacements $V \to \ell$ and $p \to -t$.

[You can read more about the thermodynamics of rubber bands in Refs. [7, 8].]

6.3 Consider a classical ideal gas which has molar specific heats at constant volume and pressure c_V and c_p (both independent of temperature). The gas is placed in a thermally insulated container and is allowed to expand quasistatically from an initial volume V_i at temperature T_i to a final volume V_f.
Use the fact that the entropy remains constant in this process in this process to find the final temperature T_f of the gas in terms of the parameters above and $\gamma = c_p/c_V$, the ratio of specific heats.

6.4 Crush the following derivatives
(a) $(\partial T/\partial V)_H$
(b) $(\partial T/\partial V)_{\bar{E}}$
and show that they vanish for an ideal gas.

6.5 Determine how the molar specific heat at constant pressure depends upon pressure. Specifically, evaluate

$$\left(\frac{\partial c_p}{\partial \bar{p}}\right)_T$$

in terms of T, v, and $(\partial \alpha/\partial T)_p$. Show that this vanishes for an ideal gas.

6.6 A substance is maintained at constant temperature T while the pressure is increased by $d\bar{p}$. Find the changes in $V, S, \bar{E}, F, G$, and H in terms of $V, T, \alpha, \kappa, \bar{p}$, and $d\bar{p}$.
Hint: Find, for example, $dV = (\partial V/\partial \bar{p})_T d\bar{p}$, and the same for the others.

6.7 Consider an ideal gas with N molecules in a container which is closed off on the top by a lid with mass m and cross-sectional area A. The entire system is thermally insulated, and you apply a pressure on the lid to initially hold it in place so the gas has a volume V_1 and a temperature T_1. When you stop holding the lid, it oscillates and after some amount of time it comes to rest in a final equilibrium situation with a final volume greater than V_1. Assume the heat capacity at constant volume C_V of the gas is independent of temperature. Additionally, neglect the heat capacities of the lid and container as well as any frictional forces between the lid and the walls of the container. Additionally, the pressure applied by the weight mg of the lid can be considered much larger than atmospheric pressure. Note: This is *not* a quasistatic process, though it is adiabatic.
(a) Does the temperature of the gas increase, decrease, or remain the same?
(b) Does the entropy of the gas increase, decrease, or remain the same?
(c) Calculate the final temperature of the gas in terms of T_1, V_1, and the other parameters mentioned above.
Hint: Conservation of energy is important here; don't forget the change in the potential energy of the lid.

6.8 A paramagnetic gas is a system with two external parameters, the volume V and the magnetization M. Thus it has three independent variables, which may be chosen to be any three of $S, \bar{p}, T, V, M$, and H. (Here H is the external magnetic field, *not* the enthalpy.) The first law of thermodynamics in this case is

$$d\bar{E} = TdS - pdV + HdM,$$

where the last term represents the increase in energy of the molecules when their magnetic dipole moments are increased in the presence of an external magnetic field.

(a) Derive the following Maxwell relations from the Helmholtz free energy:

$$\left(\frac{\partial S}{\partial V}\right)_{T,M} = \left(\frac{\partial \bar{p}}{\partial T}\right)_{V,M} \quad \text{and}$$

$$\left(\frac{\partial S}{\partial M}\right)_{T,V} = -\left(\frac{\partial H}{\partial T}\right)_{V,M}.$$

(b) Suppose there is 1 mol of the gas and that it obeys the equations of state,

$$M = \frac{aH}{T}, \quad C_{V,M} = \frac{3}{2}R, \quad \text{and} \quad \bar{p}V = RT,$$

where a is a constant and $C_{V,M}$ is the heat capacity at fixed V and M. Think of $\bar{E} = \bar{E}(T, V, M)$, and show that

$$\left(\frac{\partial \bar{E}}{\partial V}\right)_{T,M} = 0 \quad \text{and} \quad \left(\frac{\partial \bar{E}}{\partial M}\right)_{T,V} = 0,$$

and hence that $d\bar{E} = C_{V,M}dT$. From this, you can write $TdS = C_{V,M}dT + \bar{p}dV - HdM$.

(c) Consider V to be a function of $\bar{p}$, T, and M. Find dV and show that

$$TdS = \frac{5}{2}RdT - \frac{RT}{\bar{p}}d\bar{p} - HdM.$$

(d) Now treat M as a function of H, T, and $\bar{p}$. Find dM and show that

$$TdS = \left(\frac{5}{2}R + \frac{aH^2}{T^2}\right)dT - \frac{RT}{\bar{p}}d\bar{p} - \frac{aH}{T}dH.$$

Use this result for the next two parts.

(e) The system starts at temperature T_0, magnetic field H_0, and pressure $\bar{p}_0$. The magnetic field is changed quasistatically, adiabatically, and isothermally to a new value H. (This requires adjusting the pressure—and hence the volume—to keep T constant without adding heat.) Find the new pressure $\bar{p}$ in terms of H, H_0, T_0, $\bar{p}_0$, and constants.

(f) Instead, the system starts at temperature T_0 and magnetic field H_0. The magnetic field is changed adiabatically, quasistatically, and at constant pressure, to a new value H. Integrate the appropriate differential equation to find an equation relating the new temperature T to H, H_0, T_0, and constants. (This equation can be solved for T as a function of H, H_0, and T_0, but you do not need to do it.)
Hint: Define a new variable $f \equiv aH^2/(2T^2)$, find df, and use it to simplify your differential equation.

6.9 A simple introductory physics lab has a student place 75 g of ice at 0°C into 300 g of water, which is in equilibrium with a 900 g copper calorimeter at a temperature of 24°C. After the ice is placed, the system is thermally isolated.

(a) After the ice has melted and equilibrium has been reached, what is the temperature of the water? (The specific heat of copper is 0.418 J/g/K. The latent heat of fusion of ice is 333 J/g.)

(b) Compute the total entropy change resulting from the process of part (a).

(c) After all the ice has melted and equilibrium has been reached, how much work, in joules,[28] must be supplied to the system to restore the system back to 24°C?

(d) What is the *water equivalent* of this calorimeter, M_c (as in Example 6.2)?

---

28 James Prescott Joule (1818–1889).

6.10 Ferromagnetic materials are composed of atoms with a net spin, such that they maintain a net magnetization (due to the atoms all aligning in a single direction) even in the absence of a magnetic field. They are often called "permanent magnets"; however at a high enough temperature, this magnetization will vanish. At these temperatures, the atoms (like those in most materials) are randomly oriented. The **Curie temperature** T_c is the temperature where we find the onset of the ferromagnetic behavior (that is, above this temperature, the material is no longer magnetic).[29]

Consider a system of N spin-1/2 atoms (thus the spins can be aligned in one of two possible directions). The spins will contribute to the heat capacity of the material but only in the range $T_c/2 < T < T_c$. Suppose you have two competing models for this contribution:

$$C_1(T) = C_{10}\left(\frac{2T - T_c}{2T_c}\right)^2 \text{ if } \frac{1}{2}T_c < T < T_c,$$

$$C_2(T) = C_{20}\left(\frac{2T}{T_c} - 1\right) \text{ if } \frac{1}{2}T_c < T < T_c,$$

where C_{10} and C_{20} are constants and $C_{1,2}(T) = 0$ outside this temperature range. Use entropy considerations to find an explicit expression for the constants C_{10} and C_{20} in each model.

Hint: Consider how the entropy must behave in the zero-temperature and infinite temperature limits.

6.11 How would your results to Problem 6.10 change if the ferromagnetic material was composed of atoms with spin s instead of spin 1/2? Recall that a spin-s system has $2s + 1$ possible spin states.

6.12 For ultra-low temperature experiments, any changes in the system must be monitored to see how it affects the temperature. Suppose you're working on an experiment on a thermally isolated system where you quasistatically change the pressure by a small amount $\Delta\bar{p}$. What is the resulting change ΔT in the temperature of the system? Put your answer in terms of the specific heat per mass at constant pressure c_p, the coefficient of thermal expansion α, the absolute temperature T, and the volume mass density of the system $\rho = m/V$.

6.13 Two cylinders, 1 and 2 (shown below), each hold 1 mol of a substance initially with temperatures T_1, T_2 and volumes V_1, V_2. The gases are allowed to interact non-quasistatically and irreversibly when the piston between the two cylinders is suddenly allowed to slide and conduct heat. The cylinders are thermally insulated from the outside world. The cylinders each have another movable piston that is exposed to atmospheric pressure throughout the interaction, so the interaction takes place at constant pressure (but the volumes V_1 and V_2 will in general change as the temperatures change). The molar specific heat (at constant pressure) of the substance is c_p, which is assumed to be independent of temperature. The final equilibrium temperature is T_f and the final volumes are V_{1f} and V_{2f}.

(a) Imagine first that the system is brought to its final state by some different *quasistatic* process. Find the work done by each part of the system (W_1 and W_2), and the change in energies of the two parts ($\Delta\bar{E}_1$ and $\Delta\bar{E}_2$), in terms of variables given in the problem. *Hint:* Very little calculation is required.

29 This was named after Pierre Curie (1859–1906), not his wife, Marie Skłodowska-Curie (1867–1934), even though this is one of the rare cases where a female scientist is more famous than her male scientist husband.

(b) Return now to the real, non-quasistatic process. Argue that $\Delta\bar{E} = \Delta\bar{E}_1 + \Delta\bar{E}_2$, the energy change of the entire system, and $W = W_1 + W_2$, the work done by the entire system, are actually the same in part (a).

(c) Use the first law to show that $T_f = \frac{1}{2}(T_1 + T_2)$ in the actual process. This was not obvious originally, why not?

(d) Show that the change in entropy of the entire system in the actual process is

$$\Delta S = 2c_p \ln\left(\frac{T_1 + T_2}{2\sqrt{T_1 T_2}}\right).$$

(e) Show that $\Delta S > 0$ if $T_1 \neq T_2$.

(f) Suppose that instead of suddenly being allowed to slide and to conduct heat, the barrier was simply removed, so the gases on the two sides could mix directly. Would your answers to parts (c) and (d) be any different? Explain.

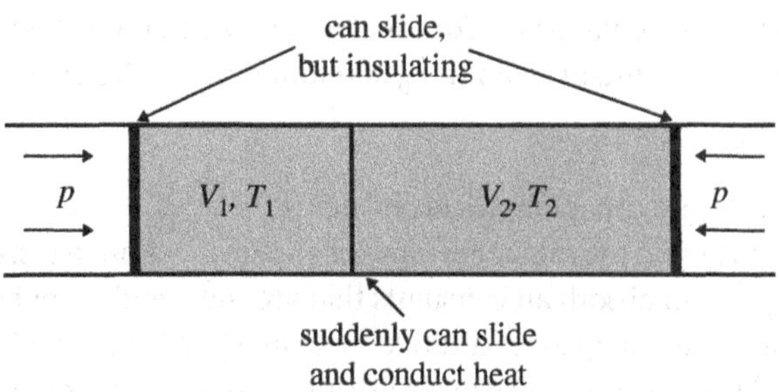

6.14 A classical system of N_1 molecules of type 1 and N_2 molecules of type 2 is confined within a box of volume V. The molecules are supposed to interact very weakly so that they constitute an ideal gas mixture.

(a) How does the total number of states $\Omega(E)$ in the range between E and $E + \delta E$ depend on the volume V of this system?

(b) Use this result to find the mean pressure $\bar{p}$ as a function of V and T.

6.15 Sound waves are pressure waves (compression and expansion) of a medium as they move through it. To a good approximation, they can be described by the one-dimensional wave equation

$$\frac{\partial^2 p}{\partial t^2} = v_s^2 \frac{\partial^2 p}{\partial x^2},$$

where v_s is the speed of sound and p is the pressure of the medium. This can be considered a quasistatic, adiabatic process: the medium relaxes to equilibrium quickly after the sound passes, and the transfer of energy from the wave to the medium occurs without heat flow. The speed of sound can be shown to be $v_s = 1/\sqrt{\rho\kappa_S}$, with ρ the equilibrium density of the medium and κ_S is the **adiabatic compressibility** of the medium,

$$\kappa_S \equiv -\frac{1}{V}\left(\frac{\partial V}{\partial p}\right)_S.$$

(a) Determine κ_S assuming the sound wave passes through an ideal gas in terms of the pressure and specific heat ratio γ.

(b) Calculate the velocity of sound in an ideal gas in terms of γ, $\mu = m/\nu$ (the molar mass), and the absolute temperature T.

(c) Calculate the velocity of sound in nitrogen (N_2) gas at 25°C and pressure, taking γ from Table 6.1.
(d) Repeat parts (a)–(c) assuming the interaction between the sound wave and the medium occurs *isothermally* instead of adiabatically. This amounts to replacing κ_S with the isothermal compressibility κ.
(e) Given your answers above, can you determine whether or not the traveling of sound waves should be considered an isothermal process or an adiabatic process?

6.16 You want to determine what the entropy and internal energy are for a gas that obeys the **Redlich–Kwong equation of state,**[30]

$$p = \frac{RT}{v-b} - \frac{a}{\sqrt{T}v(v+b)},$$

where a and b are constants (different than those in the Van der Waals equation) that would depend on the particular gas. Additionally, suppose that you have measured the molar specific heat for some (molar) volume v_0 and found that it depends upon the temperature as

$$c_V(T, v_0) = c_0 + c_1 T,$$

where c_0 and c_1 are other empirical constants.
(a) Determine what the molar entropy s is (up to an additive constant s_0) as a function of T and v. You can absorb all constants that are independent of both v and T into s_0.
(b) Determine (again up to an additive constant ε_0) what the molar energy $\bar{\varepsilon}$ is as a function of T and v. As before, you can absorb all constants that are independent of both v and T into ε_0.
(c) Show that when $a \to 0$, $b \to 0$, and $c_1 \to 0$, your results for s and $\bar{\varepsilon}$ agree with our results for the ideal gas.
(d) What must c_0 be explicitly (according to the laws of thermodynamics) and why? Simplify your results with this determination.
(e) (OPTIONAL) Use Mathematica or another computer program to create a parametric plot of s and $\bar{\varepsilon}$, so that you can visualize s vs. $\bar{\varepsilon}$ (its natural variable) for a given volume.

6.17 Return to Problem 6.2, and suppose we found the equation of state (how the tension depends on the temperature and length) of the rubber band to be

$$t = bT\frac{\ell - \ell_0}{\ell_1 - \ell_0}, \quad \ell_0 < \ell < \ell_1,$$

where ℓ_0 is the unstretched length of the rubber band, ℓ_1 is the **elastic limit** (beyond which the rubber band would break), and b is some constant. Additionally, suppose that the heat capacity at constant length C_ℓ is a constant. In this case, the first law can be written as

$$d\bar{E} = C_\ell dT + t d\ell.$$

(a) Determine the entropy of the rubber band as a function of the temperature and length, in terms of the variables given in the problem.
(b) Determine the energy of the rubber band as a function of the temperature and length, in terms of the variables given in the problem.

30 Otto Redlich (1896–1978) and Joseph Neng Shun Kwong (1916–1998).

(c) Show explicitly with your results here that the results of Problem 6.2 hold here:
  i. $\bar{E}$ is independent of ℓ,
  ii. S decreases when the length increases at constant temperature, and
  iii. T increases when ℓ increase adiabatically.
  (The last result from that problem follows immediately from our equation of state.)

6.18 A common way to expand upon the ideal gas equation is to use the **virial expansion**, as an expansion of the form

$$\bar{p} = nk_BT\sum_{i=1}^{\infty} B_i(T)n^{i-1},$$

where $n = N/V$ is the number density and $B_i(T)$ is the i**th virial coefficient** ($B_1 = 1$, to be consistent with the ideal gas equation). Let's truncate this series at $i = 2$, so that we have

$$\bar{p} = nk_BT\left[1 + B_2(T)n\right].$$

Assume that $B_2(T)$ is an increasing function of temperature (this can be made by a simple argument) and determine how the mean internal energy $\bar{E}$ of this gas depends on its volume V, i.e., find an expression for $(\partial E/\partial V)_T$. Is it positive or negative?

References

1 F. Reif. *Fundamentals of Statistical and Thermal Physics*. McGraw Hill, Tokyo, 1965.

2 R. K. P. Zia, E. F. Redish, and S. R. McKay. Making sense of the Legendre transform. *American Journal of Physics*, 77(7):614–622, 2009.

3 H. B. Callen. *Thermodynamics and An Introduction to Thermostatistics*. Wiley, New York, NY, 2nd edition, 1985.

4 K. F. Riley, M. P. Hobson, and S. J. Bence. *Mathematical Methods for Physics and Engineering: A Comprehensive Guide*. Cambridge University Press, 2nd edition, 2006.

5 Y. A. Çengel and A. J. Ghajar. *Heat and Mass Transfer: Fundamentals & Applications*. McGraw-Hill Education, 5th edition, 2019.

6 J. R. Rumble. *CRC Handbook of Chemistry and Physics*. CRC Handbook of Chemistry and Physics. CRC Press, 2022.

7 J. B. Brown. Thermodynamics of a rubber band. *American Journal of Physics*, 31(5):397–397, 1963.

8 D. Roundy and M. Rogers. Exploring the thermodynamics of a rubber band. *American Journal of Physics*, 81(1):20–23, 2013.

7

Applications of Thermodynamics

In Chapter 6, we worked through the mathematics required to study thermodynamic problems, setting us up with the tools to be able to study physical processes. In this chapter, we will work through various examples, calculating observables that can then allow us to test our theory. Recall we made some pretty bold assumptions (such as the fundamental postulate), and now it is time to see that physical observations are consistent with them. In most of our examples, the volume will be the only external parameter, although we will also see some other examples where different external parameters will come into play.

After finishing this chapter, you should be able to

- understand the adiabatic expansion of a system, specifically an ideal gas,
- learn about simple processes used to cool gases (such as free expansion and the Joule–Thomson process), and
- understand the basics of heat engines and refrigerators.

7.1 Adiabatic Expansion

The first process we will study is a thermally isolated system that is allowed to do mechanical work. Such a process has first been discussed in Section 4.6.2 (shown in Figure 4.10), and now we will consider it in more detail. Because our system is thermally isolated, there can be no heat flow into or out of the system, so

$$đQ = 0, \tag{7.1}$$

and thus via the first law,

$$d\bar{E} = -\bar{p}dV. \tag{7.2}$$

We would like to determine how the pressure changes with the volume for an adiabatic process.

In order to do this generally, let us assume the adiabatic process is also *quasistatic*. For a quasistatic process, we can relate the heat flow to the entropy change, $đQ = TdS$, so in this case,

$$dS = 0, \qquad \text{or} \qquad S = \text{constant}. \tag{7.3}$$

Statistical Thermodynamics: An Information Theory Approach, First Edition. Christopher Aubin.

Companion website: www.wiley.com/go/Aubin/StatisticalThermodynamics

This way we can determine how the pressure depends upon the volume by considering the pressure as a function of volume and entropy, and we can write

$$d\overline{p} = \left(\frac{\partial \overline{p}}{\partial V}\right)_S dV + \left(\frac{\partial \overline{p}}{\partial S}\right)_V dS = \left(\frac{\partial \overline{p}}{\partial V}\right)_S dV, \tag{7.4}$$

where the second equality is true for an adiabatic process. Now we just crush the derivative to obtain

$$\left(\frac{\partial \overline{p}}{\partial V}\right)_S = -\frac{C_p}{C_V}\frac{1}{V\kappa} = -\frac{\gamma}{V\kappa}, \tag{7.5}$$

with γ the specific heat ratio defined in Eq. (6.101).

Exercise 7.1 Show that Eq. (7.5) follows from our derivative crushing procedure.

We see in Eq. (7.5) that this derivative is negative, so the pressure decreases as the volume increases, which is something we would expect. This particular result is also valid for any system—any gas, liquid, or solid. For an ideal gas, we have $\kappa = 1/\overline{p}$, so putting Eq. (7.5) into Eq. (7.4), we get the equation

$$\frac{d\overline{p}}{\overline{p}} = -\gamma\frac{dV}{V}, \tag{7.6}$$

which we can integrate to obtain $\overline{p}V^\gamma$ = constant. This constant can be determined by the initial pressure and volume. We first saw this in Eq. (4.70) from Exercise 4.9.

As an alternate approach, if we have a specific equation of state, we can solve this more directly. For now again assuming from the beginning, we have a classical ideal gas. From Eq. (6.97), $d\overline{E} = \nu c_V dT$, and Eq. (7.2) becomes

$$\nu c_V dT = -\overline{p}dV. \tag{7.7}$$

Exercise 7.2 Show that you can write

$$d\overline{p}V + \overline{p}dV = \nu R dT. \tag{7.8}$$

by differentiating the ideal gas equation.

Putting Eq. (7.8) into Eq. (7.7), we get

$$\frac{c_V}{R}\left(d\overline{p}V + \overline{p}dV\right) = -\overline{p}dV. \tag{7.9}$$

Exercise 7.3 Assuming c_V is constant, integrate Eq. (7.9) to show that

$$pV^\gamma = \text{constant} \tag{7.10}$$

for an adiabatic process involving an ideal gas, the same as above.

For an ideal gas, the pressure behaves like $\overline{p} \propto V^{-\gamma}$. Given that $\gamma > 1$, we see that the pressure decreases faster for an adiabatic expansion than it does for an isothermal expansion, as shown in Figure 7.1. In the latter case, we have T = constant and $p \propto V^{-1}$ via the ideal gas equation. Additionally, we see that the pressure changes in the same way for all ideal gases for an isothermal

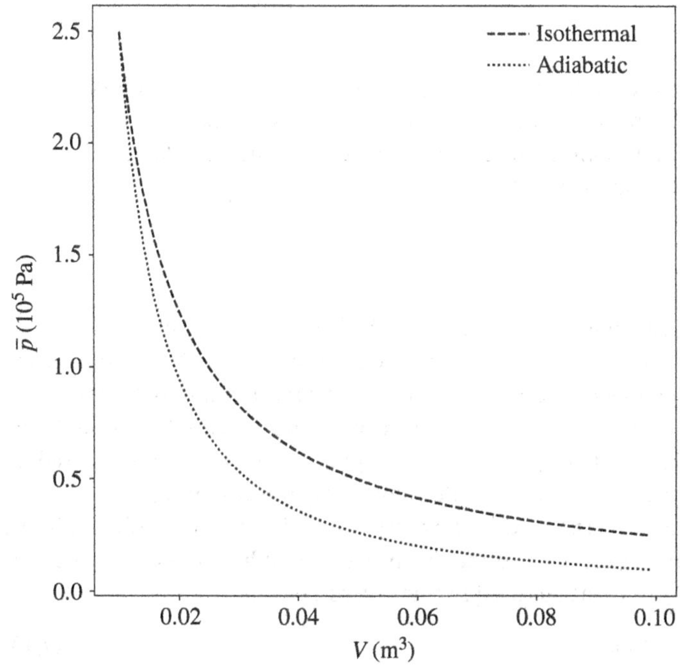

Figure 7.1 Comparing $\overline{p}$ vs. V for a diatomic ideal gas undergoing isothermal and adiabatic expansions.

process, but given that γ is different for each gas, the dependence $\overline{p}$ has on V differs for monatomic, diatomic, etc. gases undergoing an adiabatic expansion.

Exercise 7.4 Using the ideal gas equation and Eq. (7.10), show that we can relate the temperature and volume for an adiabatic process as

$$V^{\gamma-1}T = \text{const}, \tag{7.11}$$

and the pressure and temperature as

$$\overline{p}^{1-\gamma}T^{\gamma} = \text{const}. \tag{7.12}$$

I want to pause to stress something: even though no heat flows into or out of a system during an adiabatic process, the temperature of the system *can* (and usually does!) change. This points to a confusion that is common: the difference between a *thermally isolated* process (thus it's adiabatic and the temperature can change), and an *isothermal* process, where heat can be added to the system but the temperature remains constant. Remember:

- Isothermal process $\Rightarrow dT = 0$, while Q could be zero or non-zero.
- Thermally isolated process = adiabatic process $\Rightarrow Q = 0$, while the temperature may or may not change.

Exercise 7.5 Return to Problem 4.12. The figure in this problem showed the adiabatic path to take the system from the initial to final state—is this a monatomic or diatomic ideal gas?

7.2 Cooling Gases

In low temperature physics, liquid nitrogen and helium are important tools to keep systems at temperatures of around 77 and 4 K, respectively. In order to liquefy these gases, we need a method with which to cool them down from room temperature. There are two methods for doing so that we can understand with the tools at our disposal.

7.2.1 Free Expansion

Let's consider the following setup, as shown in Figure 7.2(a): We have a gas in equilibrium at some volume V_1 (the left side of the container in the figure) with a temperature T_1, and the entire system (including the right side of the container) is thermally isolated from its surroundings. We then open a valve and the gas is allowed to move to the right side of the container (Figure 7.2(b)) such that eventually it fills the entire volume $V_2 > V_1$ (Figure 7.2(c)). This process, known as **free expansion**, will bring the system to a temperature T_2 which we will show is not greater than T_1.

As the system is thermally isolated, this is an adiabatic process, and thus $Q = 0$. Additionally, the gas does no work (there is nothing for the gas to do work on in the right half of the container), so $W = 0$. By the first law, the energy remains constant in this process, or

$$\Delta \overline{E} = 0 \Rightarrow \overline{E}(T_2, V_2) = \overline{E}(T_1, V_1), \tag{7.13}$$

which is true even if this is not a quasistatic process. If the point of this procedure were to cool the gas, this would be a bad approach if the gas were ideal because we know the energy is independent of the volume, so

$$T_2 = T_1 \text{ (ideal gas)}.$$

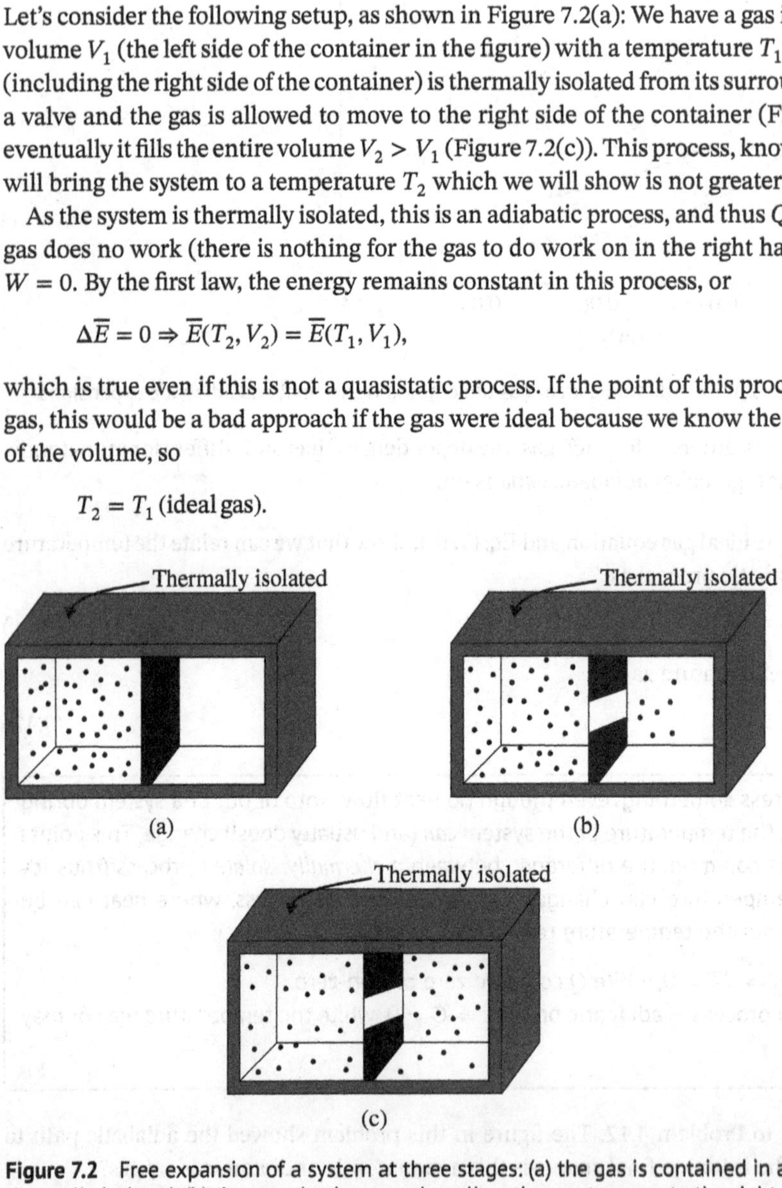

Figure 7.2 Free expansion of a system at three stages: (a) the gas is contained in a volume V_1 and is thermally isolated, (b) then a valve is opened to allow the gas to move to the right side, and (c) finally the gas is in equilibrium in some new volume $V_2 > V_1$.

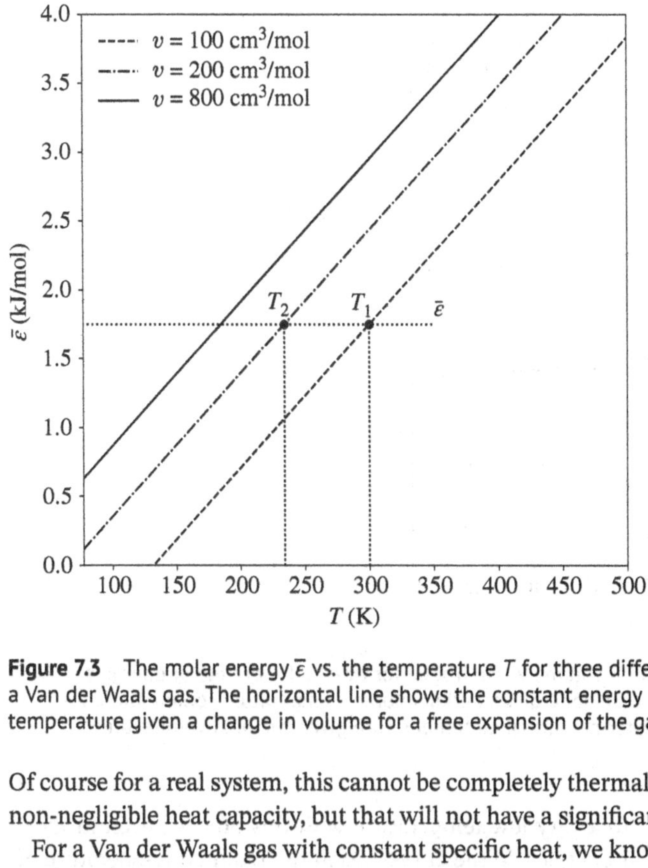

Figure 7.3 The molar energy $\bar{\varepsilon}$ vs. the temperature T for three different volumes of nitrogen, treating it as a Van der Waals gas. The horizontal line shows the constant energy line to determine the change in temperature given a change in volume for a free expansion of the gas.

Of course for a real system, this cannot be completely thermally isolated; the container will have a non-negligible heat capacity, but that will not have a significant effect on this result.

For a Van der Waals gas with constant specific heat, we know the molar energy difference, from Eq. (6.126), is

$$\bar{\varepsilon}(T_2, v_2) - \bar{\varepsilon}(T_1, v_1) = c_V(T_2 - T_1) - \left(\frac{a}{v_2} - \frac{a}{v_1}\right).$$

For the free expansion process, this vanishes, so

$$T_2 = T_1 + \frac{a}{c_V}\left(\frac{1}{v_2} - \frac{1}{v_1}\right) \quad \text{(Van der Waals gas)}, \tag{7.14}$$

and since $v_2 > v_1$, the second term is negative, which allows us to say

$$T_2 - T_1 < 0,$$

which is what we were hoping for a free expansion process: to cool a gas.

In Figure 7.3, I show the molar energy vs. temperature for various fixed molar volumes (using parameters relevant to nitrogen gas, see Table 6.2).[1] Drawing a horizontal line for a given energy, we can determine what volume we would have to expand the gas to in order to decrease the temperature to a particular value. Note it requires doubling the molar volume just to bring the gas from room temperature to about 234 K. Similarly, in Figure 7.4, I show the temperature vs. the molar volume for various molar energies. In this form, we start at a given (T, v) combination and move along a given curve to determine the final temperature for a given final volume. This graph makes

1 Strictly speaking, the specific heat is not constant as a function of temperature for nitrogen, but in the range of 175–400 K, c_V is constant within roughly 0.2%.

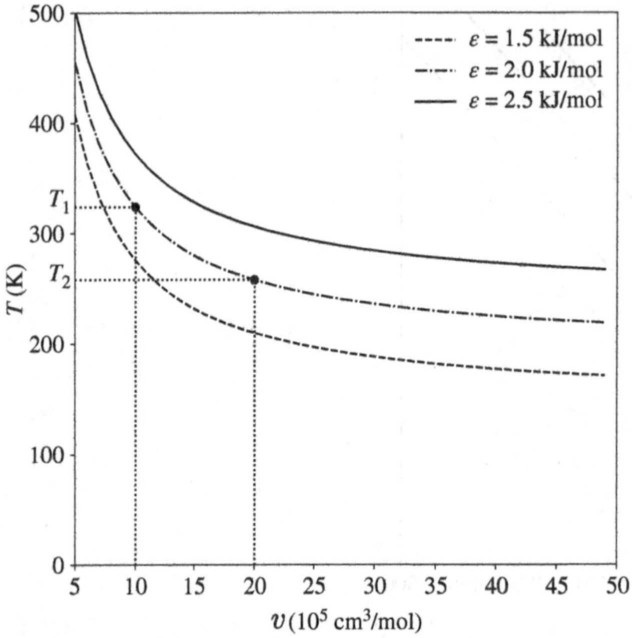

Figure 7.4 The temperature T vs. the molar volume v of nitrogen for three different molar energies, treating it as a Van der Waals gas. In this case, for a given initial volume–temperature combination, we move along a constant-energy curve to determine the final temperature given the final volume after undergoing free expansion.

it more clear how difficult it is to get to a very low temperature with this process, even in ideal circumstances (for none of these molar energies do the curves reach the required temperature of 77 K to liquefy nitrogen for these volumes!).

To consider this more generally, the relevant partial derivative we are interested in for a free expansion is

$$\left(\frac{\partial T}{\partial V}\right)_E, \tag{7.15}$$

which you showed in Problem 4 to be

$$\left(\frac{\partial T}{\partial V}\right)_E = \frac{\overline{p}\kappa - T\alpha}{C_V\kappa}.$$

Given $\kappa = 1/\overline{p}$ and $\alpha = 1/T$ for an ideal gas, this vanishes as expected.

Exercise 7.6 Show that the result for $(\partial T/\partial V)_E$ gives the expected result for the Van der Waals gas (for constant heat capacity).

Exercise 7.7 We can also calculate the entropy change,

$$\left(\frac{\partial S}{\partial V}\right)_E. \tag{7.16}$$

Crush this derivative and show that it is positive, which is expected as the number of states clearly increases.

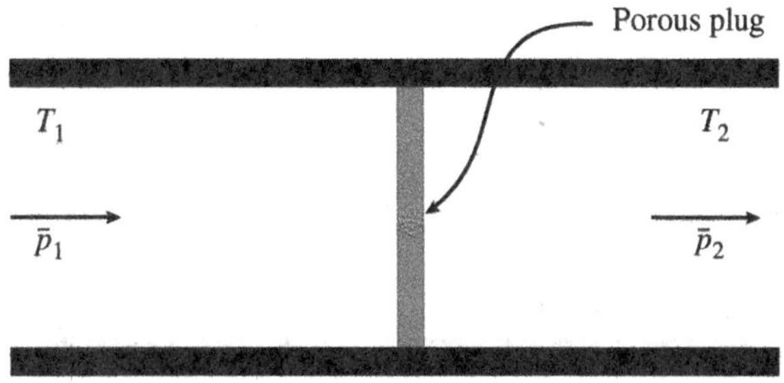

Figure 7.5 The setup for the throttling process. On the left is a gas at temperature T_1 which is pushed through a porous plug with a pressure $\overline{p}_1$. The gas emerges from the plug at a lower pressure $\overline{p}_2$ and lower temperature T_2.

While the free expansion of a gas will cool it, we can see that it cannot lower the temperature very much, and when we start to take into account the finite heat capacity of the container, then the internal energy change of the container makes a given volume change have a noticeably smaller effect on the temperature change of the gas. As such we will consider another method that can be used to achieve the goal we are after much more efficiently.

7.2.2 Throttling (Joule–Thomson) Process

A more efficient method for cooling a gas is known as the **throttling process** (also known as the **Joule–Thomson** or **Joule–Kelvin process**), the setup for which is shown in Figure 7.5. We consider a thermally insulated pipe divided into two sections by a porous plug, which impedes the flow of gas through the pipe. The gas is at temperature T_1 on the left of the plug and a pressure $\overline{p}_1$ is applied to it to force it through the plug, where it emerges with a pressure $\overline{p}_2$ and temperature T_2. The final pressure is less than the initial pressure, which is easy to see if you imagine taking a kitchen sponge and blowing through it: You may blow with a large pressure but the air has a lower pressure as it passes through the sponge.[2]

To determine the final temperature of the gas, consider ν moles of the gas in a volume V_1 as shown by the shaded portion on the left of the plug in Figure 7.6(a). We assume the size of this volume is large enough so that the volume of the plug can be neglected. After some time, because of the pressure $\overline{p}_1$ applied to it, this volume of gas moves to the right of the plug so that it is contained within some volume V_2 as shown in Figure 7.6(b).

Just as in the free expansion process, there is no heat flow ($Q = 0$), but unlike that case, the gas does work, so the internal energy changes by an amount

$$\Delta\overline{E} = \overline{E}_2 - \overline{E}_1 = \overline{E}(T_2, \overline{p}_2) - \overline{E}(T_1, \overline{p}_1) = -W \neq 0. \tag{7.17}$$

Whatever is providing the (constant) pressure $\overline{p}_1$ does an amount of work $\overline{p}_1 V_1$ on the gas, so

$$W_1 = -\overline{p}_1 V_1$$

is the amount of work done *by* the gas. The gas itself, as it passes through the plug, does some work W_2 by applying a constant pressure $\overline{p}_2$ on the rest of the gas on the right, so

$$W_2 = +\overline{p}_2 V_2.$$

Thus the total work done is

$$W = \overline{p}_2 V_2 - \overline{p}_1 V_1, \tag{7.18}$$

2 When I was teaching this during the COVID-19 pandemic, when we all had masks on during class, I could easily have everyone demonstrate this in real time. Go find a mask, you know you still have one, and try it out for yourself.

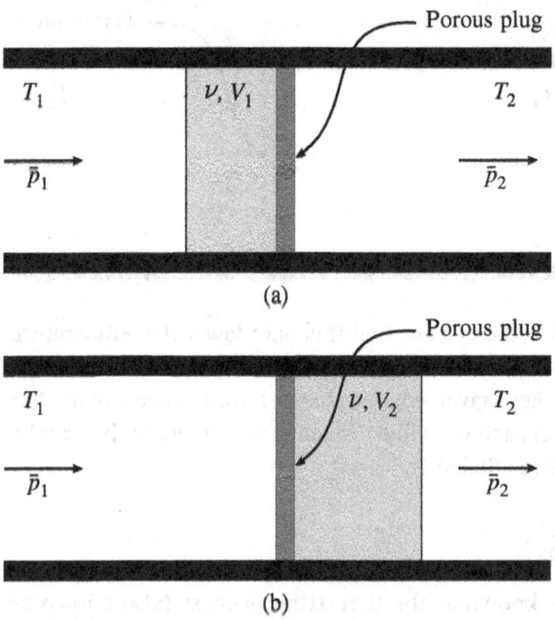

Figure 7.6 The setup for the throttling process, this time focusing on ν moles of gas with a volume V_1 before passing through the plug. The same amount of gas has a volume V_2 after passing through it.

and the change in the energy is

$$\overline{E}_2 - \overline{E}_1 = \overline{p}_1 V_1 - \overline{p}_2 V_2. \tag{7.19}$$

If we move all of the variables pertaining to the same side of the gas to one side, we have

$$\overline{E}_1 + \overline{p}_1 V_1 = \overline{E}_2 + \overline{p}_2 V_2. \tag{7.20}$$

On each side of this equation is just the *enthalpy* $H = \overline{E} + \overline{p}V$ of the gas before or after passing through the plug, and we see this is constant (not the internal energy) during the throttling process.

If the gas being passed through the plug is ideal, using Exercise 6.25 we can write

$$H(T, \overline{p}) = \overline{E} + \overline{p}V = c_V T + \nu RT = c_p T, \tag{7.21}$$

so $H(T, \overline{p}) = H(T)$ only, just like the energy. Therefore, as we saw in the free expansion process, the temperature is constant after the throttling process for an ideal gas: We can't even use the throttling process to cool an ideal gas! We are seeing more limitations to the ideal gas approximation; while it is quite useful for an introduction to thermodynamics, it quickly shows its lack of utility in general.

Considering a more general gas, we can show that if we know $H(T, \overline{p})$ and the equation of state for a gas, then we can plot T vs. $\overline{p}$ for constant H (just as we did by studying the T vs. V graphs for lines of constant $\overline{E}$ in Figure 7.4 for the free expansion). Choosing an initial temperature and pressure, we can move along this curve to a final, smaller pressure to determine the final temperature T_2. In general, it is difficult to determine H in terms of its natural variables for real gases—for example, for a Van der Waals gas, we would have to solve a cubic equation if we wish to eliminate the volume.

Treating T as a function of $\overline{p}$ and H, but keeping H constant, the relevant expression for the throttling process is

$$dT = \left(\frac{\partial T}{\partial \overline{p}}\right)_H d\overline{p} \equiv \mu_{JT} d\overline{p}, \tag{7.22}$$

which defines what is known as the **Joule–Thomson coefficient**, μ_{JT}. For the throttling process, because $d\overline{p} < 0$, we have three possibilities:

$$\text{if } \mu_{\text{JT}} \begin{cases} > 0 & \text{then } T \text{ decreases} \\ = 0 & \text{then } T \text{ is unchanged} \\ < 0 & \text{then } T \text{ increases} \end{cases} . \tag{7.23}$$

Thus, in order to ensure our gas is cooled by this process, we require the Joule–Thomson coefficient to be positive.

Exercise 7.8 Crush the derivative in μ_{JT} to obtain

$$\mu_{\text{JT}} = \frac{V}{C_p}(T\alpha - 1) = \frac{v}{c_p}(T\alpha - 1), \tag{7.24}$$

where we write this in two forms: in terms of the volume and heat capacity, and in terms of the molar volume and specific heat.

For an ideal gas, the Joule–Thomson coefficient vanishes for all temperatures and pressures because $\alpha = 1/T$. For other gases this will not always vanish, and at different values of T and $\overline{p}$, any of the possibilities in Eq. (7.23) are possible, so the point where it *does* vanish is important. The values of T and $\overline{p}$ where $\mu_{\text{JT}} = 0$ define the **inversion curve**. The inversion curve separates the two regions where we are able to cool a gas ($\mu_{\text{JT}} > 0$) or not ($\mu_{\text{JT}} \leq 0$).

In order to liquefy a gas, this process is often still used (see for example, Ref. [1]) for the initial cooling stage. To do so, the gas must initially be at a temperature and pressure such that $\mu_{\text{JT}} > 0$. I show in Figure 7.7 the inversion curve for nitrogen gas, treating it as a Van der Waals gas. We can

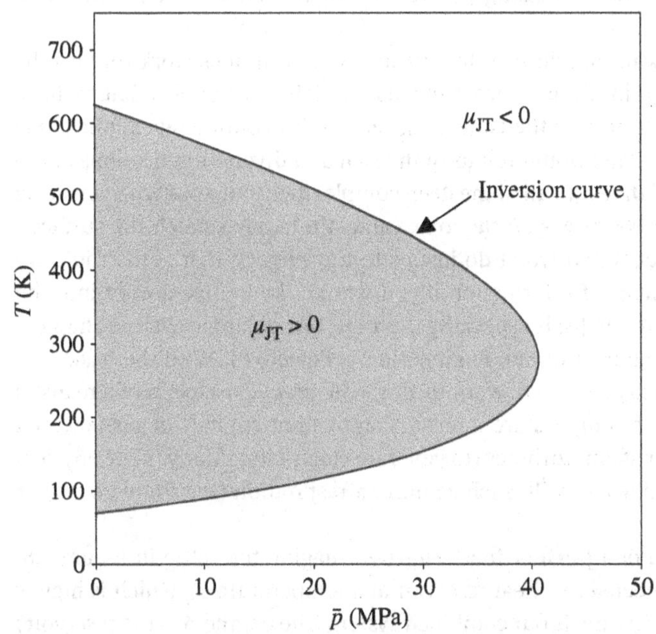

Figure 7.7 The inversion curve for nitrogen gas, where the solid curve denotes the points where $\mu_{\text{JT}} = 0$, the unshaded region is where $\mu_{\text{JT}} < 0$, and the shaded region is where $\mu_{\text{JT}} > 0$ and is the region we are interested in.

see that there is a maximum temperature the gas can have if we want to even consider using this process to cool a gas (for nitrogen, this is 625 K). Conveniently, at room temperature, we see that if we have nitrogen at a fairly high pressure initially, it is quite simple to reduce the temperature by a significant amount. Helium, however, has a maximum inversion temperature of around 34 K, so in order to liquefy helium (which happens around 4 K), one would have to use another process to cool the gas to at most 34 K before using the Joule–Thomson process.

As mentioned, the Joule–Thomson process has long been used to liquefy gases [1, 2], but at the same time, it is a phenomenon that often needs to be avoided. An example of trying to minimize the throttling effect is in pipelines used to transport fluids like natural gas. As the gas passes through pipelines, it can cool down to such a degree such that the valves used to control the flow of the gas can freeze. If this were to happen, it could clog the pipe or worse, cause it to burst. You can read the recent study in Ref. [3] as an example of estimating such effects.

7.3 Heat Engines

One of the earliest applications of thermodynamics, which led to the advent of the industrial revolution, was the construction of a **heat engine**. This is a broad term to describe the process of using heat from some source to provide useful work on another system. The reverse is something that is common in our everyday life: often work is done on a system and heat is generated as a byproduct (often considered *waste heat* because it is not used for a practical purpose). The simplest example of this in our lives is friction: When rubbing two objects together, the work done by friction is released in the form of heat (which I would not always call waste heat, for example, if you're standing outside on a frigid winter day, you can rub your hands together to warm them up). Electrical work is in common use in many households: Passing a current through a resistor releases heat that can be used for cooking.

In the case of a heat engine, we want to add heat to a system so that it does work that can be useful in some way. This subject is so simple at its heart and also so rich and complex that we have two options here. First, we could just discuss the concept at its simplest, with a few applications without getting into too much detail. This would rob us of the rich and diverse applications of the subject. On the other hand, we could dive into all of the deep complexities that arise when creating heat engines that we spend another 300 pages on the topic (and still barely scratch the surface). While the first option is less than ideal and won't do justice to the subject, that is the route we will take. There are many other sources of information if you would like to dive deeper into the subject of heat engines. Chapter 13 of Ref. [4] is especially useful in this regard as well as chapters 8 and 9 of Ref. [5] (which covers engines from an engineering perspective). With the basics we will cover in this section, you will easily be able to work through the practical aspects of a realistic engine. Several real-world examples of engines are internal combustion engines in gas-powered automobiles, jet or rocket engines, or steam turbines (to generate electricity). Many other physical systems can be modeled as heat engines as well, such as human respiration (see Problem 7.9) or the global climate [6].

To begin, let us discuss the notion of a **perfect heat engine**. Imagine the setup in Figure 7.8, which shows a system A that is connected to a heat reservoir at a temperature T_h which is higher than that of the system. We assume not only is our combined system (the engine A + the reservoir) isolated, but that the system undergoes a **cyclic process**. That is, since we don't want to have to reset the system ourselves on a regular basis, we would like the system to return to the same state periodically. Every quantity we discuss in this section will be that of a single cycle.

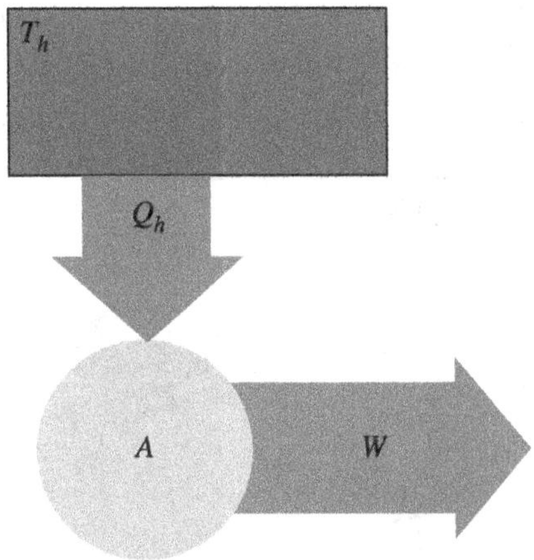

Figure 7.8 A schematic of a perfect heat engine: heat Q_h flows from a reservoir at temperature T_h into a system A which converts all of it to useful work W.

After absorbing some heat $Q_h > 0$, our system A does some work $W > 0$ on something else during one cycle. Because the system returns to its original state after this happens, it must be true that $\Delta\overline{E} = 0$, so we can say

$$W = Q_h. \tag{7.25}$$

We will now show that this is impossible by applying the second law of thermodynamics. For any spontaneous process, recall the total entropy of an isolated system (here, A + reservoir) must increase, or

$$\Delta S^{(0)} = \Delta S + \Delta S_h \geq 0$$

during each cycle (ΔS is the entropy change for the engine A and ΔS_h is the entropy change for the reservoir). As with the internal energy, the entropy of A must be unchanged (it returns to its original state), so $\Delta S = 0$, while the entropy of the heat reservoir is given by Eq. (6.3),

$$\Delta S_h = -\frac{Q_h}{T_h}.$$

We include the negative sign because we defined Q_h to be positive and the energy is leaving the reservoir. Putting this into the second law we have

$$-\frac{Q_h}{T_h} \geq 0,$$

or given that $Q_h = W$,

$$\frac{W}{T_h} \leq 0.$$

For positive absolute temperatures, this contradicts our assertion that work is being done *by* the system (that it is positive), which is what we want for a useful heat engine. Thus **it is impossible to construct a perfect heat engine**, which was Kelvin's formulation of the second law of thermodynamics before the concept of entropy was understood. All this means is that we must have some energy that is wasted—some of the heat that comes from our reservoir must be lost as *waste heat*. In other words, a real heat engine requires an exhaust, which makes sense when we think of various real-world heat engines such as the internal combustion engine in a car.

For a real engine we have two reservoirs at different temperatures, as shown in Figure 7.9. An amount of heat Q_h is transferred from the hot reservoir (with temperature T_h) to the system A, and

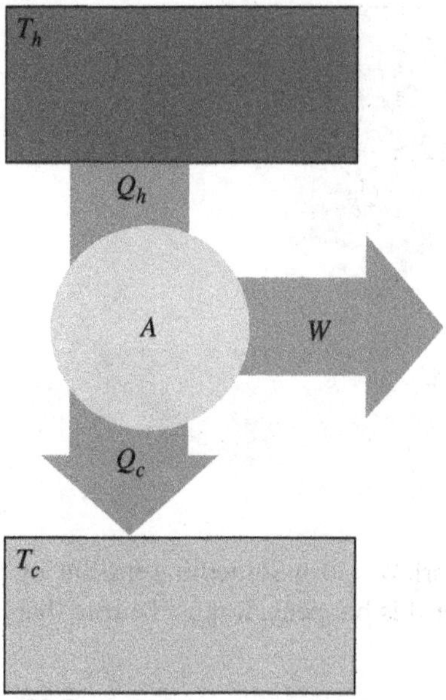

Figure 7.9 A schematic of a real heat engine. In this case, energy Q_h is transferred from the hot reservoir at temperature T_h to the system, which produces useful work W as well as waste heat Q_c, which is released to the cold reservoir at temperature $T_c < T_h$.

subsequently an amount of heat $Q_c > 0$ is transferred from the system to the cold reservoir (with temperature $T_c < T_h$). Additionally, and most importantly, the system does some positive work W on something else. We still insist that the process is a cyclic, so that $\Delta \overline{E} = 0$, and the first law implies

$$0 = Q_h - Q_c - W \Rightarrow Q_h = Q_c + W. \tag{7.26}$$

(It is conventional for all quantities in discussions of heat engines to be positive, so minus signs enter into such equations explicitly to denote if a system loses energy.)

The entropies of the two reservoirs change via Eq. (6.3). Considering the total entropy change of the combined system (and requiring it never to be negative to comply with the second law), we have

$$\Delta S^{(0)} = -\frac{Q_h}{T_h} + \frac{Q_c}{T_c} \geq 0, \tag{7.27}$$

where we have used the fact that $\Delta S = 0$ for the cyclic system A. Eliminating the waste heat Q_c and some simple algebra allows us to relate the work done by the engine to the heat absorbed from the hot reservoir. We obtain

$$\eta \equiv \frac{Q_h}{W} \leq 1 - \frac{T_c}{T_h}, \tag{7.28}$$

where this quantity is so important we define it as the **heat engine efficiency**. This is the fraction of useful work from our input heat we can get from running our engine.

Exercise 7.9 Derive the inequality in Eq. (7.28) from Eqs. (7.26) and (7.27).

If we could have a perfect engine, then $\eta = 1$, which, from this expression, would imply that either the hot reservoir were at infinite temperature or the cold reservoir were at absolute zero, both of which are impossible. If the engine were *reversible*, then the equality in Eq. (7.28) would be satisfied, and we get the **Carnot efficiency** of a heat engine,

$$\eta_C \equiv 1 - \frac{T_c}{T_h}. \tag{7.29}$$

Such an engine is known as an **ideal heat engine** (not a perfect one!), but be warned: This is *not the efficiency of a general heat engine*! The efficiency is best calculated using the definition $\eta \equiv W/Q_h$—this can always be shown to satisfy the inequality in Eq. (7.28). However, the Carnot efficiency is only applicable to a very unique situation which we discuss next. It is very common to want to use Eq. (7.29) in calculations, but don't—always return to the equality of Eq. (7.28). Related to this is that in many cases, the two reservoirs aren't always so obvious—often the cold reservoir is just the surrounding environment, but there could also be multiple reservoirs in the engine. This only accents the point that we should use $\eta = W/Q_h$ and not the inequality in Eq. (7.28).

As seen in Problem 4.11, the work done in a cyclic process is given by the area of the closed loop as shown in a $\overline{p}$–V diagram. The heat flow can be determined in a T–S diagram, or as we will see, if we have well-defined stages of the engine, often the input heat is clearly discernible. We will work through some examples to make this clear.

7.3.1 Carnot Cycle

Let's work through the details of the reversible engine designed by the French physicist Nicolas Léonard Sadi Carnot (1796–1832) in 1824 that has the maximum possible efficiency in Eq. (7.29). It is a **four-stage heat engine**: There are four separate processes that we can describe as the engine operates through a cycle. Each of these stages has well-defined heat entering or leaving the system, and the work done can be readily calculated.

The $\overline{p}$–V diagram is shown on Figure 7.10(A): the curves correspond to the stages which move the engine between the four labeled states. On Figure 7.10(B), I show a diagram of a Carnot engine, where the different points are labeled as well. Each of the processes described are quasistatic. The four stages which take our engine from $a \to b \to c \to d \to a$ are:

Stage 1, isothermal expansion from $a \to b$: With the gas at the temperature of the hot reservoir T_h, it absorbs heat Q_h as it expands from $V_a \to V_b$ *isothermally*. In the process, the pressure increases from $\overline{p}_a \to \overline{p}_b$.

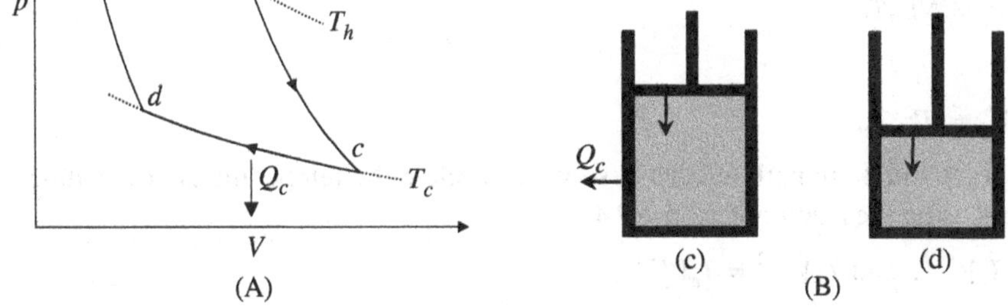

Figure 7.10 (A) The $\overline{p}$–V diagram for the Carnot cycle, where Stage 1 is from $a \to b$, Stage 2 is from $b \to c$, Stage 3 is from $c \to d$, and Stage 4 is from $d \to a$, returning our system to its original state. (B) A diagram of a Carnot engine, with the states labeled to correspond to the points on the diagram.

Stage 2, adiabatic expansion from b → c: Upon reaching the desired pressure and volume, the gas is now *thermally insulated* and allowed to expand from $V_b \to V_c$ (and the pressure continues to decrease, from $\bar{p}_b \to \bar{p}_c$) while its temperature decreases from $T_h \to T_c$.

Stage 3, isothermal compression from c → d: Upon reaching T_c, it is now put into thermal contact with the cold reservoir and is compressed isothermally from $V_c \to V_d$. During this stage, the system loses heat Q_c, and the pressure increases from $\bar{p}_c \to \bar{p}_d$.

Stage 4, adiabatic compression from d → a: Finally, we again thermally insulate the system as it is compressed from $V_d \to V_a$ while the temperature increases from $T_c \to T_h$ and the pressure increases from $\bar{p}_d \to \bar{p}_a$, thus returning the system to its original state.

If an ideal gas is the substance in the engine which does the work, Carnot showed that the efficiency of this engine is exactly that of Eq. (7.29), which I will show now. This is the simplest possible engine to consider, because at each stage it is very clear how to calculate the work done as well as the heat flow, and specifically it is clear that the added heat Q_h and the waste heat Q_c can be calculated during Stages 1 and 3, respectively.

In fact, because the heat is so easy to calculate, and we know $W = Q_h - Q_c$, we can calculate the efficiency as

$$\eta = 1 - \frac{Q_c}{Q_h}.$$

During Stage 1, we know $Q_h = \int_a^b đQ$, and for an ideal gas, if T is constant so is $\bar{E}$, and

$$đQ = d\bar{E} + \bar{p}dV = \bar{p}dV.$$

Thus,

$$\begin{aligned} Q_h &= \int_a^b đQ \\ &= \int_a^b \frac{Nk_BT_h}{V}dV \\ &= Nk_BT_h \ln\left(\frac{V_b}{V_a}\right) > 0. \end{aligned}$$

Similarly, we can calculate the heat flowing out of the system during Stage 3,

$$Q_c = -Nk_BT_c \ln\left(\frac{V_d}{V_c}\right) = Nk_BT_c \ln\left(\frac{V_c}{V_d}\right) > 0, \tag{7.30}$$

(where we put in the minus sign to ensure Q_c is a positive quantity, as we know that $\ln\left(V_d/V_c\right) < 0$). Using the ideal gas equation, we have

$$\bar{p}_aV_a = \bar{p}_bV_b = Nk_BT_h,$$

and

$$\bar{p}_cV_c = \bar{p}_dV_d = Nk_BT_c.$$

Additionally, Stages 2 and 4 are both adiabatic, so we can relate the temperatures and volumes during those stages using the result of Exercise 7.4,

$$T_hV_a^{\gamma-1} = T_cV_d^{\gamma-1}, \text{ and } T_hV_b^{\gamma-1} = T_cV_c^{\gamma-1}.$$

These allow us to write

$$\ln\left(\frac{V_b}{V_a}\right) = \frac{1}{\gamma-1}\ln\left(\frac{V_b^{\gamma-1}}{V_a^{\gamma-1}}\right) = \ln\left(\frac{V_c}{V_d}\right), \tag{7.31}$$

so then it is quite simple to see

$$\begin{aligned}\eta &= 1 - \frac{Q_c}{Q_h} \\ &= 1 - \frac{Nk_BT_c \ln\left(V_c/V_d\right)}{Nk_BT_h \ln\left(V_b/V_a\right)} \\ &= 1 - \frac{T_c}{T_h}.\end{aligned}$$

Exercise 7.10 Calculate the total work done during the Carnot cycle (by calculating the area of the $\bar{p}$–V diagram in Figure 7.10) and verify that $W = Q_h - Q_c$.

There are a couple of examples in the problems at the end of the chapter regarding heat engines, but notice that even for the simplest engines, the calculation to determine the efficiency gets tedious fairly quickly.

7.4 Refrigerators

Suppose we wish to cool a system, by transferring heat Q_c from a cold reservoir that is at a temperature T_c to a system, and then expelling heat Q_h to some hot reservoir $T_h > T_c$. Such a device is known as a **refrigerator**, something that we are all very familiar with in our kitchens. The *perfect* refrigerator, shown in Figure 7.11, cannot be realized in nature, as we could easily expect from our discussion of heat engines. That is, we must do work on the system to extract heat from the cold reservoir and reject it to the hot reservoir (shown in Figure 7.12). This is Clausius's statement of the second law of thermodynamics,[3] that is, **it is impossible to construct a perfect refrigerator.**[4]

Figure 7.11 A schematic of a perfect refrigerator, where energy Q_c is "transferred" from a cold reservoir to a system, expelling heat Q_h to a hot reservoir, thus cooling the system.

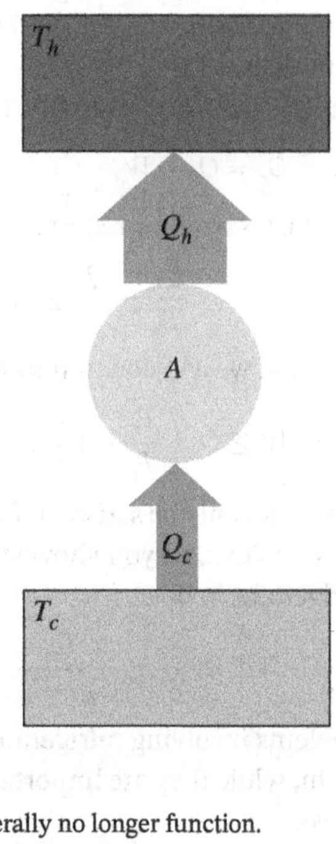

3 Rudolf Clausius (1822–1888).

4 Go ahead and test this: Unplug your refrigerator and you'll see that it will generally no longer function.

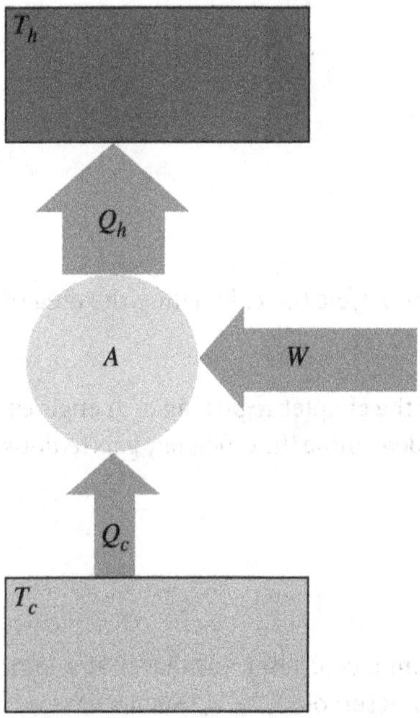

Figure 7.12 A schematic of a real refrigerator, where work must be done on the system in order to transfer the energy from the cold reservoir to the system (while still expelling heat to a hot reservoir) to cool it.

Exercise 7.11 Prove Clausius's statement for a perfect refrigerator. That is, using the first and second laws, show that you arrive at a contradiction as we saw with heat engines.

I will now work through the same steps for a real refrigerator that we did for a real heat engine. Again we assume Q_h, Q_c, and W are all positive (and will put signs as needed to denote the direction of the energy flow). We do work W on our system A so that it takes in some heat Q_c and emits some heat Q_h to the hot reservoir. The first law tells us

$$Q_c = Q_h - W, \tag{7.32}$$

and the second law requires the change of the combined system to never be negative,

$$\Delta S^{(0)} = \frac{Q_h}{T_h} - \frac{Q_c}{T_c} \geq 0. \tag{7.33}$$

As usual, we consider our system to be cyclic just as we did the engine, so $\Delta S = 0$, or

$$W \geq Q_c \left(\frac{T_h}{T_c} - 1 \right), \tag{7.34}$$

which can only be satisfied if the work done is non-zero because the right-hand side of the inequality is positive, as you showed in Exercise 7.11. We define the **coefficient of performance** for a refrigerator to be

$$\text{COP} \equiv \frac{Q_c}{W} \leq \frac{T_c}{T_h - T_c}. \tag{7.35}$$

Problems involving refrigerators are very similar (and similarly tedious) to those for heat engines. Again, while they are important for many applications, we will limit our discussion to what we did above.

7.5 Summary

- The pressure and volume of an ideal gas which undergoes an adiabatic expansion are related by $\overline{p}V^{\gamma} = \text{constant}$, where $\gamma = C_p/C_V$.
- One can cool gases, an important tool when liquefying gases, by allowing the gas to undergo free expansion ($\overline{E} = \text{constant}$) or the Joule–Thomson process ($H = \text{constant}$), so long as the gas is not ideal.
- Variations of the second law of thermodynamics can be formulated with heat engines or refrigerators. These state that one cannot construct a perfect device, where all input heat is converted to useful work (for the engine) or all heat is extracted from a system without needing to do work (for the refrigerator).

Problems

7.1 (a) Crush the following derivatives for adiabatic processes, and simplify them for an ideal gas. (The first should be in terms of $\overline{p}$ and C_V, while the second should be in terms of C_p and V.)

$$\left(\frac{\partial T}{\partial V}\right)_S, \text{ and } \left(\frac{\partial \overline{p}}{\partial T}\right)_S.$$

(b) Use these results to derive the relationships between T and V and between $\overline{p}$ and T for an ideal gas expanding adiabatically. Your results should agree with Exercise 7.4.

7.2 The earth's atmosphere can be treated to a first approximation as an ideal gas of molar mass μ in a uniform gravitational field (with g the acceleration due to gravity).

(a) Consider a small volume of air given by $dV = Adz$, where A is some cross-sectional area and the bottom of this volume is at a height z above sea level while the top is at a height $z + dz$. The pressure at any given height comes from the air above that height. Use this to determine the pressure difference $d\overline{p} = \overline{p}(z + dz) - \overline{p}(z)$ between these two heights and show that you can write

$$\frac{d\overline{p}}{\overline{p}} = -\frac{\mu g}{RT}dz,$$

where T is the absolute temperature at height z.

(b) It's a good approximation (to first order) to assume that the decrease of pressure occurs adiabatically. In this case, show that

$$\frac{d\overline{p}}{\overline{p}} = \frac{\gamma}{\gamma - 1}\frac{dT}{T}.$$

(c) From (a) and (b) calculate dT/dz in kelvin per kilometer. Assume the atmosphere to consist only of nitrogen N_2, which you can treat as a diatomic ideal gas.

7.3 Let us work through determining the inversion curve for a Van der Waals gas.

(a) First determine the **critical pressure** $\overline{p}_c$, **critical temperature** T_c, and **critical volume** v_c. These are found by solving the equations

$$\left(\frac{\partial \overline{p}}{\partial v}\right)_T = 0 \quad \text{and} \quad \left(\frac{\partial^2 \overline{p}}{\partial v^2}\right)_T = 0.$$

(The importance of these values will be discussed in Chapter 10.)

(b) Define dimensionless variables $\overline{p}' = \overline{p}/\overline{p}_c$ (and similarly for T' and v'), to eliminate a and b so that the Van der Waals equation can be written as

$$\left(\overline{p}' + \frac{3}{v'^2}\right)(3v' - 1) = 8T'. \tag{7.36}$$

This is now a semi-universal equation—it is the same for all gases, even though the critical parameters will be different for each.

(c) Implicitly differentiate Eq. (7.36) to determine $\alpha = (1/v)(\partial v/\partial T)_p$ (or use your result from Exercise 6.34).

(d) The vanishing of the Joule–Thomson coefficient implies that

$$T\alpha = 1.$$

With your result to part (c), solve for $\overline{p}'$ as a function of T'. You should find

$$\overline{p}' = 9 - 12\left(\sqrt{T'} - \sqrt{3}\right)^2. \tag{7.37}$$

Invert this and plot T' vs. $\overline{p}'$. You should see something that looks like Figure 7.7.

(e) Find the maximum inversion temperature for helium and nitrogen by setting $\overline{p}' = 0$ (using the values for a and b in Table 6.2), and convert the maximum temperature to kelvin. How does it compare to the experimentally determined values, 34 and 625 K, respectively?

7.4 Consider a Van der Waals gas such that $v \gg b$ and b is of the same order of magnitude as a/RT.

(a) By working to second order in the small quantities, show that

$$v \simeq \frac{RT}{\overline{p}} + b - \frac{a}{RT} + \frac{2ab\overline{p}}{R^2T^2} - \frac{a^2\overline{p}}{R^3T^3}.$$

Hint: To do this, start by setting $a = b = 0$ and find v (call it v_0). Then set $v = v_0 + xb + y\left[a/(RT)\right]$, and expand the Van der Waals equation to linear order in b and $[a/(RT)]$, and solve for x and y. Repeat with quadratic terms in the expansion to then obtain the result above.

(b) Using your result in (a), find an approximate expression, to first order in small quantities, for T as a function of $\overline{p}$ on the inversion curve for the Joule–Thomson process. You should find

$$T \simeq \frac{2a}{Rb} - \frac{2b\overline{p}}{R}.$$

(c) The inversion curve for a Van der Waals gas is shown in Figure 7.7 Argue that, *in the region where the approximation here is likely to be valid*, the inversion curve found in (b) has roughly the same shape as the complete inversion curve.

(d) Compare your expression for the inversion curve from Problem 7.3 to what you obtained here. Expand that expression to show that your results are the same.

7.5 A gasoline engine can be approximately represented by the $\overline{p}$–V diagram shown below. It has four stages like the Carnot engine, but in this case, the two stages from $a \to b$ and $c \to d$ represent the adiabatic compression and expansion of the air–gasoline mixture and the stages from $b \to c$ and $d \to a$ correspond to isochoric rises and falls in the pressure. Assume this cycle to be carried out quasistatically for a fixed amount of ideal gas having a constant specific heat. Calculate the efficiency η for this process, expressing your answer in terms of V_1, V_2, and $\gamma = C_p/C_V$.

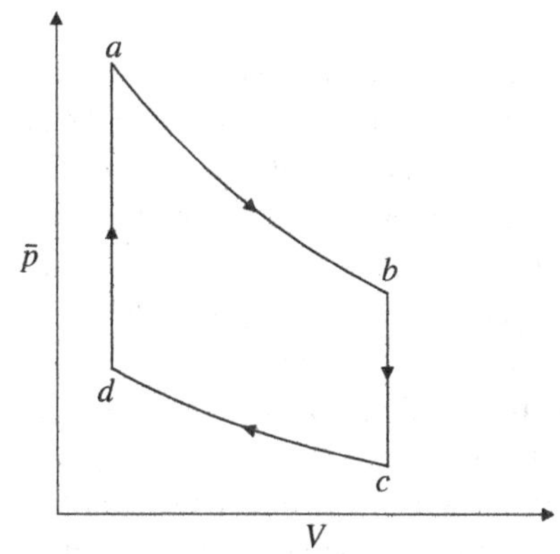

7.6 Low temperature research is often performed on samples whose temperatures are on the order of one millikelvin (1 mK).

(a) The price of work (in the form, say, of electrical energy) is roughly 25 cents per kilowatt-hour, at least in New York City. What would be the minimum cost of extracting 1 calorie of heat from a system at 1 mK if the surrounding atmosphere is at 290 K?

(b) The cost of extracting one calorie in part (a) may not seem very high, but imagine that it cost you that much *for each calorie* you extract from inside your refrigerator. Estimate, very roughly, what your monthly electrical bill would be in that case. You might start by estimating the total amount of liquids (soft drinks, juice, milk, beer, etc.) you cool down from room temperature to ~ 40°F in a month, and how much ice you freeze in a month. This should convince you that low temperature experiments must be done on small, well-insulated samples.

7.7 An ideal gas undergoes a circular cycle as shown on the $\bar{p}$–V diagram below [7]. What is the Carnot efficiency (that is, the maximum possible efficiency of a real engine) which operates between the same high and low temperatures as the gas in this circular cycle? Your answer should be a pure number, not dependent on p_0 and V_0.

Hint: Find the maximum and minimum temperatures by maximizing and minimizing the product $\bar{p}V$ on the circle.

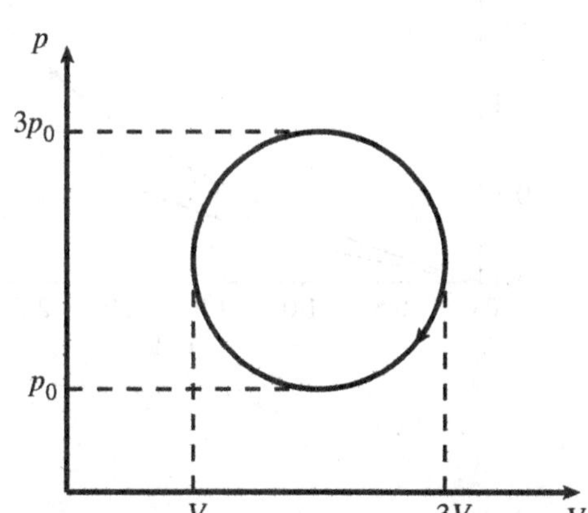

7.8 In classical thermodynamics, one can prove that the efficiency of a reversible heat engine is greater than or equal to that of an irreversible engine—without ever mentioning entropy. You are asked to reproduce the argument by using the procedure below. Your only assumptions should be the first law of thermodynamics and Kelvin's statement of the second law.

(a) Consider two cyclic heat engines, A and A', both of which operate from heat reservoirs at temperatures T_h and T_c ($T_h > T_c$). Let Q_h, Q_c, and W be the heat absorbed from a hot reservoir at temperature T_h, the heat given off to a cold reservoir at temperature T_c, and the work done, respectively, in one cycle of engine A. Define Q'_h, Q'_c, and W' similarly for A'. Write down the efficiency η of engine A in terms of Q_h and Q_c only, and write a similar expression for η'.

(b) Let $Q_c/Q'_c = m'/m$, where m and m' are integers.[5] Assume A is a reversible engine, and consider the combined heat engine $A^{(0)}$ that could be made by running A' for m' cycles, and running A *backward* for m cycles. Relate $Q_h^{(0)}$, the total heat absorbed by $A^{(0)}$ from the hot reservoir, to $W^{(0)}$, the total work done by $A^{(0)}$.

(c) Show that Kelvin's statement of the second law implies that $Q_h^{(0)} \leq 0$. (This combined "engine" is actually what is known as a **heat pump**.)

(d) Use (c) and (a) to show that $\eta' \leq \eta$, proving the theorem.

(e) A corollary to the theorem is that "all reversible engines have the same efficiency." Show that if A' is reversible also, then $\eta' = \eta$.

7.9 In Ref. [8], it is shown that human respiration can be modeled as a heat engine, with the $\bar{p}$–V diagram for each breath shown in the figure below. The solid line corresponds to inhalation and the dashed line corresponds to exhalation.

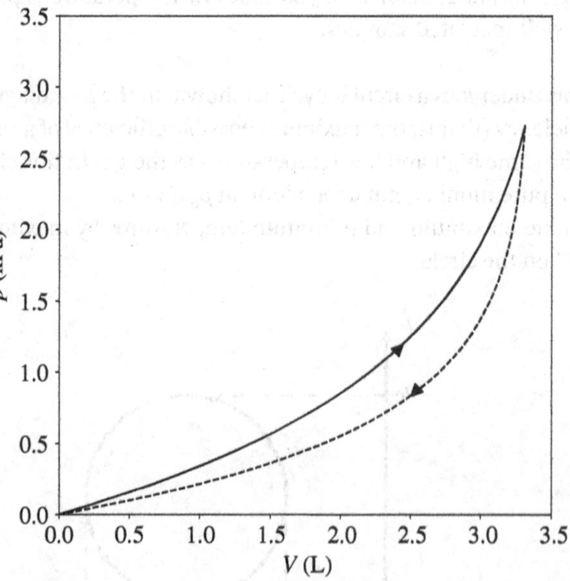

5 Recall in our discussion when deriving our expression for missing information, any real number can be approximated by a rational number to any desired level of accuracy via a Diophantine approximation.

We can consider the air we breathe to a good approximation as a diatomic ideal gas (it's mostly nitrogen and oxygen, after all), so we will use the following relationship for the internal energy,

$$\overline{E} = \frac{5}{2}\nu RT = \frac{5}{2}\overline{p}V.$$

However, for the relationship between $\overline{p}$ and V, we will use an equation of state that eliminates these and incorporates the elastic nature of the lungs (considering them as rubber balloons),

$$V = V_{\text{max}}\left(1 - e^{-\kappa\overline{p}}\right), \tag{7.38}$$

instead of the ideal gas equation.[6] Here V_{max} is the maximum lung capacity, and κ is the isothermal compressibility, and with this model we'll treat them both as constants that differ for inhalation and exhalation.

(a) Invert Eq. (7.38) to get $\overline{p}$ as a function of V.

(b) We have for inhalation and exhalation, respectively,

$$V_{\text{max},i} = 3.57 \times 10^{-3}\ \text{m}^3,\ \kappa_i = 0.964 \times 10^{-3}\ \text{Pa}^{-1}$$
$$V_{\text{max},e} = 3.35 \times 10^{-3}\ \text{m}^3,\ \kappa_e = 1.644 \times 10^{-3}\ \text{Pa}^{-1}$$

The equation of state is defined so that the initial pressure is 0 Pa and the initial volume of the lungs is 0 m ^{3}. The final pressure and volume are determined by the fact that they must be the same for both inhalation and exhalation. Using this, determine the final pressure and volume. You will need to solve these numerically using Mathematica or some other software.

(c) Calculate the total work done during one cycle, W_{net}.

(d) What is the heat input Q_h during inhalation?

(e) What is the efficiency of this "heat engine?"

(f) Compare this to the Carnot efficiency if you assume the hot reservoir is the body at 37°C and the cold reservoir is room temperature at 20°C.

7.10 The Carnot cycle is central to meteorology, as many weather patterns can be described by assuming parcels of air correspond to ideal heat engines. The stages of such an "engine" correspond to the four stages of a Carnot engine from Section 7.3.1. Stage 1 (isothermal expansion) occurs near the surface of the Earth, which acts as our hot reservoir. The adiabatic expansion of Stage 2, the parcel that recently expanded now rises adiabatically to the tropopause (the cold reservoir), which marks the beginning of the atmosphere, and is roughly 17 km above the surface of the ground, expanding some more. Once in thermal equilibrium with the tropopause, the parcel then contracts isothermally in Stage 3, and finally sinks back to the surface adiabatically during Stage 4, contracting more to its original volume.

Suppose we have a parcel of air with a mass $M = 8$ kg, and while in contact with the surface of the Earth, it expands isothermally by 30 m ^{3} before rising to the tropopause (and expanding some more). When in contact with the tropopause, it contracts isothermally by the same amount before sinking back to the surface. Consider the surface to be about 15°C at a pressure of 1.0×10^5 Pa, and at the tropopause the temperature is −60°C and the pressure is 1.5×10^4 Pa. With the density of air on the surface being 1.293 kg/m ^{3}, calculate

6 Note you can look at other models of the $\overline{p}$–V equation of state in the lungs in Ref. [9].

(a) The efficiency of this engine,
(b) the entropy change of the environment, and
(c) the total entropy change of the parcel of air plus the environment.

References

1 A. Alekseev. Basics of low-temperature refrigeration, 2014.

2 R. D. Gunn, P. L. Chueh, and J. M. Prausnitz. Inversion temperatures and pressures for cryogenic gases and their mixtures. *Cryogenics*, 6(6):324–329, 1966.

3 S. N. Shoghl, A. Naderifar, F. Farhadi, and G. Pazuki. Prediction of Joule-Thomson coefficient and inversion curve for natural gas and its components using CFD modeling. *Journal of Natural Gas Science and Engineering*, 83:103570, 2020.

4 K. Stowe. *An Introduction to Thermodynamics and Statistical Mechanics*. Cambridge University Press, 2007.

5 M. J. Moran, H. N. Shapiro, D. D. Boettner, and M. B. Bailey. *Fundamentals of Engineering Thermodynamics*. Wiley, 2018.

6 M. S. Singh and M. E O'Neill. Thermodynamics of the climate system. *Physics Today*, 75(7):30–37, July 2022.

7 B. Korsunsky. Round and round. *The Physics Teacher*, 49(1):61–61, Jan 2011.

8 T. C. Lipscombe and C. E. Mungan. Breathtaking physics: human respiration as a heat engine. *The Physics Teacher*, 58(3):150–151, 2020.

9 J. G. Venegas, R. S. Harris, and B. A. Simon. A comprehensive equation for the pulmonary pressure-volume curve. *Journal of Applied Physiology*, 84(1):389–395, 1998.

Further Reading

If you're interested in more about the applications of thermodynamics to meteorology, you should consider

- S. Miller. *Applied Thermodynamics for Meteorologists*. Cambridge University Press, 2015.

8

The Canonical Distribution

At this point we have everything we need to study statistical mechanics and thermodynamics; however, there is a different approach (from the microscopic level) we can take that is easier to work with, known as the **canonical ensemble**. It is both simpler to perform first-principles calculations with, and often easier to understand physically than the microcanonical ensemble. You might be wondering why I didn't start with this approach from the beginning, and the reason is simple: I like to avoid introducing temperature until we discuss equilibrium, and with this approach, we fix the temperature of the system of interest instead of the energy.

From an introductory standpoint, fixing the energy is conceptually simple, even though it is not necessarily clear how, experimentally, to fix the energy of the system. The microcanonical ensemble is a good way to introduce the ideas of statistical mechanics because it requires just understanding the counting of states. Because of our everyday understanding of temperature, fixing this quantity makes a little more sense (to me at least), and as we will see, many times calculations will be easier with this new approach, even if it might seem tougher at first.

After finishing this chapter, you should be able to

- learn how to describe systems in thermal contact with a heat reservoir,
- understand how to calculate the partition function and relate it to the entropy,
- apply this to the classical ideal gas and non-ideal gases, and
- understand how the classical assumption gives an incorrect expression for the entropy while also learning how to correct an error we have inadvertently been making so far.

8.1 Restarting Our Study of Systems

To set up the statistical problem in a new way, let us first revisit how we set up the microcanonical ensemble. As before, we start with a system A in thermodynamic equilibrium and we assume we have been able to enumerate all of the microstates. Each microstate r has an energy E_r and a probability P_r, and the entropy of the system is defined by the Shannon entropy in Eq. (4.8),

$$S = -k_B \sum_r P_r \ln P_r.$$

We first consider this system to be isolated so that it can be described by the microcanonical ensemble, and then we consider it in contact with a heat reservoir, which will be our introduction to the canonical ensemble.

Statistical Thermodynamics: An Information Theory Approach, First Edition. Christopher Aubin.

Companion website: www.wiley.com/go/Aubin/StatisticalThermodynamics

8.1.1 A as an Isolated System

First we consider A to be isolated from the surroundings—then the energy of the system is constant, and we require the energy of a microstate to be in the range $E < E_r < E + \delta E$ with $\delta E \ll E$. Such a setup allows us to require that the system A is in a state which maximizes the missing information, or the entropy, so we saw in Chapter 3 (specifically you proved it in Problem 3.8) that this gives us the probability distribution

$$P_r = \begin{cases} \frac{1}{\Omega} & \text{if } E < E_r < E + \delta E \\ 0 & \text{otherwise} \end{cases}, \tag{8.1}$$

where $\Omega(E)$ is constant and is the total number of accessible states in this energy range. We can write this as

$$P_r = \begin{cases} C\Omega_r & \text{if } E < E_r < E + \delta E \\ 0 & \text{otherwise} \end{cases}, \tag{8.2}$$

where $C = 1/\sum_r \Omega_r = 1/\Omega$, and Ω_r is the number of microstates corresponding to the macrostate r (which would be greater than one if there is a degeneracy in the energy E_r).

Other than mechanical constraints (such as fixing the volume), we place no other requirements on the system. Theoretically this is easy to understand, as we have a solid idea of what energy is from classical mechanics; however, in practice this can be hard to realize. That is, how do you hold the energy of the system fixed? One way, to keep beverages hot or cold for example, is to use an insulating container such as a thermos. This minimizes heat entering the system so $Q = 0$ (at least approximately), and if we do work on the system (while keeping it sealed) we can change other thermodynamic parameters at constant energy. But while energy is familiar with us now (after years of studying physics), it is still a somewhat abstract concept.

In everyday life, we are more accustomed to understanding temperature (even if it too is an abstract idea at its core), so it is easier to consider fixing this instead of the energy. Generally if we consider a system (say a cup of coffee) in the room you're sitting in, the temperature of the room (and often the pressure) is very close to being fixed. And even though the microcanonical ensemble is merely a setup that allows us to relate the internal energy of the system to other thermodynamic parameters, it seems odd to study problems with an isolated system in mind. In our everyday lives, it's not always easy to completely isolate a system to maintain a constant energy, but if we put a system in contact with a heat reservoir, it is straightforward to keep it at a fixed temperature. The only reason we didn't start with this approach was because I wanted to keep temperature out of the picture until it was defined in Eq. (5.14). This way we could look at temperature first in the abstract sense, and then connect it with our everyday physical understanding of it. As such, using a different viewpoint from the beginning will be helpful.

8.1.2 System in Contact with a Heat Reservoir

Now let's put the system A in thermal equilibrium with a heat reservoir A', as shown in Figure 8.1 such that the combined system, $A^{(0)} = A + A'$, is thermally isolated from the outside world. The reservoir will have a fixed temperature T, and as they are in equilibrium, A will also have this temperature (this is why we use T for reservoir temperature and not T' as before, I would rather not bother with the extra notation). For now we will assume the container which holds the system A is rigid, so the volume cannot change. The energy of our system A in a state r, E_r, is clearly not fixed in this case—we can only fix the energy *or* the temperature, not both.[1]

1 At least, for the purposes of setting up the problem.

Figure 8.1 A system A with rigid boundaries in thermal equilibrium with a heat reservoir, A'. The combined system is thermally isolated from the outside world. This is the same as Figure 6.1 but with $T = T'$.

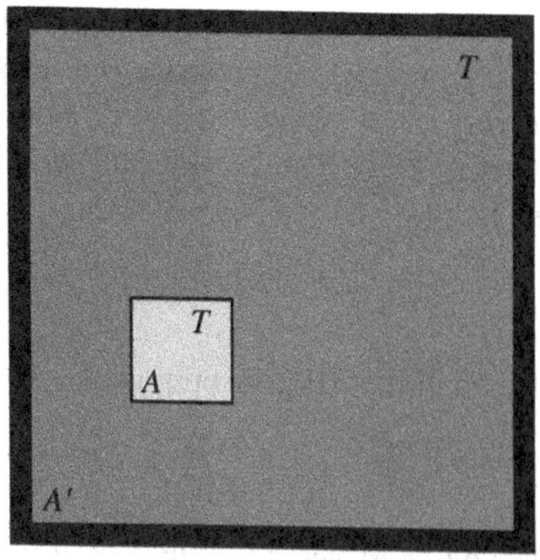

Because the *combined* system is thermally isolated, we know that the total energy is fixed, and when A is in a given microstate r, it is given by[2]

$$E^{(0)} = E_r + E'.$$

In this setup, we assume we only know E_r (in principle), but E' and thus $E^{(0)}$ are both unknown. All we will require is that A' is a heat reservoir when compared with A, so we can say that $E_r \ll E'$ and as such $E_r \ll E^{(0)}$. Because the total energy $E^{(0)}$ is fixed, if we increase the energy of our system, then the energy of the reservoir decreases, and vice versa, or

$$E' = E^{(0)} - E_r,$$

and we know that for general systems, the number of states accessible to a system increases with energy. We can write the number of states accessible to A' in terms of the energy of system A as

$$\Omega'(E') = \Omega'(E^{(0)} - E_r).$$

But remember, for the microcanonical ensemble, we know that all possible states (of the combined system), according to the fundamental postulate of statistical mechanics, are equally likely. For the system we are interested in, on the other hand, the probability that A is in a given state r is proportional to the number of states that exist in A', or

$$P_r = C\Omega'(E^{(0)} - E_r). \tag{8.3}$$

The constant C comes from normalization, ensuring $\sum_r P_r = 1$, which we will explicitly calculate shortly.

Exercise 8.1 Consider rolling five fair six-sided dice, one of which is red, and the others are white. Suppose that when you roll these dice, you end up with a sum of 11. Show that the probability to roll a 1–6 on the red die is given by

$$P_1 = \frac{80}{205},\ P_2 = \frac{56}{205},\ P_3 = \frac{35}{205},\ P_4 = \frac{20}{205},\ P_5 = \frac{10}{205},\ P_6 = \frac{4}{205}.$$

In this exercise, the white dice act as a "heat reservoir" for our single red die, and we see the probability of a given roll of the red die is proportional to the number of states accessible to the other dice.

2 In reality, of course, we still treat the total system within some energy range $E^{(0)} + \delta E$ with $\delta E \ll E^{(0)}$, since as usual we will never be able to know with absolute precision what the total energy is, but we are really only focused on the system A, so I will omit this detail.

To turn Eq. (8.3) into a more useful expression, let's recall that $A \ll A'$, that is, A' is a heat reservoir compared with A so we can say $E_r \ll E' \sim E^{(0)}$. The question is how does Ω' behave when E_r is very small (or equivalently $E' = E^{(0)}$)? This calls for a Taylor expansion (about $E_r = 0$), but as we already know that Ω' is a rapidly changing function of the energy, while $\ln \Omega'$ changes more slowly, so let's expand the latter of these,

$$\ln \Omega'(E^{(0)} - E_r) \approx \ln \Omega'(E^{(0)}) + \frac{\partial \ln \Omega'}{\partial E_r} E_r + \cdots .$$

Given that $E^{(0)}$ = constant, we know $dE_r = -dE'$, so

$$\frac{\partial \ln \Omega'}{\partial E_r} = -\frac{\partial \ln \Omega'}{\partial E'} = -\beta.$$

From this we can write the logarithm of the probability of the system being in a state r to be

$$\ln P_r = \ln C + \ln \Omega'(E^{(0)} - E_r) = \ln C + \ln \Omega'(E^{(0)}) - \beta E_r, \tag{8.4}$$

or

$$P_r = C' e^{-\beta E_r}.$$

The constant C' can be easily determined by requiring $\sum_r P_r = 1$, so we get

$$\frac{1}{C'} \equiv Z \equiv \sum_r e^{-\beta E_r}. \tag{8.5}$$

Even though Z only enters as (one over) the normalization constant, it has a special name: the **partition function**. The notation Z comes from the German, *Zustandssumme*, which means "state sum." This is quite literal: We sum this exponential factor, known as the **Boltzmann factor,** over all states of the system.

The probability distribution

$$P_r = \frac{e^{-\beta E_r}}{Z} \tag{8.6}$$

is known as the **Boltzmann distribution** or **canonical distribution**, and we say that such a system is described by the **canonical ensemble**. It is very different from the microcanonical ensemble which has P_r = constant for all microstates. Not only is the probability in Eq. (8.6) far from constant, but the microstates accessible to the system vary wildly between the two ensembles: For the microcanonical ensemble, only those within a small range of energies are allowed, but for the canonical ensemble, *all* microstates are allowed.

Note that there is one limitation to our system, in that we cannot have $E_r > E^{(0)}$, because of course that would violate the conservation of energy. For all practical purposes, the energy of the reservoir is so much greater than that of the system, that when E_r gets anywhere close to E', the exponential in Eq. (8.6) is so damped that it is completely negligible. Thus, we will always perform sums over *all* states, because the high energy states do not contribute significantly to the sum, and ironically if we cut out those high energy states the summations become much more intractable.

At first glance, this seems to contradict the idea that the probability of a system increases rapidly with the energy because the number of states does. However, recall that unlike the isolated system, we have an additional constraint on our system now—if we raise the energy of A, then we lower that of the reservoir (and this was seen explicitly in Exercise 8.1). Unlike the microcanonical ensemble, our system is not completely independent, and any change in the system A requires a corresponding change in the reservoir. This constraint causes the probability of a given state of A to drop rapidly with increasing energy, since the number of accessible states of the reservoir drops rapidly.

Once we calculate Z for a given system, then we can easily calculate mean values for some quantity y,

$$\bar{y} = \sum_r P_r y_r = \frac{\sum_r y_r e^{-\beta E_r}}{\sum_r e^{-\beta E_r}} = \frac{1}{Z}\sum_r y_r e^{-\beta E_r}. \tag{8.7}$$

Although this seems, on the surface, to be more difficult to calculate than Ω, it turns out that the partition function is often simpler to calculate as a function of the parameters of the system than the number of states within a given energy range. This is simply due to the sum no longer being restricted in calculating Z as it is for Ω (the complexity of the sum is explicit in the exponential, if you will). Additionally, as we will see, often if we can calculate Z as a function of the temperature and external parameters (such as the volume), then explicitly evaluating Eq. (8.7) will not be necessary. Of course, regardless, for many problems, the partition function is just as difficult to calculate as the number of states and often can only be done numerically. Nevertheless this is a cleaner way to approach our problems and we will see later that there is a wonderful equivalence between the two different formulations, at least for very large systems.

Let's revisit a few examples from previous problems and exercises.

Example 8.1 Consider a system of N molecules as in Problem 4.6 and Problem 5.5. Each spin only interacts with some background magnetic field $\mathbf{H}$, and can be either spin up or spin down relative to the field. The potential energy of one of the molecules is given by

$$U = -\boldsymbol{\mu} \cdot \mathbf{H}, \tag{8.8}$$

where $\boldsymbol{\mu}$ is the magnetic dipole moment. For now we will write this as (in Section 9.3 we will be more explicit with what μ_0 is)

$$\mu = \pm\mu_0,$$

where the upper (lower) sign indicates the spin is aligned (anti-aligned) with the magnetic field. With this same convention, we can write the energy as

$$U = \pm\mu_0 H. \tag{8.9}$$

If the kinetic degrees of freedom are negligible, then the partition function for this system is easy to calculate, especially if the spins don't interact (strongly) with each other. We can write the partition function as

$$Z = \sum_r e^{-\beta E_r} = \sum e^{-\beta(U_1+U_2+\cdots+U_N)}, \tag{8.10}$$

where the sum is over the two possible states of all N spins (U_1 has two possible values, U_2 has two possible values, etc.). However, because the energies are not dependent on each other,[3] we can split this up as

3 This is an approximation of course, because each spin produces its own magnetic field, which would modify the field felt by the other spins. This approximation is often called the **mean field approximation**, where the field H that a given spin feels can be thought of as the external field plus the average of the contributions to the field from the other $N-1$ spins. It is straightforward, if not simple, to systematically improve upon this approximation, but we will stick with this simple assumption for now.

$$Z = \left(\sum e^{-\beta U_1}\right)\left(\sum e^{-\beta U_2}\right)\cdots\left(\sum e^{-\beta U_N}\right),$$

and we can define the **single-particle partition function** ζ, to write our full partition function as

$$Z = \zeta_1\zeta_2\cdots\zeta_N = \zeta^N, \tag{8.11}$$

where the second equality follows because the spins are identical. We can easily calculate

$$\zeta = e^{\beta\mu_0 H} + e^{-\beta\mu_0 H} = 2\cosh\left(\beta\mu_0 H\right). \tag{8.12}$$

If the spins are independent, then we can write the average energy as N times the average energy per spin $\bar{E} = N\bar{\varepsilon}$, which was determined in the aforementioned problems. We can easily calculate this with

$$\bar{\varepsilon} = \frac{-\mu_0 H e^{\beta\mu_0 H} + \mu_0 H e^{-\beta\mu_0 H}}{Z} = -\mu_0 H \frac{2\sinh\left(\beta\mu_0 H\right)}{2\cosh\left(\beta\mu_0 H\right)}. \tag{8.13}$$

Thus the mean energy is

$$\bar{E} = -N\mu_0 H \tanh\left(\frac{\mu_0 H}{k_B T}\right), \tag{8.14}$$

which was the result of Problem 5.5. That the two different calculations obtained exactly the same result is not something that is obvious, given how different the two setups are. Later we will discuss this equivalency and when we should expect it to hold.

Exercise 8.2 Using $\mu = \pm\mu_0$, calculate the mean value of the magnetic moment of the system. You should get for a single spin,

$$\bar{\mu} = \mu_0 \tanh\left(\frac{\mu_0 H}{k_B T}\right), \tag{8.15}$$

so the magnetization (total magnetic moment per unit volume) is

$$\overline{M} = \frac{N\mu_0}{V}\tanh\left(\frac{\mu_0 H}{k_B T}\right), \tag{8.16}$$

also found in Problem 5.5!

Let's look at the results from Example 8.1 and Exercise 8.2 for extreme temperatures (very high and very low temperatures). In this case, "high (low) temperature" means that the thermal energy $k_B T$ is much larger (smaller) than the relevant energy scale of the problem; for this case, this scale is $\mu_0 H$.

In the low temperature limit, $k_B T \ll \mu_0 H$, then any thermal fluctuations effectively go away,[4] and the energy provided by the external magnetic field is dominant. This means that we expect all of the spins in the system to be in the ground state, so $\bar{E} \to -N\mu_0 H$. This is true in this case, as $\tanh\left[\mu_0 H/(k_B T)\right] \to 1$ as $\mu_0 H/(k_B T) \to \infty$. Additionally we see $\overline{M} \to N\mu_0/V$, so all of the spins are aligned with the field, corresponding to the ground state.

In the high temperature limit ($k_B T \gg \mu_0 H$), then the energy associated with the magnetic field cannot significantly affect each spin, and the thermal energy allows fluctuations to occur quite

4 While imprecise, many refer to the system as "freezing out" as this occurs. It's not a great term given that for most materials, the temperature where the system is frozen is often much higher than absolute zero, but I do find it useful to envision what is going on.

readily. (There's so much thermal energy in the system that flipping spins from up to down or vice versa doesn't really cost any energy.) Thus, we find each spin equally likely to be in either of its possible states, so both the energy and the magnetization will average to zero, which can be seen in the above expressions.

We will examine more general spin systems in more detail later (including specifically *how* the magnetization goes to zero in this limit) in Section 9.3.

Example 8.2 Let's now look at a case that is a bit more general but can be used to understand previous spin system. Assume we have a system of N molecules confined to a fixed volume V which can be in one of n states with energies $\epsilon_1 < \epsilon_2 < \cdots < \epsilon_n$. We also assume the system is ideal so the individual particles are very weakly interacting (as we also assumed in Example 8.1).

Let us first look at the mean energy per molecule $\overline{\epsilon} = \bar{E}/N$ qualitatively as a function of T. As before we can look at the two limits as did above: low and high temperature, and this time be a little more explicit in what occurs in these two limits.

- As $T \to 0$ (or more precisely, $k_BT \ll \epsilon_1, \epsilon_2, \dots, \epsilon_n$), the system gradually approaches its ground state, and thus we expect $\overline{\epsilon} \to \epsilon_1$, assuming ϵ_1 is the lowest energy state.
- As $T \to \infty$ ($k_BT \gg \epsilon_1, \epsilon_2, \dots, \epsilon_n$), the thermal fluctuations are so great that we would be equally likely to be in any state so that $\overline{\epsilon} \to (\epsilon_1 + \epsilon_2 + \cdots + \epsilon_n)/n$,
- We expect $\overline{\epsilon}$ to be a smooth function between these two points.

Additionally we can look at the specific heat (per molecule) which we first saw in Exercise 6.10,

$$c_V = \left(\frac{\partial \overline{\epsilon}}{\partial T}\right)_V, \tag{8.17}$$

and also consider this qualitatively as a function of temperature. c_V is positive for ordinary systems, and we expect $\overline{\epsilon}$ to go to zero smoothly as a function of T. Thus we expect $c_V \to 0$ as $T \to 0$, which is consistent with the third law of thermodynamics. Similarly, if the mean energy approaches its high-T asymptotic value smoothly, then the heat capacity should also vanish for high temperatures. (After all, at high energy it shouldn't require much energy to change the temperature of the system.) Thus, there must be some turnover of the function; the specific heat cannot be zero for all temperatures, so we expect at least one maximum for c_V.

Exercise 8.3 Of course it is easy to check the statements in the previous example with the canonical ensemble. Following the procedure from Example 8.1, show that for the $n = 2$ case,

$$\overline{\epsilon} = \frac{\epsilon_1 + \epsilon_2 e^{-\beta(\epsilon_2 - \epsilon_1)}}{1 + e^{-\beta(\epsilon_2 - \epsilon_1)}}. \tag{8.18}$$

Additionally, show you obtain the result in Eq. (8.14) if you set $\epsilon_1 = -\mu H = -\epsilon_2$.

It is easy to see from Eq. (8.18) that if $T \to 0, \overline{\epsilon} \to \epsilon_1$ (the ground state) as expected and as $T \to \infty$, then $\overline{\epsilon} \to (\epsilon_1 + \epsilon_2)/2$. This can be seen explicitly in Figure 8.2; the far right point is a temperature given by $k_BT = 10\epsilon_1$, where the energy is about 3.3% from the limiting value.

Exercise 8.4 Use Eq. (8.17) to evaluate the specific heat per molecule for this two-state system. Note that

$$\frac{d\beta}{dT} = -\frac{1}{k_BT^2},$$

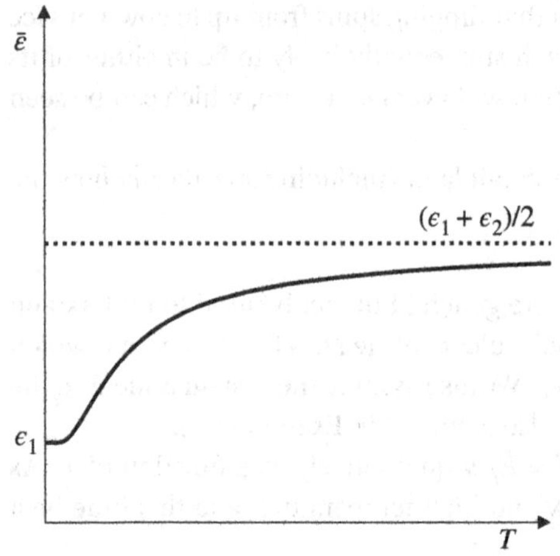

Figure 8.2 $\bar{\epsilon}$ vs. T for our simple two-state system in Example 8.2.

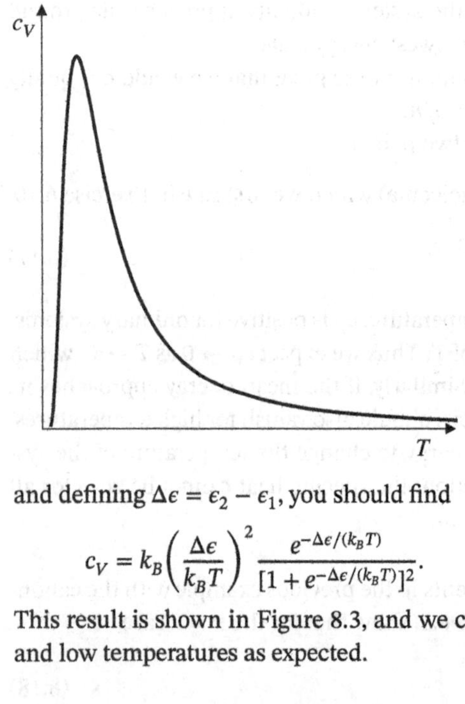

Figure 8.3 c_V vs. T for our simple two-state system in Example 8.2.

and defining $\Delta\epsilon = \epsilon_2 - \epsilon_1$, you should find

$$c_V = k_B\left(\frac{\Delta\epsilon}{k_B T}\right)^2 \frac{e^{-\Delta\epsilon/(k_B T)}}{[1 + e^{-\Delta\epsilon/(k_B T)}]^2}. \tag{8.19}$$

This result is shown in Figure 8.3, and we can see it is always positive, and vanishes at both high and low temperatures as expected.

Up to now our two examples involve systems with a finite number of discrete degrees of freedom. As is the case with the microcanonical ensemble, our sum will change to an integral when we have continuous degrees of freedom. We'll discuss this in detail when we return to the classical ideal gas in Section 8.4.

8.2 Connecting to the Microcanonical Ensemble

With Eq. (8.7), we can calculate any mean values we would like. However, applying that expression directly can become tedious rather quickly, so we will derive some expressions to make some of these calculations much simpler (and will guide us to other ways to simplify later expressions readily). Additionally, this will allow us to compare our results to those from the microcanonical ensemble and allow us to determine when our results should agree when studying a problem with either ensemble.

8.2.1 Mean Energy

Let's start with the usual starting point, the mean internal energy,

$$\bar{E} = \frac{1}{Z}\sum_r E_r e^{-\beta E_r}. \tag{8.20}$$

If we are able to calculate Z as a function of β for a given system (which we have already done above for a couple of examples), then we can use a simple trick to extract $\bar{E}$ directly. We can see that

$$\frac{\partial}{\partial\beta}\left(e^{-\beta E_r}\right) = -E_r e^{-\beta E_r},$$

so we can write

$$\begin{aligned}\bar{E} &= -\frac{1}{Z}\sum_r \frac{\partial}{\partial\beta}\left(e^{-\beta E_r}\right)\\ &= -\frac{1}{Z}\frac{\partial}{\partial\beta}\left(\sum_r e^{-\beta E_r}\right)\\ &= -\frac{1}{Z}\frac{\partial Z}{\partial\beta},\end{aligned}$$

or, remembering the chain rule,

$$\bar{E} = -\frac{\partial \ln Z}{\partial\beta}. \tag{8.21}$$

Exercise 8.5 Show that the results of Eqs. (8.14) and (8.18) can be found by applying Eq. (8.21).

8.2.2 Variance in $\bar{E}$

In the microcanonical ensemble, the mean energy is simply the constant $\bar{E} = E$, because the energy is fixed to be in the range E to $E + \delta E$ and $\delta E \ll E$.[5] The spread of energies about this mean value is treated as zero because δE is so small compared with the energies of interest. With the canonical ensemble, on the other hand, given that *all* energies are possible, there might be a much wider spread of possible energies, so to study this, we wish to examine the variance about the mean,

$$\begin{aligned}\overline{(\Delta E)^2} &= \overline{\left(E - \bar{E}\right)^2}\\ &= \overline{E^2} - \bar{E}^2.\end{aligned} \tag{8.22}$$

The second term on the right-hand side we have already written in terms of Z, and the first term can also be determined in terms of the partition function by applying the derivative trick above twice,

$$\begin{aligned}\overline{E^2} &= \frac{1}{Z}\sum_r E_r^2 e^{-\beta E_r}\\ &= \frac{1}{Z}\left(-\frac{\partial}{\partial\beta}\right)^2\sum_r e^{-\beta E_r}\end{aligned}$$

5 If we wanted to be precise, $\bar{E} = E + \delta E/2$, but again δE is so small that this second term can be neglected.

$$= \frac{1}{Z}\frac{\partial^2 Z}{\partial \beta^2}. \tag{8.23}$$

Exercise 8.6 Show that this can be written in terms of the mean energy,

$$\overline{E^2} = \frac{1}{Z}\frac{\partial^2 Z}{\partial \beta^2} = \frac{\partial^2 \ln Z}{\partial \beta^2} + \bar{E}^2. \tag{8.24}$$

To do so, use the fact that

$$\frac{1}{Z}\frac{\partial f(\beta)}{\partial \beta} = \frac{\partial}{\partial \beta}\left[\frac{f(\beta)}{Z}\right] - \left(\frac{\partial}{\partial \beta}\frac{1}{Z}\right)f(\beta).$$

We can use Eq. (8.24) to write the variance in the energy as

$$\overline{(\Delta E)^2} = \frac{\partial^2 \ln Z}{\partial \beta^2} = -\frac{\partial \bar{E}}{\partial \beta}. \tag{8.25}$$

The first result from this is that

$$-\frac{\partial \bar{E}}{\partial \beta} \geq 0$$

or

$$\frac{\partial \bar{E}}{\partial T} \geq 0,$$

because $\overline{(\Delta E)^2} \geq 0$. This tells us that the mean energy is an increasing function of temperature, as we would expect (and saw before).

Exercise 8.7 Show that the heat capacity at constant volume can be written in terms of the variance as

$$C_V = \frac{\overline{(\Delta E)^2}}{k_B T^2}. \tag{8.26}$$

If we expect the results for the mean energy from our two ensembles to be the same, then not only would we want

$$\bar{E}_{\text{canonical}} = \bar{E}_{\text{microcanonical}},$$

as a function of temperature and volume, but we would want to ensure the variance in the energy in the canonical ensemble is negligible. That is, we require

$$\frac{\overline{(\Delta E)^2}}{\bar{E}^2} \ll 1. \tag{8.27}$$

We will see this will be valid in large systems, those where $N \gg 1$, which is the (thankfully!) primary focus for our studies. There will be examples where the equality between the two ensembles will not exist, but always when we consider a small system—when it matters these two can be considered equivalent.

8.2.3 Mean Pressure

Now that we have a way to calculate $\bar{E}(T, V)$ (we ignored the volume above, but it was there implicitly, being held constant), let's now consider the mean pressure, $\bar{p}$.

The infinitesimal work done by changing the volume is given by (explicitly holding β constant)

$$\begin{aligned} dW &= \sum_r \left(-\frac{\partial E_r}{\partial V}\right)_\beta P_r dV \\ &= \frac{1}{Z}\sum_r \left(-\frac{\partial E_r}{\partial V}\right)_\beta e^{-\beta E_r} dV. \end{aligned}$$

Using the same derivative trick as above, now combined with the chain rule, we have

$$\begin{aligned} dW &= \frac{1}{Z}\sum_r \left(-\frac{\partial E_r}{\partial V}\right)_\beta e^{-\beta E_r} dV \\ &= \frac{1}{Z}\frac{1}{\beta}\left(\frac{\partial}{\partial V}\right)_\beta \sum_r e^{-\beta E_r} dV \\ &= \frac{1}{Z}\frac{1}{\beta}\left(\frac{\partial Z}{\partial V}\right)_\beta dV \\ &= \frac{1}{\beta}\left(\frac{\partial \ln Z}{\partial V}\right)_\beta dV. \end{aligned}$$

However, writing the work done in terms of the mean pressure,

$$dW = \bar{p}dV = -\overline{\left(\frac{\partial E_r}{\partial V}\right)_\beta} dV,$$

we can calculate the mean pressure directly from the logarithm of the partition function,

$$\bar{p} = \frac{1}{\beta}\left(\frac{\partial \ln Z}{\partial V}\right)_\beta. \tag{8.28}$$

Of course if we had different external parameters, then we would have similar expressions for the generalized forces coming from changing those parameters.

As we can see from these few examples, mean values of quantities can often be obtained from derivatives of the logarithm of the partition function. This tells us that whenever a new problem enters the picture, we should not immediately calculate *any* mean value without first seeing if we can evaluate it by some derivative.

8.3 Thermodynamics and the Canonical Ensemble

Our development of thermodynamics from the microcanonical ensemble allowed us to obtain any thermodynamic quantity we were interested in from the entropy, which we could (in principle) calculate as a function of the energy and external parameters from the microscopic statistical nature of the system. Specifically we found $S(\bar{E}, V)$, the fundamental relation, and from that we found the equations of state, which could give us everything we need to know about the system. In other words, "If we know the entropy, then we know everything."

If we are able to calculate the partition function, then can we say the same thing? More specifically, if we can find a way to relate the partition function to the entropy, then we know can connect the canonical ensemble to all of the other thermodynamic functions we discussed in Chapter 6. Based upon my statement above, if we can get the entropy from the partition function, then we can also say, "If we know the partition function, then we know everything."

Given what we derived from information theory, we start from the Shannon entropy,

$$S = -k_B \sum_r P_r \ln P_r, \tag{8.29}$$

and then simply substitute the Boltzmann distribution, $P_r = e^{-\beta E_r}/Z$, to find for the entropy,

$$\begin{aligned} S &= -k_B \sum_r P_r \ln\left(\frac{e^{-\beta E_r}}{Z}\right) \\ &= k_B\beta \sum_r P_r E_r + k_B \ln Z \sum_r P_r \\ &= k_B \beta \bar{E} + k_B \ln Z. \end{aligned}$$

To simplify this between the second and third lines, I used $\sum_r P_r = 1$. We can then write

$$S = k_B \left(\ln Z + \beta \bar{E}\right). \tag{8.30}$$

Therefore a connection between the partition function and the entropy comes straight from our understanding of the origin of the entropy from information theory. This means that if we can calculate Z, we can obtain both $\bar{E}$ and thus S, and everything from Chapter 6 will follow.

Exercise 8.8 We can connect the entropy to the partition function another way. Do this by considering $\ln Z$ to be naturally a function of β and V, or $\ln Z(\beta, V)$, and calculate the differential $d \ln Z$. The partial derivatives you find here can be written in terms of the mean energy and the pressure, and Eq. (8.30) quickly follows.

When working with the canonical ensemble, the Helmholtz free energy is used more frequently than in the microcanonical ensemble, and this can be seen in Eq. (8.30). If we multiply both sides by the temperature and bring $\bar{E}$ over to the same side as TS, we can write the free energy as[6]

$$F = -k_B T \ln Z = \bar{E} - TS. \tag{8.31}$$

Because we can obtain everything of interest from F just as easily as we can from $\ln Z$, and it also has a physical interpretation as discussion in Section 6.3.3, you will see this commonly appear as what some physicists calculate.

How does the partition function behave in the zero temperature limit? All of the molecules in the system will collapse to their ground state, and thus we see that

$$Z \to Z_0 = \Omega_0 e^{-\beta E_0},$$

where Ω_0 is the number of states with the same (ground state) energy E_0. The entropy is easily shown to be

$$S = k_B \ln \Omega_0 = \text{constant}, \tag{8.32}$$

consistent with the third law of thermodynamics.

Exercise 8.9 Show that S is in fact $k_B \ln \Omega_0$ as $T \to 0$.

To complete our introduction to the partition function, let us ask: is it extensive or intensive? Let us look at two systems, A_1 and A_2, which are both in equilibrium with the same reservoir. In general, the energy of the two systems is

$$E_{rs} = E_r^{(1)} + E_s^{(2)} + U_{rs}^{(12)},$$

6 This makes a lot of sense that F is so simply related to the partition function, given that the natural variables of the Helmholtz free energy are T and V, which are the natural variables of Z (instead of $\bar{E}$ and S).

where the energies from the first two terms are those of the systems A_1 and A_2 in states r and s, respectively. The third term is the interaction energy of the two systems. The partition function is rather tricky to evaluate for a general setup, so let's also assume they are weakly interacting so we can say $U_{rs}^{(12)} \approx 0$.

The combined partition function is easy to write down,

$$\begin{aligned} Z &= \sum_{rs} e^{-\beta E_{rs}} \\ &= \sum_{rs} e^{-\beta E_r^{(1)}} e^{-\beta E_s^{(2)}} \\ &= \left(\sum_r e^{-\beta E_r^{(1)}}\right)\left(\sum_s e^{-\beta E_s^{(2)}}\right) \\ &= Z_1 Z_2. \end{aligned}$$

That is, the combined partition function is the product of the individual partition functions for the different systems—so it is neither extensive nor intensive! However, the logarithm,

$$\ln Z = \ln Z^{(1)} + \ln Z^{(2)},$$

is *extensive*. The mean energies are related in the same way as above, $\bar{E} = \bar{E}_1 + \bar{E}_2$, and thus the total entropy is just the sum of the individual entropies, $S = S_1 + S_2$. The extensive properties of the entropy and energy thus match with what we expect from before. Thus the connections between the canonical ensemble and thermodynamics are the same as those between the microcanonical ensemble and thermodynamics.

8.4 Classical Ideal Gas (Yet Again)

To see how these ideas apply to a system we know very well, let's return to the classical monatomic ideal gas.[7] As before, we have N identical molecules of mass m in a volume V. The position and momentum of the ith molecule are $\mathbf{r}_i$ and $\mathbf{p}_i$ and the total energy of a state r is given by

$$E_r = \sum_{i=1}^{N} \frac{\mathbf{p}_i^2}{2m} + U(\mathbf{r}_1, \dots, \mathbf{r}_N),$$

with r corresponding to the positions and momenta of all N molecules. Because our degrees of freedom are continuous, the partition function sum will become an integral, in the same way as it did when we calculated ϕ in the microcanonical ensemble. That is, we consider our classical phase space such that

$$\frac{\delta x_i \delta p_{xi}}{h_0} = \frac{\delta y_i \delta p_{yi}}{h_0} = \frac{\delta z_i \delta p_{zi}}{h_0} = 1,$$

for $i = 1, \dots, N$ and h_0 is the parameter we use to define the volume in phase space. The sum over states is trivially

$$\sum_r e^{-\beta E_r} = \sum_r \prod_{i=1}^{N} \left[\left(\frac{\delta x_i \delta p_{xi}}{h_0}\right)\left(\frac{\delta y_i \delta p_{yi}}{h_0}\right)\left(\frac{\delta z_i \delta p_{zi}}{h_0}\right)\right] e^{-\beta E_r}.$$

7 We will study a diatomic gas in Section 11.7.

As usual we assume our energy is large enough such that our δx_i and δp_{xi} can be treated as infinitesimals for all i (also for the y- and z-directions), so

$$\sum_r e^{-\beta E_r} \to \int \frac{d^3\mathbf{r}_1 \cdots d^3\mathbf{r}_N d^3\mathbf{p}_1 \cdots d^3\mathbf{p}_N}{h_0^{3N}} e^{-\beta E_r}.$$

Unlike in the microcanonical ensemble, the integrals here run over *all* available values of position and momentum. The positions are restricted to the volume V, and each momentum component can run from $-\infty$ to ∞.[8] Putting E_r in, our partition function can be written as

$$Z = \frac{1}{h_0^{3N}} \int \prod_{i=1}^{N} (d^3\mathbf{r}_i d^3\mathbf{p}_i) \exp\left\{-\beta\left[\frac{1}{2m}(\mathbf{p}_1^2 + \cdots + \mathbf{p}_N^2) + U(\mathbf{r}_1, \ldots, \mathbf{r}_N)\right]\right\}.$$

This is quite terrifying at first glance, but we will see that the lack of restrictions on the integration limits will make this relatively simple to evaluate. In this case, because the potential energy only depends upon position and the kinetic energy only on momentum, we can factor the integral,

$$Z = \frac{1}{h_0^{3N}} \left(\int e^{-\beta \mathbf{p}_1^2/(2m)} d^3\mathbf{p}_1\right) \cdots \left(\int e^{-\beta \mathbf{p}_N^2/(2m)} d^3\mathbf{p}_N\right) \times \left(\int e^{-\beta U(\mathbf{r}_1,\ldots,\mathbf{r}_N)} d^3\mathbf{r}_1 \cdots d^3\mathbf{r}_N\right). \tag{8.33}$$

The momentum integrals are easy to do, and we will do those in a moment. The position integral is not so simple as it stands, but remember for an ideal gas, we set $U \approx 0$, so $e^{-\beta U(\mathbf{r}_1,\ldots,\mathbf{r}_N)} \approx 1$. This allows us to factor the position integrals, so again

$$Z = \frac{1}{h_0^{3N}} \left(\int e^{-\beta \mathbf{p}_1^2/(2m)} d^3\mathbf{p}_1 d^3\mathbf{r}_1\right) \cdots \left(\int e^{-\beta \mathbf{p}_N^2/(2m)} d^3\mathbf{p}_N d^3\mathbf{r}_N\right) = \zeta^N, \tag{8.34}$$

with ζ the single-particle partition function,

$$\zeta = \frac{V}{h_0^3} \int e^{-\beta \mathbf{p}^2/(2m)} d^3\mathbf{p} = V\left(\frac{2m\pi}{h_0^2\beta}\right)^{3/2}. \tag{8.35}$$

Exercise 8.10 Fill in the steps required to obtain Eq. (8.35). The results in Appendix C will help here.

Equation (8.35) allows us to write the (logarithm of the) partition function for a monatomic ideal gas as

$$\begin{aligned}\ln Z &= N \ln \zeta \\ &= N \ln\left[V\left(\frac{2m\pi}{h_0^2\beta}\right)^{3/2}\right] \\ &= N\left[\ln V - \frac{3}{2}\ln\beta + \frac{3}{2}\ln\left(\frac{2m\pi}{h_0^2}\right)\right],\end{aligned}$$

8 In the microcanonical ensemble, the integration limits were restricted by the requirement that $E_r < E$ to calculate ϕ.

where we have separated the temperature (or β) and volume dependence, and we see the final term is independent of thermodynamic parameters. The mean pressure easily follows,

$$\bar{p} = \frac{1}{\beta}\frac{\partial \ln Z}{\partial V} = \frac{N}{\beta V}, \tag{8.36}$$

which is just the ideal gas equation,

$$\bar{p}V = Nk_BT.$$

The mean energy easily is determined to match what we saw before,

$$\bar{E} = -\frac{\partial \ln Z}{\partial \beta} = \frac{3}{2}Nk_BT = N\bar{\varepsilon},$$

where $\bar{\varepsilon}$ is the mean energy per molecule. Finally, the heat capacity at constant volume is given by

$$C_V = \left(\frac{\partial \bar{E}}{\partial T}\right)_V = \frac{3}{2}Nk_B.$$

All of these results match what we found from the microcanonical ensemble, so we could say that the two ensembles are consistent. However, in the microcanonical ensemble, the energy was fixed to one value—that value determined the temperature of the system—even though we often wrote it as $\bar{E}$. With the canonical ensemble, the temperature is fixed, and this is truly the *mean* energy of the system—the actual energy of the system can have any value. So as discussed earlier, to truly see if the two ensembles match, we should consider the variance in the energy, Eq. (8.25),

$$\overline{(\Delta E)^2} = -\frac{\partial \bar{E}}{\partial \beta} = \frac{3}{2}Nk_B^2T^2, \tag{8.37}$$

so the spread of energies about the mean is $\sqrt{\overline{(\Delta E)^2}} = \sqrt{\frac{3}{2}N}k_BT$. Compared with the mean energy,

$$\frac{\sqrt{\overline{(\Delta E)^2}}}{\bar{E}} = \sqrt{\frac{2}{3N}}. \tag{8.38}$$

This is on the order of 10^{-12} for a system with a mole of molecules in our system. Thus, we can say safely that the fluctuations in the energy are small enough that the mean energy is equal to the energy of the system. Note of course this only holds for very large systems—if we considered a system with one or two particles in it, such a statement wouldn't hold, and there would be no equivalency between the two ensembles.

None of this is new, but now let's calculate the entropy from our expression for $\ln Z$ for the ideal gas. We didn't do this before because we didn't do the explicit integral over the $3N$-dimensional hypersphere in Eq. (4.36) to get the prefactor A in Eq. (4.38) (although you can use the results in Appendix D to do so, see Problem 8.13). However, the canonical formulation makes this much simpler.

Plugging our partition function into Eq. (8.30), we obtain

$$S = Nk_B\left\{\ln V + \frac{3}{2}\ln T + \sigma\right\}, \tag{8.39}$$

where σ is a constant that is independent of T and V. In Section 8.5, we will show that this expression for the entropy is in fact incorrect!

Exercise 8.11 Derive Eq. (8.39), and show that

$$\sigma = \frac{3}{2}\ln\left(\frac{2m\pi k_B}{h_0^2}\right) + \frac{3}{2}. \tag{8.40}$$

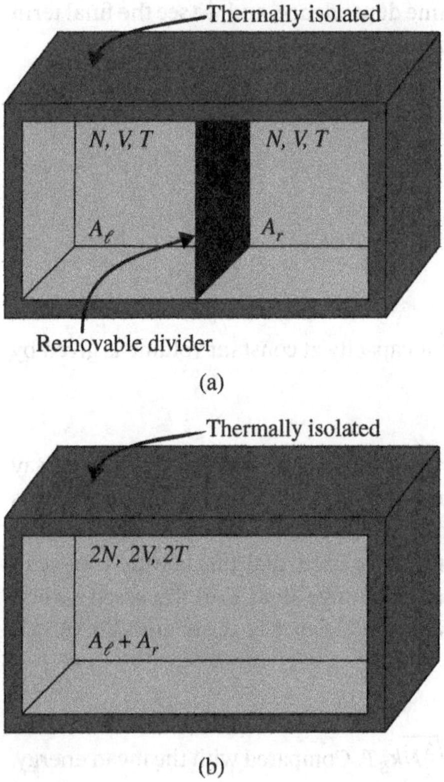

Figure 8.4 (a) Two identical systems, both with N molecules of a classical ideal gas at temperature T and volume V. (b) The same setup with the divider removed, so the system is double in size. This is a reversible process.

8.5 Fudged Classical Statistics

The problem with Eq. (8.39) is apparent when we recall that the entropy is an extensive quantity. If we scale our system by some factor a, then all extensive variables should scale by that same factor. However, this is clearly not the case for the entropy in Eq. (8.39). With $V \to aV$ and $N \to aN$, we see that

$$S \to aNk\left[\ln(aV) + \frac{3}{2}\ln T + \sigma\right] = aNk\ln(a) + aS \neq aS. \tag{8.41}$$

Gibbs formulated this problem as follows. Suppose we have a container as shown in Figure 8.4, with two systems: A_ℓ on the left and A_r on the right. They both have the same type and number N of molecules, the same volume V, the same temperature T, and of course the same pressure. In this situation, you can write the entropies of each side as

$$S_\ell = S_r = Nk_B\left(\ln V + \frac{3}{2}\ln T + \sigma\right). \tag{8.42}$$

Suppose now that the divider between the two systems is removed without disturbing the system, doubling doubled the system. The new entropy of the combined system is given by

$$S_{\text{combined}} = 2Nk_B\left[\ln(2V) + \frac{3}{2}\ln T + \sigma\right], \tag{8.43}$$

and we can calculate the entropy change during this (spontaneous) process as

$$\Delta S = S_{\text{combined}} - (S_\ell + S_r) = 2Nk_B \ln 2. \tag{8.44}$$

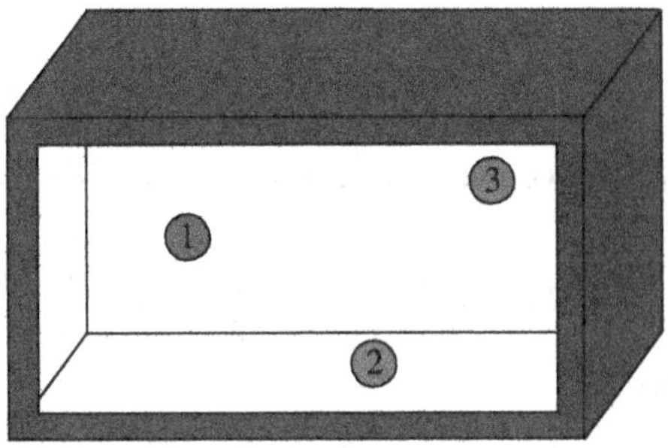

Figure 8.5 Three identical molecules; if we interchange any of them we have the same state as shown. The six distinct classical states should really be considered the same state given that we can't distinguish any of them.

However, this is a *reversible process*, because the distribution of the molecules hasn't changed. The molecules were equally distributed throughout the container before and after removing the divider, so we should have found $\Delta S = 0$. This is known as the **Gibbs paradox**,[9] but the solution to this paradox is actually quite easy to understand.

The problem is that we took the classical approximation too seriously: In the classical world, every identical molecule is distinct, but in reality we can't actually freely interchange them. That is, two classical oxygen molecules could be labeled (perhaps molecules 1 and 2), so that interchanging any two of them in a system of such molecules would result in a distinct state, such as the three molecules shown in Figure 8.5. However, in reality, interchanging two oxygen atoms does not yield a different state in any meaningful way. (I only care about whether or not oxygen molecules are near me so that I can inhale them! The specific molecules entering my lungs are irrelevant.) In other words, while we can *in principle* label each molecule as distinct, for all practical purposes, these oxygen molecules should be considered as indistinguishable.

This is something that will get taken into consideration automatically when we treat the system quantum mechanically. We don't want to consider this more complicated setup until Chapter 11, so instead we need to come up with a way to treat our system classically while still taking into account the fact that the particles in our system are indistinguishable. Figure 8.5 has three identical molecules, and so there are six ways to arrange them so that we get the same state. However, the sum we perform to obtain the partition function includes each of these six states as different—and thus would be six times larger than what we would expect if these were all considered a single state. In this case, to account for this overcounting, we just need to divide our partition function by six.

In general, for a system with N identical particles, then we should modify our partition function,

$$Z \to Z_f = \frac{Z}{N!}. \tag{8.45}$$

The $1/N!$ factor accounts for the $N!$ ways to arrange our molecules that each result in the same state. For each different species in our system, we would include a similar factor corresponding to the number of molecules of each species. For example, if we had a mixture of oxygen and nitrogen gas, we would divide our partition function by $N_{O_2}!N_{N_2}!$. The notation f in Eq. (8.45) is used to denote what we'll call the **fudged partition function**, and this corresponds to **fudged classical statistics.**[10] I use this term because we have introduced a fudge factor to make our classical problem make physical sense when looking at thermodynamics, most specifically the entropy (as we will see shortly). However, it is *not* true classical statistical mechanics! We will see in Chapter 11 that fudged classical statistics are however the proper classical limit of the quantum mechanical problem.

---

9 Physicists love the term *paradox* for any situation that doesn't make sense after a first study.

10 Claude Bernard gets credit again for yet another term I'm using in this book; he used it in his undergraduate statistical mechanics course and I like to stick with it.

With this fudge factor, we find

$$\ln Z_f = N \ln \zeta - \ln N! \sim N \ln \zeta - N \ln N + N, \tag{8.46}$$

where we have used Stirling's formula in its simplest form. We see that the entropy now has the form

$$S = Nk_B \left[\ln \left(\frac{V}{N} \right) + \frac{3}{2} \ln T + \sigma_0 \right], \tag{8.47}$$

where

$$\sigma_0 = \sigma + 1 = \frac{3}{2} \ln \left(\frac{2\pi m k_B}{h_0^2} \right) + \frac{5}{2}. \tag{8.48}$$

The entropy in Eq. (8.47) is, as it should be, an extensive quantity, because V now appears in the ratio V/N. If we scale the system by some factor a we will see $S \to aS$ as expected.

Exercise 8.12 Show that the Gibbs paradox is avoided in his specific example by explicitly calculating the change in the entropy using Eq. (8.47).

One final note: Quantities that arise from derivatives of $\ln Z$ or the entropy with respect to β and V (or other external parameters) are unchanged here—the extra fudge factor is not needed to obtain the correct expressions for the equations of state (this is, after all, why we only just now introduced it). We need fudged statistics for the entropy, and as we'll see in Chapter 10, for quantities that involve derivatives of the entropy with respect to the number of molecules (when we allow that to change). Thus, sometimes we will not bother to include it if we don't need it.

8.6 Non-ideal Gases

It's very clear that not all systems can be treated as ideal gases. For one, as we will show in Chapter 10, the ideal gas equation won't be able to properly describe other phases of matter such as solids and liquids. Even without that issue, not all gases are accurately treated as ideal. We have seen the Van der Waals equation of state which was of course determined from an empirical standpoint, but how can we derive it from a more first-principles approach? This is what we plan to do in this section.

For now we will just consider classical a non-ideal dilute gas, so that the number density $n = N/V$ can be treated as small. If the gas is dilute, then the interactions between different molecules can be simplified, although not neglected.

Again we consider N identical molecules (for now still monatomic) in a volume V in thermal contact with a heat reservoir at temperature T. The energy of a state is again given by

$$E_r = \sum_{i=1}^{N} \frac{\mathbf{p}_i^2}{2m} + U(\mathbf{r}_1, \ldots, \mathbf{r}_N),$$

as before, and the fudged partition function is

$$Z_f = \frac{1}{N!} \int \prod_{i=1}^{N} \left(\frac{d^3\mathbf{r}_i d^3\mathbf{p}_i}{h_0^3} \right) e^{-\beta E_r}.$$

The momentum integrals are trivial here; they are the same as for the ideal gas, so we can simplify this to

$$Z_f = \frac{1}{N!}\left(\frac{2\pi m}{h_0^2\beta}\right)^{3N/2} \int \prod_{i=1}^{N} d^3\mathbf{r}_i e^{-\beta U(\mathbf{r}_1,\ldots,\mathbf{r}_N)}.$$

This is where we get stuck, because if U cannot be neglected, we cannot easily evaluate this integral in general. Also, for most realistic situations, each molecule will interact with every other molecule, so we cannot just factor this into separate integrals over the individual molecular positions. We need some more assumptions to make any headway, especially without an explicit expression for U.

For a dilute gas, we realize that it will be rare for two molecules to come close to each other, and practically impossible for three molecules (which is why we can neglect the interactions for the ideal gas case), and thus we will assume we only have two-body interactions. As is generally the case, we also assume that the potential energy only depends on the distance between two molecules, so we can write

$$U(\mathbf{r}_1,\ldots,\mathbf{r}_N) = \sum_{i=1}^{N}\sum_{j=i+1}^{N} U_{ij}(|\mathbf{r}_i - \mathbf{r}_j|) = \sum_{i<j} U_{ij}, \tag{8.49}$$

where we will keep the argument $|\mathbf{r}_i - \mathbf{r}_j|$ implicit in the potential energy from now on. This allows us to factor the exponential at least, if not the integral, so

$$Z_f = \frac{1}{N!}\left(\frac{2\pi m}{h_0^2\beta}\right)^{3N/2} \int d^{3N}\mathbf{r} \prod_{i<j} e^{-\beta U_{ij}}.$$

Generally for gases, such potentials vanish as $|\mathbf{r}_i - \mathbf{r}_j| \to \infty$ (because of course the interaction should go away as the molecules are infinitely far apart), and they tend to become large and positive as $|\mathbf{r}_i - \mathbf{r}_j| \to 0$, because of the electrostatic repulsion between the outer shell electrons. An example of such a potential is shown in Figure 8.6.

We can rewrite the exponential in terms of a new parameter which will be easy to work with, as

$$e^{-\beta U_{ij}} = 1 + \left(e^{-\beta U_{ij}} - 1\right) \equiv 1 + \xi_{ij}. \tag{8.50}$$

Figure 8.6 An example of a realistic intermolecular potential $U(r)$ as a function of r.

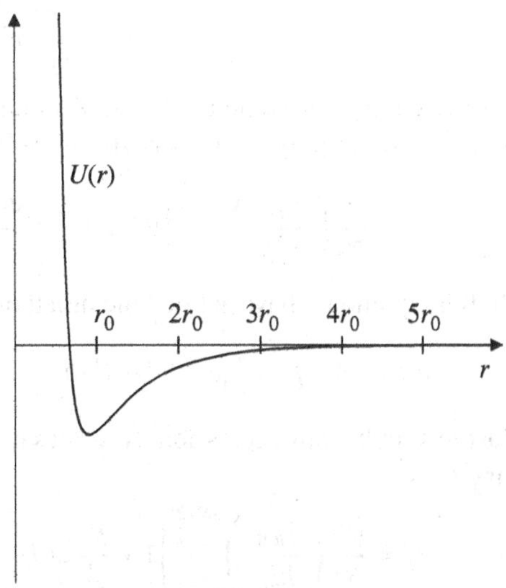

This is simply a trivial rewriting with $\xi_{ij} = e^{-\beta U_{ij}} - 1$. We can then consider the integrand that goes into the partition function and expand it as

$$\prod_{i<j}(1+\xi_{ij}) = 1 + \sum_{i<j}\xi_{ij} + \sum_{i,j,k,\ell}\xi_{ij}\xi_{k\ell} + \cdots, \tag{8.51}$$

where the second sum has many constraints in it, which you can explore in the next exercise.

Exercise 8.13 Consider the case where $N = 4$, and explicitly expand the product on the left-hand side of Eq. (8.51), and determine what the constraints on the second sum on the right-hand side are (that is, what combinations of i, j, k, and ℓ will appear?).

For now let us just consider the first two terms in Eq. (8.51), which we will see lead to an expansion in powers of the density n. The partition function in this case is

$$\begin{aligned} Z_f &\approx \frac{1}{N!}\left(\frac{2\pi m}{h_0^2\beta}\right)^{3N/2} \int d^{3N}\mathbf{r}\left(1+\sum_{i<j}\xi_{ij}\right) \\ &= \frac{1}{N!}\left(\frac{2\pi m}{h_0^2\beta}\right)^{3N/2} \left[V^N + \sum_{i<j}\int d^{3N}\mathbf{r}\,(e^{-\beta U_{ij}}-1)\right]. \end{aligned}$$

The integral is over $3N$ positions but only two ($\mathbf{r}_i$ and $\mathbf{r}_j$) are in U_{ij}, so we can evaluate $N-2$ of the position integrals immediately,

$$\sum_{i<j}\int d^{3N}\mathbf{r}\,(e^{-\beta U_{ij}}-1) = V^{N-2}\sum_{i<j}\int d^3\mathbf{r}_i d^3\mathbf{r}_j\,(e^{-\beta U_{ij}}-1).$$

We can simplify this second set of integrals by using two-body coordinates, with the center-of-mass coordinate $\mathbf{r}_{\text{cm}} \equiv (\mathbf{r}_i + \mathbf{r}_j)/2$ and the relative coordinate $\mathbf{r} \equiv \mathbf{r}_i - \mathbf{r}_j$, so that (using spherical coordinates in the last equality)

$$\begin{aligned} \int d^3\mathbf{r}_i d^3\mathbf{r}_j\,(e^{-\beta U_{ij}}-1) &= \int d^3\mathbf{r}_{\text{cm}} d^3\mathbf{r}\left[e^{-\beta U(r)}-1\right] \\ &= V\int d^3\mathbf{r}\left[e^{-\beta U(r)}-1\right] \\ &= 4\pi V\int_0^\infty r^2\left[e^{-\beta U(r)}-1\right]dr. \end{aligned}$$

As these integrals would be the same for all i and j, and with $i<j$, there are $N(N-1)/2$ integrals that we can simplify in this way, so we are left with

$$Z_f = \frac{1}{N!}\left(\frac{2\pi m}{h_0^2\beta}\right)^{3N/2}\left[V^N + V^{N-1}\frac{N(N-1)}{2}a(\beta)\right].$$

In this equation, I have defined the function

$$a(\beta) \equiv 4\pi\int_0^\infty r^2\left[e^{-\beta U(r)}-1\right]dr, \tag{8.52}$$

just to simplify our expression. Now let's use the fact that we have a large system, so that $N \gg 1$, and thus

$$Z_f = \frac{V^N}{N!}\left(\frac{2\pi m}{h_0^2\beta}\right)^{3N/2}\left[1+\frac{N^2}{2V}a(\beta)\right]. \tag{8.53}$$

Exercise 8.14 Use Eq. (8.53) to show that

$$\beta\bar{p} = \frac{N}{V} - \frac{\frac{N^2}{2V^2}a(\beta)}{1 + \frac{N^2}{2V}a(\beta)}.$$

and for small N/V,

$$\bar{p} \approx \frac{Nk_BT}{V}\left[1 - \frac{a(\beta)}{2}\frac{N}{V}\right]. \tag{8.54}$$

The second term on the right-hand side of Eq. (8.54) is the first correction to the ideal gas for a dilute system. It shows us that we can write, with $n = N/V$,

$$\frac{\bar{p}}{k_BT} = n + \sum_{i=2}^{\infty} B_i(T)n^i, \tag{8.55}$$

which is known as the **virial expansion** which you first saw in Problem 6.18, and the B_i's are known as **virial coefficients.** One could determine these coefficients from the partition function if the potential is known (we will do this for a special case shortly), or determine them empirically.

A realistic example of a potential between the two molecules is the **Lennard–Jones potential,**

$$U(r) = U_0\left[\left(\frac{r_0}{r}\right)^{12} - 2\left(\frac{r_0}{r}\right)^{6}\right], \tag{8.56}$$

with a minimum at r_0, $U(r_0) = -U_0$. This is shown in Figure 8.7, and from this we could determine $a(\beta)$. However, it is quite difficult to evaluate the integral in Eq. (8.52) analytically, so let us examine a similar but much simpler potential, the **Sutherland potential,** which nicely approximates the important features,

$$U(r) = \begin{cases} +\infty & r < r_0 \\ -U_0\left(\frac{r_0}{r}\right)^6 & r \geq r_0 \end{cases}, \tag{8.57}$$

as shown in comparison with the Lennard–Jones potential in Figure 8.8. This has the three most important features of the Lennard–Jones potential: first that it increases rapidly below $r = r_0$, second that it has a minimum at $r = r_0$, and third it vanishes like r^{-6} as $r \to \infty$.

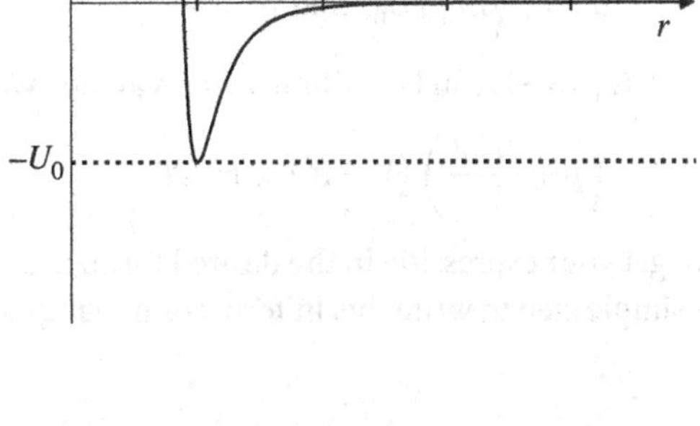

Figure 8.7 The Lennard-Jones potential, Eq. (8.56), as a function of r.

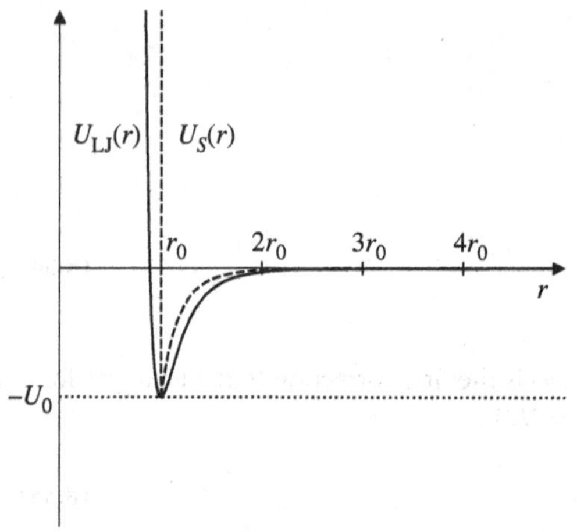

Figure 8.8 A comparison of the Sutherland potential $U_s(r)$ with the Lennard–Jones potential $U_{LJ}(r)$.

This potential assumes the molecules are all hard spheres of radius $r_0/2$, which creates the infinite wall at $r = r_0$: The centers of any two molecules can never be closer than this. We will assume βU_0 is small enough so we can expand for $r \geq r_0$,

$$e^{-\beta U(r)} \approx 1 + \beta U_0 \left(\frac{r_0}{r}\right)^6 + \cdots,$$

and for $r < r_0$, $e^{-\beta U(r)} = 0$. This allows us to explicitly evaluate $a(\beta)$,

$$\begin{aligned} a(\beta) &= 4\pi \int_0^\infty r^2 \left[e^{-\beta U(r)} - 1\right] dr \\ &= 4\pi \int_0^{r_0} r^2 \left[e^{-\beta U(r)} - 1\right] dr + 4\pi \int_{r_0}^\infty r^2 \left[e^{-\beta U(r)} - 1\right] dr \\ &= -\frac{4}{3}\pi r_0^3 + 4\pi\beta U_0 r_0^6 \int_{r_0}^\infty \frac{dr}{r^4} \\ &= \frac{4}{3}\pi r_0^3 \left(\beta U_0 - 1\right). \end{aligned} \tag{8.58}$$

With a little bit of algebra, we can write

$$\overline{p} = \frac{Nk_BT}{V}\left[1 - \frac{2\pi r_0^3}{3}\frac{N}{V}\left(\frac{U_0}{k_BT} - 1\right)\right] \tag{8.59}$$

Exercise 8.15 Show that our result for $a(\beta)$ in Eq. (8.58), when substituted into Eq. (8.54), gives Eq. (8.59).

Exercise 8.16 Show that if you define

$$a' = \left(\frac{2}{3}\pi r_0^3\right) U_0, \text{ and } b' = \frac{2}{3}\pi r_0^3,$$

then Eq. (8.59) can be written as the Van der Waals equation of state,

$$\left(\overline{p} + \frac{N^2 a'}{V^2}\right)(V - b') = Nk_BT.$$

To get your expression in the desired form, use the fact that for small x, $1 + x \approx 1/(1 - x)$. It is then a simple step to write this in terms of molar quantities to get Eq. (6.121).

Exercise 8.17 Evaluate $\bar{E}$ for the Van der Waals gas, and show that if you keep only the first term in expanding in N/V, you obtain the result we found in Eq. (6.126).

Not only does this show us how to obtain the Van der Waals equation of state from the microscopic system, it gives us a systematic way to improve our approximations if we wish to take a theoretical approach to determining the equations of state. So from now on we needn't worry about apologizing anymore: If our results don't match experimental observations, we can now just improve our theory bit by bit to make it more realistic.

8.7 Specified Mean Energy

We have seen that the spread of the energy about the mean energy in the canonical ensemble is small enough for large systems that we can often treat the same system using either ensembles we have studied thus far (explicitly in Eq. (8.38) for the ideal gas). This implies that we have a well-specified *mean energy*, and if that is the case, why can't we just assume this first to derive the canonical distribution? That is, in the microcanonical ensemble, we specified the *total* energy, but we would like to formulate our theory such that we can specify the *mean* energy instead. We will see in this section that we can do this, and will just have to determine the temperature from $\bar{E}$ (although determining the temperature this way is not as simple as we would like).

We start with a system with mean energy $\bar{E}$ and a total number of states accessible to the system given by Ω (but in this case not with a specific energy). Suppose also that there are Ω_s states with some energy E_s, so that we can write the mean energy as

$$\bar{E} = \frac{1}{\Omega}\sum_s \Omega_s E_s, \tag{8.60}$$

and we can define a "total energy,"

$$E^{(0)} \equiv \Omega\bar{E} = \sum_s \Omega_s E_s. \tag{8.61}$$

This is the total energy of an *ensemble* of systems, not a single system, but we can consider it to be such if we treat the ensemble of systems as our combined system $A^{(0)}$ from when we first introduced the canonical ensemble. From now on we will consider a *system* in our ensemble as a *state* of the system as before.

Let's now focus on just one of these systems in a state r, with energy E_r and we'll separate it from the rest as shown in Figure 8.9. With the total set of systems "having" energy $E^{(0)} = \Omega\bar{E}$, then the system we pick out will be equally likely to be any of the possible microstates. The rest of the states will have energy

$$\Omega\bar{E} - E_r \equiv E', \tag{8.62}$$

which, with our definition of $E^{(0)}$, is mathematically identical to the setup of the canonical ensemble. Our system A is the one we picked out with energy E_r, and the energy it can have depends upon the combined energy of the remaining states, which together act as our heat reservoir. This will only be a valid comparison if we have a very large number of states, so that $\Omega \gg 1$, and as such $E_r \ll \Omega\bar{E}$ and $E_r \ll E'$.

With this mathematical equivalence, we can say that the probability of our chosen state having energy E_r is given by the Boltzmann factor,

$$P_r \propto e^{-\beta E_r}.$$

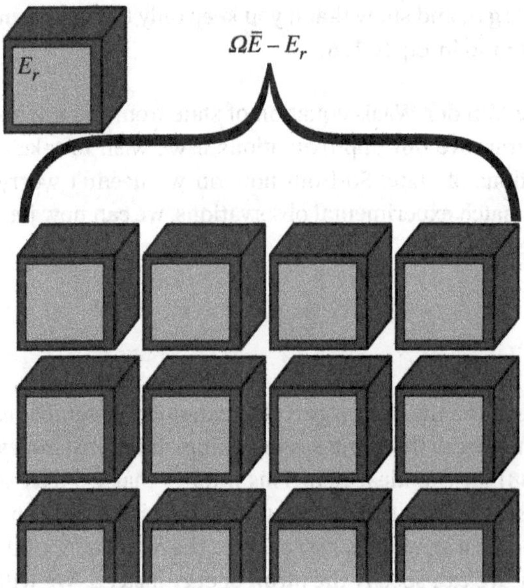

Figure 8.9 An ensemble of systems with one system in a state with energy E_r separated from the remaining ones. The remaining systems with total energy $\Omega\bar{E} - E_r$ act as a heat reservoir.

Therefore we again have the same definition of the partition function Z as before in Eq. (8.5). The difference here is that the derivative $(\partial \ln \Omega/\partial E)$ doesn't directly relate to the temperature of a physical heat bath, but is instead determined implicitly by the relationship

$$\frac{1}{Z}\sum_r E_r e^{-\beta E_r} = \bar{E}. \tag{8.63}$$

To summarize, if a system is in contact with a heat reservoir at temperature T, then we use this relation to determine the mean energy $\bar{E}$. On the other hand, if we know the mean energy of the system, then this same expression allows us to determine the temperature. This allows us yet again another connection between our two ensembles: As long as the system is large, then specifying the total energy *or* the mean energy allows for us to derive the same relationship between the (mean) energy and the temperature. However, for small systems, then our results will in general be different with the canonical and microcanonical ensembles.

For small systems, such as the two-state system where the mean energy was given by Eq. (8.18), solving for β is straightforward. However, for larger systems this can be difficult (see Problem 8.4). Because the primary goal here was to justify the correspondence between the microcanonical and canonical ensembles, this is not serious concern.

Exercise 8.18 Solve Eq. (8.18) for the temperature, assuming we knew the mean energy.

8.8 Summary

- In the canonical ensemble, a system is in thermal contact with a heat reservoir with temperature $T = 1/(k_B\beta)$, and the probability it is in a state with energy E_r is proportional to $e^{-\beta E_r}$, the Boltzmann factor. Such a system can be thought of as having a well-defined temperature or a well-defined mean energy.
- All thermodynamic quantities can be determined from the partition function, $Z = \sum_r e^{-\beta E_r}$.

- For large systems, the results from calculations in the microcanonical ensemble and canonical ensemble will be consistent with each other.
- When one calculates the entropy for a classical monatomic ideal gas without treating the molecules in the system as indistinguishable (a true classical system), one runs into the Gibbs paradox, which is solved by using fudged classical statistics.
- Using the partition function, it is straightforward to determine the equations of state for non-ideal gases, the most simple of these is the Van der Waals equation. Using the virial expansion we can systematically improve upon the naive ideal gas assumptions.

Problems

8.1 There is yet another way to derive the canonical ensemble starting from the initial assumption that we maximize the missing information in our theory, consistent with any constraints. We wish to maximize the Shannon entropy,

$$S = -k_B \sum_r P_r \ln P_r,$$

with the constraint that the probabilities are normalized, $\sum_r P_r = 1$. This leads to the microcanonical ensemble, as seen in Problem 3.8 in the context of information theory. For the canonical ensemble, we fix the temperature of the system, which we can argue fixes the average internal energy of the system (based upon our interpretation of temperature). Thus, we can impose a second constraint that

$$\bar{E} = \sum_r P_r E_r = \text{fixed}.$$

(a) Use the method of Lagrange multipliers to show that maximizing the Shannon entropy subject to these two constraints leads to the Boltzmann distribution.
(b) Use the relationship between $\ln Z$ and S to show that the second Lagrange multiplier you must introduce should be identified with $1/T$ (see Problem 8.6).

8.2 Consider a one-dimensional quantum mechanical simple harmonic oscillator with angular frequency ω which is in thermal contact with a heat reservoir at temperature T low enough so that $k_BT/(\hbar\omega) \ll 1$.
(a) Find the ratio of the probability of the oscillator being in the first excited state to the probability of its being in the ground state.
(b) Assuming that only the ground state and first excited state are appreciably occupied, find the mean energy of the oscillator as a function of the temperature T.
(c) For a hydrogen iodide (HI) molecule, the vibrations have a natural frequency of $f = 6.69 \times 10^{13}$ Hz. What is the spacing between energy levels of this quantum harmonic oscillator, $\hbar\omega$?
(d) What temperature is "low enough" here? That is, find the temperature where your result to part (a) is about 0.01 (so that it is 100 times more likely to be in the ground state than the first excited state).

8.3 Consider a three-dimensional harmonic oscillator, with the same spring constant in each direction so we can write

$$\varepsilon_{n_x,n_y,n_z} = \left(n_x + n_y + n_z + \frac{3}{2}\right)\hbar\omega, \; n_x, n_y, n_z = 0, 1, 2, \ldots. \tag{8.64}$$

Show using the canonical ensemble that the mean energy of one oscillator is

$$\bar{\epsilon} = 3\hbar\omega\left(\frac{1}{2} + \frac{1}{e^{\beta\hbar\omega} - 1}\right). \tag{8.65}$$

8.4 One billion ordinary six-sided dice, one of which is green, are thrown, and the total roll turns out to be exactly two billion (unlikely, but assume it occurs). Use the canonical distribution to calculate the probability for each face 1–6 of the green die. This is a case where $\bar{E}$, rather than β, is known, and you'll have to find β, or equivalently $x = e^{-\beta}$, implicitly. Remember when we talk of rolling dice, the roll of a die is its "energy."
Hint: This needs to be done via trial and error, or perhaps by numerically solving for x with Mathematica or other software.

8.5 Redo Problem 8.4 for the case where the result of the roll is three billion, four billion, or five billion.

8.6 Consider a system whose external parameters are fixed and return to our original form for the Shannon entropy,

$$S = -k_B \sum_r P_r \ln P_r,$$

where P_r is the probability of being in microstate r. Only assume that P_r is a function of β, but do not assume that it is proportional to the Boltzmann factor.

(a) Show that

$$\frac{\partial S}{\partial \beta} = -k_B \sum_r \left(\frac{\partial P_r}{\partial \beta}\right) \ln P_r.$$

Hint: You will have to show that the other term in the derivative (where $\partial/\partial\beta$ acts on $\ln P_r$) vanishes after summing over r.

(b) Assuming E_r is independent of the parameter β, write down a similar expression for $\partial\bar{E}/\partial\beta$.

(c) Suppose you have proved that $P_r = e^{-\beta E_r}/Z$ for a system with known mean energy but have not yet figured out the thermodynamic meaning of the parameter β. Use the results above plus the fact that $dS = đQ/T$ (for a quasistatic thermal interaction) to show that $\beta = 1/(k_B T)$.

8.7 Suppose you have a rigid rotor with moment of inertia I and angular momentum L such that the only relevant energy is the rotational kinetic energy

$$K_{\text{rot}} = \frac{L^2}{2I}.$$

(a) Assume this is a classical system, so that $0 \le L < \infty$ is a continuous variable. Calculate the partition function for a single rigid rotor, and from that calculate the mean energy.

(b) Now assume this is a quantum system, so that

$$L^2 = \ell(\ell+1)\hbar^2, \ell = 0, 1, 2, \ldots.$$

For each value of ℓ, there are $2\ell + 1$ states with a given energy. Write down the partition function for the quantum case.

(c) Assume that the temperature is large enough for the quantum system that you can treat ℓ as a continuous variable. Convert your result from (b) into an integral, and evaluate it.

(d) From (c) calculate the mean energy of the quantum rigid rotor.

(e) The results above were for a single rotor for either the quantum or classical case. Imagine you had a system of N such rotors, where only this energy contributes. Is there a way that you could perform a measurement on this system to be able to distinguish a classical system from a quantum system? Explain.
Hint: Consider the heat capacities for the two cases.

8.8 Suppose you have a nonmagnetic liquid in a container of height h and cross-sectional area A. You mix in N spin-1/2 molecules with magnetic dipole moment μ_0 and apply an external magnetic field,

$$H(z) = H_0 \frac{z}{h},$$

along the axis of the cylinder (the z-direction). What is the number density (number per unit volume) of the magnetic molecules as a function of z, if the temperature of the system is T?

8.9 Semiconductors are created by *doping* a material with impurities to have an effective positive or negative net charge. For example, silicon has two electrons in its outer shell while boron has only one electron in its outer shell. By doping silicon with boron atoms, one creates a **p-type** semiconductor, because the boron atoms effectively have a positive charge due to the unfilled outer shell (known as a **hole** in condensed matter physics), which can propagate through the material. Such materials are crucial for modern electronics, and we can understand some of the basics of this setup as a thermodynamic system.

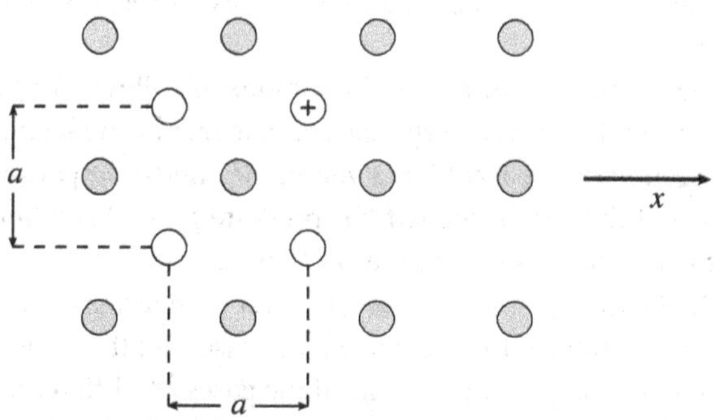

If we take the "positive" charge of the boron atoms seriously (the material is still electrically neutral), we can imagine it can be equally spaced between silicon atoms (which are on average a distance a from each other) as shown—it can be in one of the four possible spaces (sticking with two dimensions for simplicity).

Consider n boron atoms per unit volume which can freely move through the lattice structure formed by the silicon atoms, and the combined system is in contact with a heat reservoir at temperature T. Now suppose a small electric field $\mathcal{E}$ is applied along the x-direction. What is the electric polarization (the mean electric dipole moment per unit volume) along the x-direction?

Hint: With just some slight differences, this is almost identical to our spin-1/2 magnetization question.

8.10 Nuclei of spin-one atoms can exist in one of three possible states, labeled by the quantum number $m = -1, 0$, and $+1$. Suppose these energies are labeled as ε_{-1}, ε_0, and ε_{+1}, respectively, and assume the lowest energy is ε_0.

(a) If these are the only relevant degrees of freedom in the system, calculate the mean molar energy as a function of the absolute temperature T.
(b) Calculate the molar entropy of the system.
(c) Using the method of Example 8.2 (using qualitative reasoning), determine the molar entropy at low and high temperatures.
(d) Show that your expression in part (b) reduces properly to these values in part (c) in the proper limits.
(e) Calculate the nuclear contribution to the molar heat capacity of the solid as a function of temperature. What is the temperature dependence for large and small values of T?
(f) Often such nuclei have a charge distribution to give them an ellipsoidal shape, so that $\varepsilon_{+1} = \varepsilon_{-1}$. Simplify your results above in this case, setting $\varepsilon_{+1} = \varepsilon$ and $\varepsilon_0 = 0$.

8.11 Extend our work on the classical monatomic ideal gas to include M different types of molecules at absolute temperature T in a container of volume V. Assume you have N_k molecules of type k, where $k = 1, \ldots, M$.

(a) Work through the steps we went through in Section 8.4 to determine $\overline{p}(V, T)$. Define $N = N_1 + N_2 + \cdots + N_M$.
(b) The **partial pressure** $\overline{p}_i$ is defined by the pressure the ith gas would have if it were the only molecule in the container at this temperature. How does $\overline{p}_i$ relate to the total mean pressure $\overline{p}$?
(c) Did you need to use fudged classical statistics for these questions? Explain.

8.12 Consider a generalization of the Gibbs paradox as follows. You have a container fixed to temperature T divided into two separate compartments, the section on the left has a volume V_ℓ, and the section on the right has a volume V_r. Both compartments have N molecules of an ideal gas, and the gas on the left has pressure $\overline{p}$. At this point, you remove the divider separating the two sides, allowing the gas to mix.

(a) Calculate the final pressure of the gas mixture in terms of $\overline{p}$.
(b) Determine the total entropy change if the gases are the same.
(c) Determine the total entropy change if the gases are different.

Hint: Your results will be different than what we found when discussing the Gibbs paradox, but should match precisely if $V_\ell = V_r$ and the gases are the same.

8.13 Return to the microcanonical ensemble but now explicitly calculate the prefactor in Eq. (4.38) using the results in Appendix D. Show that you get the same result in the large-N limit as we found in Eq. (8.39).

Hint: Remember that the Γ function is just the factorial function, so $\Gamma(n) = (n-1)!$ for positive integer n, which you can apply Stirling's formula to simplify. (Note: this has a similar problem as in the canonical ensemble, that we would have to add a fudge factor to avoid the Gibbs paradox, but we won't worry about this here.)

8.14 Consider the virial expansion with the following **hard sphere** free-particle model,

$$U(r) = \begin{cases} \infty & r < r_0 \\ 0 & r \geq r_0 \end{cases}.$$

Calculate the second virial coefficient $B_2 = -a(\beta)/2$ for this model and the corresponding equation of state. Compare this to both the ideal gas equation and the Van der Waals equation.

9

Applications of the Canonical Distribution

In Chapter 8 we formulated our statistical problem with the canonical ensemble and showed that for large systems physical results were the same as what we obtained with the microcanonical ensemble. Additionally, the canonical ensemble lends itself to a simpler method for calculating thermodynamic quantities, so it is generally the preferred approach for most problems.

In this chapter we will apply the canonical ensemble to several examples to show the sheer power of it, as we'll be able to not only cover more complicated examples that were more difficult before, but also set up the basics behind the kinetic theory of gases.

After finishing this chapter, you should be able to

- derive the equipartition theorem using the canonical ensemble,
- calculate the specific heat of solids with both the Einstein model and the Debye model,
- understand paramagnetism for a general spin system, and
- learn the basics of kinetic theory, namely the Maxwell velocity distribution.

9.1 Equipartition Theorem

Our first application of the canonical ensemble is to one of the most often used (and more often misused) theorems in thermodynamics. Before proving it, I'll begin by stating the theorem, which is known as the

Equipartition theorem:
For a classical system, any degree of freedom that appears only *quadratically* in the energy of the system will contribute $\frac{1}{2}k_BT$ to the mean internal energy of the system.

These are fairly strict assumptions and while this means it has limited use, this theorem is useful for many calculations. If nothing else, this theorem has given us our qualitative understanding of the meaning of absolute temperature. Additionally, its derivation also shows the simplicity of performing calculations in the canonical ensemble and thus gives yet another reason to use this approach.

We begin with the energy of a microstate of a classical system of N molecules in terms of the positions and momenta,

$$E_r = E_r\left(x_1, \ldots, z_N, p_{x1}, \ldots, p_{zN}\right) = E_r\left(q_i\right),$$

Statistical Thermodynamics: An Information Theory Approach, First Edition. Christopher Aubin.

Companion website: www.wiley.com/go/Aubin/StatisticalThermodynamics

where q_i is a shorthand for any of the position or momentum coordinates (so it could correspond to x_4, z_{10}, p_{y8}, etc.). In general this is a complicated function of the coordinates and momenta (such as in our discussion of the non-ideal gas in Section 8.6); however, it is often the case that one (or more) of these degrees of freedom can be separated out, allowing us to write E_r as the sum of two functions,

$$E_r = \varepsilon_i(q_i) + E'_r\left(q_{j\neq i}\right),$$

where the first term is only a function of the degree of freedom q_i, and the second term includes all of the remaining terms in E_r, none of which depend on q_i. Because calculating the mean value of a quantity is a linear operation, we will first show that we can calculate the mean value of ε_i—the contribution to the mean energy coming from this single degree of freedom q_i without needing to know E'_r. And, if every degree of freedom appears separately in E_r, then we calculate $\overline{\varepsilon}_i$ for each i, and simply add them up to obtain the total mean energy $\overline{E}$. To calculate this, we ensure our system is in contact with a heat reservoir at temperature T, and evaluate

$$\overline{\varepsilon}_i = \frac{\int dx_1 \ldots dp_{zN}\varepsilon_i(q_i)e^{-\beta E_r}}{\int dx_1 \ldots dp_{zN}e^{-\beta E_r}} = \frac{\int d^{6N}\mathbf{q}\varepsilon_i(q_i)e^{-\beta E_r}}{\int d^{6N}\mathbf{q}e^{-\beta E_r}}. \tag{9.1}$$

In the second equality, we are using a shorthand for the position and momentum variables, putting them in a 6N-dimensional "vector," $\mathbf{q}$. This is purely for a mathematical shorthand, and we'll see it won't make much of a difference shortly. I won't include the limits of integration—but will point out what they will be when its needed.

Because we can separate out ε_i and E'_r is independent of q_i, we can write

$$\overline{\varepsilon}_i = \frac{\left[\int dq_i\varepsilon_i(q_i)e^{-\beta\varepsilon_i(q_i)}\right]\int d^{6N-1}\mathbf{q}e^{-\beta E'_r}}{\left[\int dq_ie^{-\beta\varepsilon_i(q_i)}\right]\int d^{6N-1}\mathbf{q}e^{-\beta E'_r}}, \tag{9.2}$$

and the integrals over the variables other than q_i cancel, leaving us with

$$\overline{\varepsilon}_i = \frac{\int dq_ie^{-\beta\varepsilon_i(q_i)}\varepsilon_i(q_i)}{\int dq_ie^{-\beta\varepsilon_i(q_i)}}. \tag{9.3}$$

We have seen this before—we can calculate the mean energy per particle for such separable problems. If we use our derivative trick from before, we can simplify this further to write

$$\overline{\varepsilon}_i = -\frac{\partial}{\partial\beta}\ln\left[\int dq_ie^{-\beta\varepsilon_i(q_i)}\right]. \tag{9.4}$$

To get to this point, we have assumed two things: that we have a classical system and the contribution to the energy from a single degree of freedom can be separated out. Now let us make the third assumption required for the theorem, that ε_i is quadratic in q_i, so $\varepsilon_i = bq_i^2$ for any constant b, which simplifies the integral quite a bit. In this case, we will assume that the integration limits are now from $-\infty$ to ∞.[1] Then the integral is straightforward (c.f. Eq. (C.2)),

$$\int_{-\infty}^{\infty} dq_ie^{-\beta bq_i^2} = \left(\frac{\pi}{\beta b}\right)^{1/2},$$

so that

$$\overline{\varepsilon}_i = -\frac{\partial}{\partial\beta}\ln\left(\frac{\pi}{\beta b}\right)^{1/2} = \frac{1}{2\beta}, \tag{9.5}$$

1 If they are not, then we can usually extend them as such, because the exponential will generally be negligible for increasing q_i.

or

$$\bar{\varepsilon}_i = \frac{1}{2}k_B T. \tag{9.6}$$

This is precisely what was stated in the equipartition theorem at the beginning of the section.

It is important to remember that this is not valid for quantum mechanical systems. We must be in the classical limit in order for this to be valid (i.e., when the temperature is high enough that quantum effects are no longer important, or when the spacing between energy levels is much less than $k_B T$).

Example 9.1 Consider the mean kinetic energy of a molecule of a gas, where

$$K = \frac{1}{2m}(p_x^2 + p_y^2 + p_z^2).$$

The equipartition theorem immediately gives

$$\overline{K} = \frac{3}{2}k_B T,$$

which is true for one molecule of a monatomic ideal gas (in that case, $\overline{E} = \overline{K}$).

This is often where the idea that "temperature is a measure of the kinetic energy of a system" arises. While in this example, it is precisely true, remember that temperature is a more complicated concept. The next example shows that the potential energy in a system also contributes to the temperature.

Example 9.2 For a simple harmonic oscillator in one dimension, we have

$$\varepsilon = \frac{p^2}{2m} + \frac{1}{2}m\omega^2 x^2,$$

where both the kinetic and potential energy are of the form required by the theorem, so

$$\bar{\varepsilon} = k_B T.$$

Exercise 9.1 Extend Example 9.2 to N oscillators in three dimensions. What would the equipartition give for the mean energy in this case? Compare your result to what you found in Problem 5.3 (or the high temperature limit of Problem 5.4).

9.2 Specific Heat of Solids

As an application of the canonical ensemble, let's study the specific heat of solids. If we start with a classical model, then this is just a straightforward application of the equipartition theorem. This will lead to several issues (most importantly to a contradiction with the laws of thermodynamics), so we will then move to a quantum mechanical study with two different models, improving upon our classical result.

9.2.1 The Classical Case

A solid can be modeled as a large number N of molecules which are bound (possibly with a crystalline structure, but for our purposes, this will not yet be necessary). The exact mechanism for the binding may not be known, but for a first approximation, we will assume the molecules can be treated as masses connected to each other with springs which obey Hooke's law[2] as shown in Figure 9.1. This figure only shows a two-dimensional system for simplicity of visualization, but we'll consider the full three-dimensional case.

Instead of assigning the coordinates $\mathbf{r}_i$ to the ith molecule, we will use a notation common in classical mechanics: We will use *generalized coordinates*, q_ℓ, which could be the x-, y,- or z-direction of any of the N molecules (so $1 \leq \ell \leq 3N$). Additionally, we consider p_ℓ a generalized momentum (conjugate to the coordinate q_ℓ). We will keep this completely general, so that the "springs" could in principle connect molecules that are not only the nearest neighbors as shown in Figure 9.1, but any other mass in the solid. This is reasonable given that the intermolecular interactions could in principle be farther reaching than just between those closest molecules.

With these assumptions, the energy of a given state is given by

$$E_r = \sum_{\ell=1}^{3N} \frac{p_\ell^2}{2m} + \frac{1}{2} \sum_{\ell,n=1}^{3N} \tilde{k}_{\ell n} q_\ell q_n, \tag{9.7}$$

where $\tilde{k}_{\ell n} = \tilde{k}_{n\ell}$ is the spring constant that connects q_ℓ with q_n and m is the mass of the molecules. This seems impossible to evaluate, but a common exercise in classical mechanics is to write this in terms of the **normal modes** of the system (see, for example, chapter 12 of Ref. [1]). There are $3N$ such normal modes, which we will label as η_ℓ, and one determines them by diagonalizing the matrix that is formed by the spring constants $\tilde{k}_{\ell n}$, resulting in

$$E_r = \sum_{\ell=1}^{3N} \left(\frac{p_\ell'^2}{2m} + \frac{1}{2} k_\ell \eta_\ell^2 \right), \tag{9.8}$$

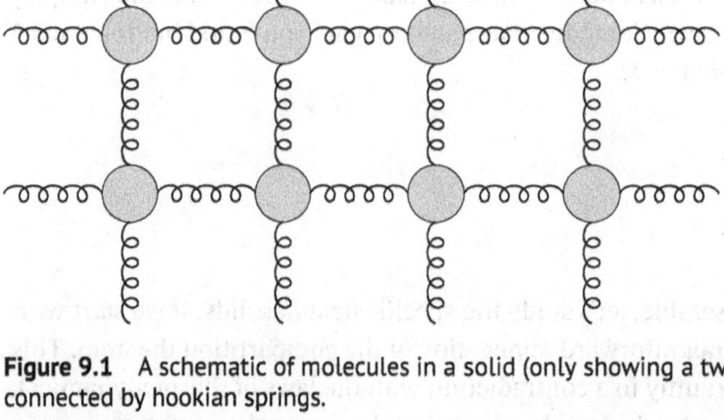

Figure 9.1 A schematic of molecules in a solid (only showing a two-dimensional system for simplicity) connected by hookian springs.

2 Robert Hooke (1635–1703).

where we are defining p'_ℓ as the momentum of the ℓth normal mode and k_ℓ the eigenvalues of the spring constant matrix.

The energy is now in the form required for us to evaluate its average value using the equipartition theorem, so we can immediately write down

$$\overline{E} = 3Nk_BT, \tag{9.9}$$

because we have two quadratic degrees of freedom for each of our $3N$ normal modes. Also note that the molecules don't even need to be the same type—all N of them could have a different mass and Eq. (9.9) would be the same.

The heat capacity at constant volume easily follows from this,

$$C_V = \left(\frac{\partial \overline{E}}{\partial T}\right)_V = 3Nk_B = 3\nu R, \tag{9.10}$$

so the molar specific heat is

$$c_V = 3R = 24.9\,\mathrm{J/mol/K}. \tag{9.11}$$

This result is independent of any specific properties of the solid in question (in a sense, this is the "ideal solid" much like we had the ideal gas). Table 9.1 lists some specific heats (both per unit mass and per mole) at constant pressure c_p (from Ref. [2]) at 298 K for some solids. Additionally I have included the molar mass of each solid, as often it is the specific heat per unit mass that is determined experimentally, so we need this to convert to the molar specific heat. While we calculated c_V, we know that c_V only differs by a small amount from c_p (around 5% or so) for solids, so that most of these are around 24.9 J/mol/K is a nice validation of the classical approximation. Just as with the specific heats of ideal gases in Table 6.1, we see the specific heats per unit mass vary wildly, but for the most part, the molar specific heats are quite similar.

There is one significant outlier: graphite, which has an interesting crystalline structure. The approximation we made above is very far from what is measured. Graphite is composed of a stack of two-dimensional layers, where in each layer the carbon atoms form a hexagonal lattice structure

Table 9.1 The molar mass and specific heats (per mass and per mole) at constant pressure for various solids [2] J.R. Rumble 2022/Taylor & Francis.

Solid	Molar mass (g/mol)	c_p (J/g/K)	c_p (J/mol/K)
Aluminum	26.98	0.897	24.20
Copper	63.55	0.385	24.440
Gold	196.97	0.129	25.42
Iron	55.84	0.449	25.10
Lead	207	0.130	26.84
Nickel	58.69	0.444	24.60
Platinum	195.08	0.133	25.86
Silver	107.87	0.235	25.35
Sulfur (rhombic)	32.07	0.708	22.70
Tin (white)	118.71	0.227	26.99
Zinc	65.4	0.388	25.39
Carbon (graphite)	12.01	0.709	8.52

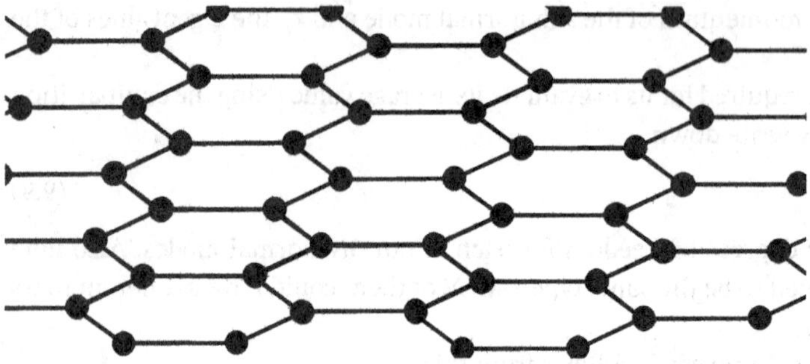

Figure 9.2 The hexagonal structure of one layer of carbon atoms. Graphite is composed of many such layers stacked on top of each other.

as shown in Figure 9.2. The layers tend to be spaced roughly 2.4 times the interatomic distance of the carbon atoms in the layers. Our approximation generally assumes a cubic lattice structure, where all atoms are equally spaced in all directions.

Exercise 9.2 Look up data for some of these solids for α and κ at room temperature (as well as their molar volumes). Use the relationship between the molar specific heats in Eq. (6.92) to determine what c_p should be from our theoretical determination of c_V. How far are your results from the various results in Table 9.1?

That the specific heat is constant is often known as the **law of Dulong & Petit**,[3] but this can only be true at high temperatures. We know this cannot be valid at all temperatures, as it would violate the third law of thermodynamics, which requires that the specific heat vanishes as $T \to 0$. Nevertheless, given the agreement with the specific heats of (many) solids at room temperature, we will use this as a check that our quantum mechanical models are valid by considering the high temperature limit.

9.2.2 The Einstein Model

The first quantum mechanical solution we will study is the **Einstein model of specific heats**, developed by Albert Einstein (1879–1955) in 1906. This is similar to the classical model in that it treats the molecules in our solid as simple harmonic oscillators; however, we will now use the discrete quantum mechanical energy levels to calculate the partition function. For a single molecule, the energy of a state is given by

$$\epsilon_{n_x,n_y,n_z} = \left(n_x + n_y + n_z + \frac{3}{2}\right)\hbar\omega, \qquad n_x, n_y, n_z = 0, 1, 2 \ldots .$$

The mean energy of a single quantum mechanical oscillator in contact with a heat reservoir at temperature T is given from Problem 8.3,

$$\overline{\epsilon} = 3\hbar\omega\left(\frac{1}{2} + \frac{1}{e^{\hbar\omega/(k_BT)} - 1}\right). \tag{9.12}$$

In general, there would be a different natural frequency ω for each of our $3N$ normal modes, but Einstein's approximation was to assume that all they are all the same (so the spring constants k_ℓ in

3 Pierre-Louis Dulong (1785–1838) and Alexis-Thérèse Petit (1791–1820).

Eq. (9.8) are the same for all ℓ). The total mean energy is then given by

$$\overline{E} = 3N\hbar\omega\left(\frac{1}{2} + \frac{1}{e^{\hbar\omega/(k_BT)} - 1}\right), \tag{9.13}$$

and we can calculate the heat capacity to be

$$C_V = 3Nk_B\left(\frac{\hbar\omega}{k_BT}\right)^2 \frac{e^{\hbar\omega/(k_BT)}}{[e^{\hbar\omega/(k_BT)} - 1]^2}. \tag{9.14}$$

Exercise 9.3 Calculate $C_V = (\partial\overline{E}/\partial T)_V$ explicitly and show you obtain Eq. (9.14).

Let's define the **Einstein temperature**

$$\Theta_E \equiv \frac{\hbar\omega}{k_B}, \tag{9.15}$$

and also divide Eq. (9.14) by the number of moles to obtain the molar specific heat,

$$c_V = 3R\left(\frac{\Theta_E}{T}\right)^2 \frac{e^{\Theta_E/T}}{(e^{\Theta_E/T} - 1)^2}. \tag{9.16}$$

I show this as a function of temperature in Figure 9.3, and we can see that when $T = 3\Theta_E$, the difference between the quantum (solid curve) and classical (dotted line) result is about 1%. For copper, $\Theta_E \sim 240$ K [3], so room temperature is only just slightly higher than the Einstein temperature. At 300 K, the deviation between these two models is around 5%.

Now let's look at the limiting behavior of the specific heat to see that it agrees with our expectations. First, for high temperatures, when $T \gg \Theta_E$ (or the thermal energy is much greater than the relevant energy scale of the theory, $k_BT \gg \hbar\omega$), then we can perform a Taylor expansion of the exponential to easily see that

$$c_V \to 3R, \tag{9.17}$$

which is the classical result. Additionally, in the low temperature limit, or $T \ll \Theta_E$ ($k_BT \ll \hbar\omega$), then this vanishes exponentially, or

$$c_V \sim 3R\left(\frac{\Theta_E}{T}\right)^2 e^{-\Theta_E/T} \to 0. \tag{9.18}$$

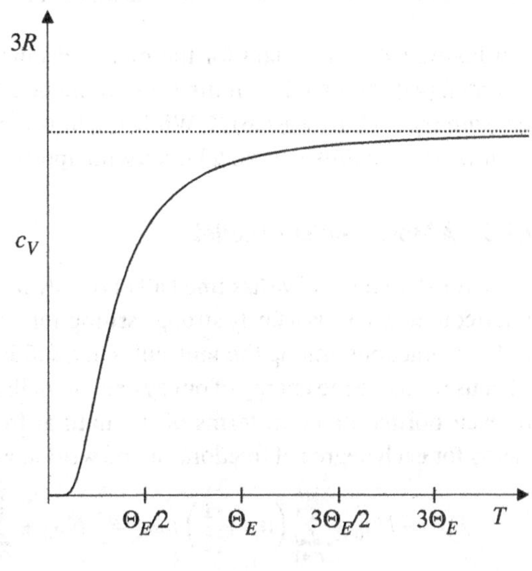

Figure 9.3 The molar specific heat as a function of temperature from the Einstein Model (solid curve) compared with the constant classical result (dotted line).

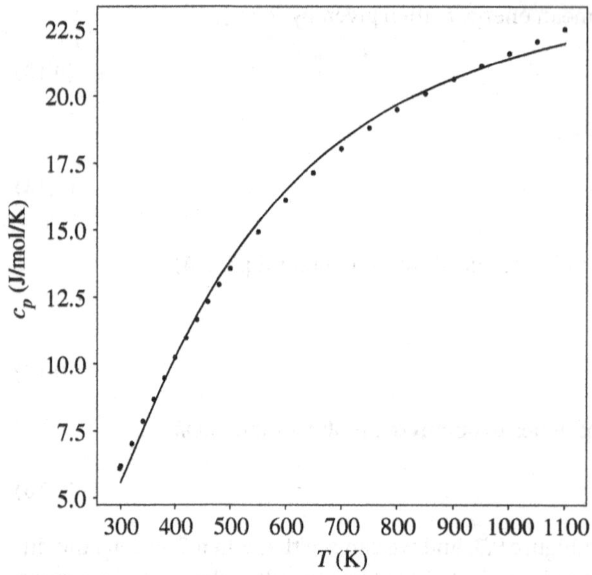

Figure 9.4 Experimental data for the specific heat of diamond from Ref. Adapted from [4] compared with the Einstein model.

While this result is consistent with the third law of thermodynamics, the specific heat vanishes too rapidly compared with what is seen experimentally. It is found that generally for small temperature,

$$c_V \propto T^3, \tag{9.19}$$

so it still vanishes, but not exponentially. This disagreement is due to our treatment of all the normal modes oscillating at the same natural frequency, which is not generally true. It is however a decent first approximation, and at the very least it agrees with both the classical result and the third law of thermodynamics in the appropriate limits. In Section 9.2.3 we will study a more realistic model which gives the correct zero-temperature behavior.

Exercise 9.4 Perform the expansions that lead to Eqs. (9.17) and (9.18).

In Figure 9.4 I show data for the molar specific heat at constant pressure for diamond (taken from Ref. [4]) compared with the Einstein model. The authors of Ref. [5] fit Eq. (9.16) to this data to obtain $\Theta_E = (1364.9 \pm 6.6)$ K. While the fit is not perfect, we can see it matches qualitatively, so again we see the Einstein model is a useful approximation in this range of temperatures.

9.2.3 A More Realistic Model

In the Einstein model, we assumed all of the $3N$ normal modes of our oscillators had the same natural frequency. This is a fairly strong assumption, useful to perform a calculation, but not realistic, as the interactions among the molecules in a solid are much more complicated.

Let us return to the energy of our system of oscillators in Eq. (9.8), and write the spring constants for each normal mode in terms of the natural frequency, $k_\ell = m\omega_\ell^2$. The quantum mechanical energy for each degree of freedom can be written as

$$E_r = -Nu_0' + \sum_{\ell=1}^{3N} \left(n_\ell + \frac{1}{2}\right) \hbar\omega_\ell = -Nu_0 + \sum_{\ell=1}^{3N} n_\ell \hbar\omega_\ell, \tag{9.20}$$

where the first term, $-Nu'_0$, is a (constant) potential energy shift that arises when making the transformation to the normal mode coordinates. This term is negative because it represents a binding energy that tends to hold the molecules in the solid together. In the second equality I have absorbed the zero-point energy $(1/2)\sum_\ell \hbar\omega_\ell$ into u_0 as it does not depend upon the quantum state.

Now the partition function is easy to calculate,[4]

$$\begin{aligned} Z &= \sum_r e^{-\beta E_r} \\ &= e^{N\beta u_0} \sum_r \exp\left(-\beta \sum_{\ell=1}^{3N} n_\ell \hbar\omega_\ell\right) \\ &= e^{N\beta u_0} \left(\sum_{n_1} e^{-\beta n_1 \hbar\omega_1}\right)\left(\sum_{n_2} e^{-\beta n_2 \hbar\omega_2}\right)\cdots\left(\sum_{n_{3N}} e^{-\beta n_{3N} \hbar\omega_{3N}}\right) \\ &= e^{N\beta u_0}\left(\frac{1}{1-e^{-\beta\hbar\omega_1}}\right)\left(\frac{1}{1-e^{-\beta\hbar\omega_2}}\right)\cdots\left(\frac{1}{1-e^{-\beta\hbar\omega_{3N}}}\right) \end{aligned}$$

so we can write the logarithm of the partition function as

$$\ln Z = N\beta u_0 - \sum_{\ell=1}^{3N} \ln(1 - e^{-\beta\hbar\omega_\ell}). \tag{9.21}$$

To make any more headway, we need to know the frequencies of the normal modes, which requires some additional information. Without losing too much generality, if the temperature is not too low, we can assume (from our classical knowledge) that the energy levels are closely spaced, so $\beta\hbar\omega_\ell \ll 1$. Let's define $\sigma(\omega)d\omega$ as the number of modes with angular frequency in the range ω to $\omega + d\omega$. Then, following our usual method of turning a finite sum into an integral,

$$\ln Z = N\beta\eta - \int_0^\infty \ln(1 - e^{-\beta\hbar\omega})\sigma(\omega)d\omega. \tag{9.22}$$

From this we can calculate the mean energy and specific heat in terms of $\sigma(\omega)$. We get

$$\overline{E} = \hbar\omega \int_0^\infty \frac{e^{-\beta\hbar\omega}}{1 - e^{-\beta\hbar\omega}} \sigma(\omega)d\omega - N\eta \tag{9.23}$$

and

$$C_V = k_B(\beta\hbar)^2 \int_0^\infty \frac{\omega^2\sigma(\omega)e^{-\beta\hbar\omega}}{\left(1 - e^{-\beta\hbar\omega}\right)^2} d\omega. \tag{9.24}$$

We are far from finished, as we still need to determine $\sigma(\omega)$, but this is something we can attempt to model. For any model, we must constrain σ such that it is normalized by the number of normal modes, in other words, σ must satisfy

$$\int_0^\infty \sigma(\omega)d\omega = 3N. \tag{9.25}$$

Much of the research of solids, at least within the context of their specific heat, relies on a detailed study of this function (or similar such functions).[5]

Exercise 9.5 Derive Eqs. (9.23) and (9.24) from Eq. (9.21).

---

4 As we are looking to calculate the mean energy and not the entropy, we do not include the fudge factor.

5 But also keep in mind that the procedure taken here is common in physics: When we cannot solve a problem, we find a way to write it in terms of a function that can hopefully be modeled. That is, we parameterize our ignorance in a way that can simplify the problem.

9.2.4 The Debye Model

Peter Debye (1884–1966) modeled the normal modes as quantized (massless) sound waves known as **phonons**. These can be treated similar to photons (the quanta of light, which we will study later), but they move with the speed of sound v_s through the solid. As we will see in Chapter 12 (in the context of photons), we can determine the density of states for phonons, assuming we restrict our system to a volume V, to be

$$\sigma(\omega)d\omega = \frac{3V}{2\pi^2 v_s^3}\omega^2 d\omega. \tag{9.26}$$

In addition to this expression, we will insist there exists a maximum frequency, the **Debye frequency** ω_D, above which $\sigma = 0$. We generally expect most physical systems to readily allow low frequency oscillations, but the higher frequency oscillations (which also correspond to higher energies) tend to be less common. In a sense, this can be understood with the canonical ensemble: Given the Boltzmann factor, $e^{-\beta E_r}$, it is less likely in general for a system to be in a higher energy state. This frequency can be found by integrating Eq. (9.25) from zero to infinity to be

$$\omega_D = v_s\left(6\pi^2\frac{N}{V}\right)^{1/3}. \tag{9.27}$$

Exercise 9.6 Use Eq. (9.25) with Eq. (9.26) to derive Eq. (9.27).

Using Eq. (9.24), with a little effort (see Problem 9.8) we can calculate the molar specific heat

$$c_V = 3R\left[3\left(\frac{T}{\Theta_D}\right)^3\int_0^{\Theta_D/T}\frac{y^4 e^y}{(e^y-1)^2}dy\right] = 3Rf_D\left(\frac{\Theta_D}{T}\right), \tag{9.28}$$

where we have defined the **Debye temperature**,

$$\Theta_D = \frac{\hbar\omega_D}{k_B} = \frac{\hbar v_s}{k_B}\left(6\pi^2\frac{N}{V}\right)^{1/3}, \tag{9.29}$$

and the **Debye function**

$$f_D(x) = \frac{3}{x^3}\int_0^x \frac{y^4 e^y}{(e^y-1)^2}dy. \tag{9.30}$$

This result for c_V is shown in Figure 9.5, and we can see from this figure that it both vanishes at low temperature and approaches the classical result (the horizontal dotted line) for high temperatures.

Again let us consider the extreme (high and low) temperature limits. For high temperature, the upper limit in the integral for f_D goes to zero, so the integration variable y is always small. Expanding the integrand in Eq. (9.30) gives us

$$f_D(x) \approx \frac{3}{x^3}\int_0^x y^2 e^y dy = 1 + \mathcal{O}(x^3).$$

Thus the Debye expression for c_V approaches $3R$, the classical result, in this limit.

For low temperature, the upper limit of the integral approaches infinity, and as such we can evaluate the integral exactly (either by using contour integration techniques, or looking it up in a table, as this is a well-known integral) to obtain

$$f_D(x) \sim \frac{3}{x^3}\frac{4\pi^4}{15}. \tag{9.31}$$

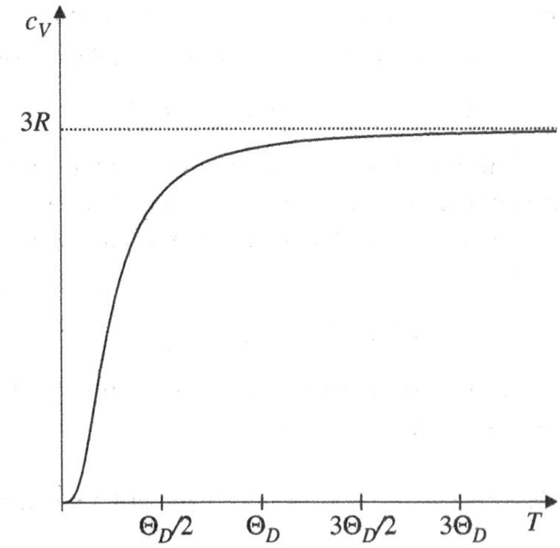

Figure 9.5 The molar specific heat as a function of temperature in the Debye model compared with the classical result (the horizontal dotted line).

We then see that c_V vanishes like T^3 as we stated earlier,

$$c_V \sim \left(\frac{12\pi^4 R}{5\Theta_D^3}\right) T^3. \tag{9.32}$$

The Debye model is a marked improvement on the Einstein model and works quite well for many solids. In fact, the Debye temperature is still a quantity that is measured and tabulated for many materials (see, for example, page 4–119 of Ref. [2]). To improve upon this model, a truly quantum mechanical calculation is required to include the effects of the conduction electrons, which we will get to when we study the Fermi gas in Chapter 12.

9.3 Paramagnetism

We have studied spin-1/2 systems both classically and quantum mechanically in previous chapters. For many reasons this system is one of the most useful spin systems to study, but let's move on to systems with general spin. As usual, we will assume we have N molecules in a volume V, and that the kinetic degrees of freedom are not relevant to the system. This will give us insight into paramagnetism, an important physical phenomenon.

A general molecule has orbital angular momentum, **L**, along with spin angular momentum **S**, for a total angular momentum of

$$\mathbf{J} = \mathbf{L} + \mathbf{S}.$$

Orbital angular momentum, due to quantum mechanical boundary conditions, is quantized such that the total angular momentum squared can only have discrete values given by

$$L^2 = \ell(\ell+1)\hbar^2, \quad \ell = 0, 1, 2, \ldots.$$

The z-component of the angular momentum, for a given ℓ, is restricted to the values

$$L_z = m_\ell \hbar,$$

where m_ℓ ranges from $-\ell$ to ℓ in integer steps. Orbital angular momentum is precisely what we would expect from a quantum mechanical study of classical angular momentum. Spin angular momentum, however, is somewhat different, as it has no classical analog, and yet behaves similar to orbital angular momentum. That is, the restrictions on S^2 and S_z are the same for a spin-s particle,

$$S^2 = s(s+1)\hbar^2 \text{ and } S_z = m_s\hbar,$$

but this time s can take on integer *and* half-integer values: 0, 1/2, 1, ... The constraint on m_s is the same: It runs from $-s$ to $+s$ in integral steps.

When we add orbital and spin angular momentum together, **J** also satisfies the same relationships,

$$J^2 = j(j+1)\hbar^2 \text{ and } J_z = m_j\hbar,$$

and j can thus take on integral (if s is an integer) or half-integral (if s is a half-integer) values, and $m_j = -j, -j+1, \dots, j-1, j$.[6]

Exercise 9.7 Show that for a given j, there are $2j+1$ allowed values of m_j.

The intrinsic magnetic dipole moment operator for a molecule with charge q, mass m, and spin **J** is given by

$$\boldsymbol{\mu} = g\frac{q}{2m}\mathbf{J}, \tag{9.33}$$

where g is imaginatively called the **g-factor**. This corrects for the fact that the spin angular momentum contribution, while somewhat like angular momentum, does not contribute in the same way to the dipole moment. For a classical calculation of the magnetic moment of a rotating charge with angular momentum, we would obtain $g = 1$, but when considering the spin angular momentum contribution, $g \neq 1$ due to quantum mechanical effects that are non-trivial to calculate.

If we now place our spin in a magnetic field with magnitude H in the z-direction, the potential energy of the system is given by

$$U = -\boldsymbol{\mu}\cdot\mathbf{H} = -g\frac{q}{2m}\mathbf{J}\cdot\mathbf{H} = -\frac{gq\hbar}{2m}m_jH$$

or, defining $\mu_0 \equiv gq\hbar/(2m)$, we can write this as

$$U_{m_j} = -m_j\mu_0 H, \tag{9.34}$$

where m_j takes on the values above. As before, we assume the spins are independent of each other (or rather, we are considering the lowest order mean-field approximation for our system).

We are interested in calculating the average **magnetization** of this system, or the magnetic dipole moment per unit volume,

$$\overline{M}_z = \frac{N\overline{\mu}}{V}. \tag{9.35}$$

Because we treat the spins as independent, we just need to calculate the mean magnetic moment for a single spin $\overline{\mu}$, then we can multiply by N/V to obtain $\overline{M}_z$.

Let's begin by calculating the single-particle partition function. We have

$$\zeta = \sum_r e^{-\beta E_r} = \sum_{m_j=-j}^{j} e^{m_j\beta\mu_0 H}.$$

6 Note there is nothing special about the z-direction; we could choose any one direction to measure the angular momentum. One strange result from quantum mechanics is that we can *only* choose one. Once we know J_z we cannot know for certain J_x or J_y. This is not needed for a thermodynamic calculation, but it's rather interesting.

We can now change the integers we sum over, by defining $n = m_j + j$, so the sum now ranges from $n = 0$ to $n = 2j$, and

$$\zeta = \sum_{n=0}^{2j} e^{(n-j)\beta\mu_0 H} = e^{-j\beta\mu_0 H} \sum_{n=0}^{2j} \left(e^{\beta\mu_0 H}\right)^n. \tag{9.36}$$

The sum is easy to determine (see the box below),

$$\sum_{n=0}^{M} x^n = \frac{1 - x^{M+1}}{1 - x}, \tag{9.37}$$

and thus the single-particle partition function is

$$\zeta = e^{-j\beta\mu_0 H} \frac{1 - e^{(2j+1)\beta\mu_0 H}}{1 - e^{\beta\mu_0 H}}. \tag{9.38}$$

The sum in Eq. (9.37) can of course be looked up, but it's a good exercise to derive it, as a reminder of some simple mathematical tools. We define the sum on the left-hand side of Eq. (9.37) as S_M and then start with the geometric series

$$S_\infty \equiv \sum_{n=0}^{\infty} x^n = \frac{1}{1 - x},$$

which can be easily derived from a Taylor expansion of the right-hand side. We can break up the sum and take some simple steps to get to the result we want,

$$\begin{aligned} S_\infty &= \sum_{n=0}^{M} x^n + \sum_{n=M+1}^{\infty} x^n \\ &= \sum_{n=0}^{M} x^n + x^{M+1} \sum_{n=M+1}^{\infty} x^{n-M-1} \\ &= \sum_{n=0}^{M} x^n + x^{M+1} \sum_{n'=0}^{\infty} x^{n'} \\ &= S_M + x^{M+1} S_\infty. \end{aligned}$$

At this point, solving for S_M gives the desired result.

Exercise 9.8 Show that with some algebra (and remembering the definitions of hyperbolic functions), you can write Eq. (9.38) as

$$\zeta = \frac{\sinh\left[(j + 1/2)\,\beta\mu_0 H\right]}{\sinh\left(\beta\mu_0 H/2\right)}. \tag{9.39}$$

The mean magnetic moment per molecule is given by

$$\overline{\mu} = \frac{\sum_{m_j} m_j \mu_0 e^{m_j \beta\mu_0 H}}{\sum_{m_j} e^{m_j \beta\mu_0 H}}, \tag{9.40}$$

but instead of evaluating the numerator and denominator directly, remember we can usually write mean values as a derivative of the partition function. In this case, we can write

$$\overline{\mu} = \frac{1}{\beta} \frac{\partial \ln \zeta}{\partial H}. \tag{9.41}$$

Exercise 9.9 Show that Eq. (9.41) follows from Eq. (9.40). Then, evaluate the derivative to obtain

$$\overline{\mu} = \mu_0 \left\{ \left(j+\frac{1}{2}\right) \coth\left[\left(j+\frac{1}{2}\right)\frac{\mu_0 H}{k_B T}\right] - \frac{1}{2}\coth\left(\frac{\mu_0 H}{2k_B T}\right) \right\}. \tag{9.42}$$

It's standard to define the **Brillouin function**[7]

$$B_j(x) = \frac{1}{j}\left\{ \left(j+\frac{1}{2}\right)\coth\left[\left(j+\frac{1}{2}\right)x\right] - \frac{1}{2}\coth\left(\frac{x}{2}\right)\right\},$$

to write the mean magnetic moment as

$$\overline{\mu} = \mu_0 j B_j\left(\frac{\mu_0 H}{k_B T}\right), \tag{9.43}$$

and finally the mean magnetization is given by

$$\overline{M} = n\mu_0 j B_j\left(\frac{\mu_0 H}{k_B T}\right), \tag{9.44}$$

where $n = N/V$ as usual.

Next we want to investigate the high and low temperature limits of the magnetization, where our two expectations should hold: As our system reaches absolute zero, the system should go to its ground state, and at high temperatures, the spins are all equally likely to be in any of their $2j+1$ possible states. To do this we'll look at the limits of the Brillouin function, where the high temperature limit corresponds to $x \ll 1$ and the low temperature limit corresponds to $x \gg 1$. For the latter, we can approximate $\coth x$ as

$$\coth x = \frac{e^x + e^{-x}}{e^x - e^{-x}} \approx 1,$$

as the e^x terms in the numerator and denominator overwhelm the e^{-x} terms. Thus for large x we have

$$B_j(x) \to \frac{1}{j}\left[\left(j+\frac{1}{2}\right) - \frac{1}{2}\right] = 1.$$

In this limit, at low temperature the magnetization becomes

$$\overline{M}_z \to n\mu_0 j, \tag{9.45}$$

so the spins are all "frozen" in the ground state, as expected.

Exercise 9.10 Expand $\coth(x)$ to show that in the high temperature limit, or small x,

$$\coth(x) = \frac{1}{x} + \frac{x}{3} + \mathcal{O}(x^3),$$

so that with some algebra, the Brillouin function becomes

$$B_j(x \ll 1) = \frac{j+1}{3}x. \tag{9.46}$$

For high temperatures ($k_B T \gg \mu_0 H$) we then obtain for the magnetization

$$\overline{M}_z \approx \frac{ng\mu_0^2 j(j+1)}{3k_B T}H. \tag{9.47}$$

7 Léon Brillouin (1889–1969).

$\overline{M}$ vanishes linearly with H, which is commonly the assumption in E&M for studying magnetic materials (linear materials are much simpler to study). The slope is called the **magnetic susceptibility,**

$$\chi_m \equiv \frac{n\mu_0^2 j(j+1)}{3k_B T}, \tag{9.48}$$

and that this is proportional to $1/T$ is known as **Curie's law**. This dependence on temperature also arises from a classical calculation of paramagnetic materials (see Problem 9.6); however, in that case the result is a factor of 3 smaller than the quantum result.

The Brillouin function is shown in Figure 9.6 for three different choices of j (1/2, 1, and 2), where the limiting cases are clear in the plot. For example, for small x, the function is linear in x and gives rise to the proper high-temperature limit, and as $x \to \infty$ we see $B_j(x)$ approaches one.

The expression we have for the magnetization is now valid for any spin system, but honestly nothing new came out of this calculation that wasn't *qualitatively* apparent in our original calculation for a spin-1/2 particle. We can use that calculation to check our result here, to ensure our expression in Eq. (9.44) reduces to that in Eq. (8.16) for $j = 1/2$. What is interesting here is that the effects here are valid for any type of spin, no matter the origin of the magnetic moment (orbital or spin angular momentum): The only difference comes in the magnitude. If the source of the paramagnetism is atomic in origin, then it generally comes from the outer electrons in the atom, so μ_0 has a scale of

$$\mu_0/g \equiv \mu_B = \frac{e}{2m_e}\hbar \approx 9.27 \times 10^{-24}\,\text{J/T},$$

which is known as the **Bohr magneton.**[8] If the paramagnetism is nuclear in origin, such as when the electrons in the atom are in a combined spin-0 state, then the relevant scale is given by the **nuclear magneton,**

$$\mu_0/g \equiv \mu_N = \frac{e}{2m_p}\hbar \approx 5.05 \times 10^{-27}\,\text{J/T}.$$

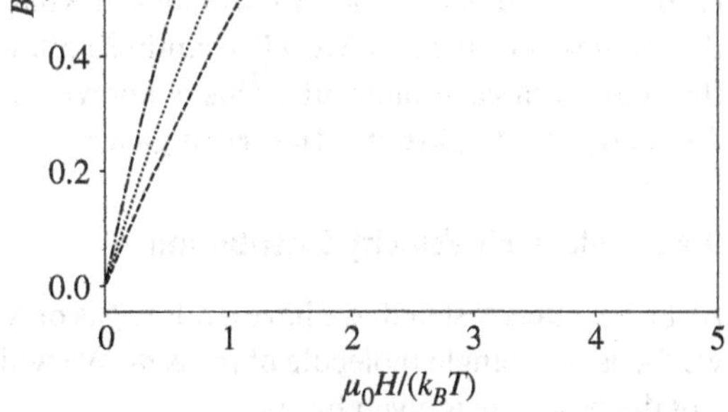

Figure 9.6 The Brillouin function for three different values of j.

8 Niels Bohr (1885–1962).

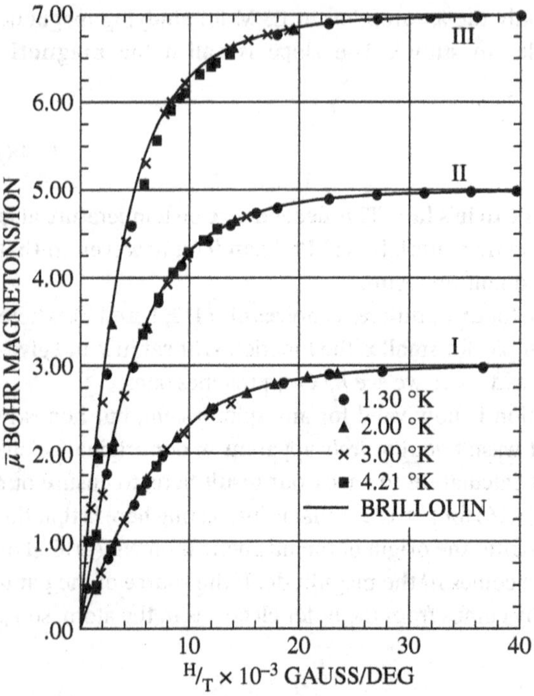

Figure 9.7 The mean magnetization per ion for three different substances (I = Potassium chromium, II = Iron ammonium, III = gadolinium sulfate octahydrate) as a function of H/T. The data points are experimental measurements and the lines are the Brillouin functions. Source: Reproduced from [6]/with permission of American Physical Society.

μ_N is roughly 2000 times smaller than μ_B, given the factor by which the proton mass is larger than the electron mass, so nuclear magnetic effects tend to be much weaker.

Figure 9.7 shows a plot of the mean magnetic moment per ion for three different substances—I = potassium chromium alum ($j = 3/2$), II = Iron ammonium alum ($j = 5/2$), III = gadolinium sulfate octahydrate ($j = 7/2$)—as a function of H/T, from Ref. [6]. The data are measurements at different (low) temperatures, and the curves are the Brillouin functions, which match the data quite well. While this reference is rather old (from 1952), it still shows a marvelous comparison of these results for different materials to the result of this section.

9.4 Introduction to Kinetic Theory

As a final application of the canonical ensemble, let's now revisit the classical ideal gas, but with a different approach. Instead of merely calculating the partition function to extract bulk properties (which we've already done), we will look at what the statistical approach can tell us about more of the microscopic features. We will examine a dilute gas and determine the probability of measuring the velocity of each molecule. This is known as the **Maxwell velocity distribution** and is the foundation for the **kinetic theory of gases**.

9.4.1 Maxwell Velocity Distribution

As I have already stated, we have a dilute gas of N molecules confined to a volume V, but this time we focus on a single molecule of mass m. We will treat the gas classically and ideal, so the energy ε of the molecule is given by

$$\varepsilon_r(\mathbf{v}) = \frac{1}{2}m\mathbf{v}^2 + \varepsilon_r^{(\text{int})},$$

where we include $\epsilon_r^{(\text{int})}$, the internal energy of the molecule; we do not insist that the gas be monatomic. This may be non-zero, but it only depends on the internal variables of the molecule (such as the relative positions and momenta of the atoms). Another change in our current treatment is that we have written the energy in terms of the velocity $\mathbf{v} = \mathbf{p}/m$, which is conventional.[9] This also helps avoid confusion between momentum and pressure as we perform calculations. We already omitted a potential energy term that could depend on $\mathbf{r}$, the position of the molecule, as we are treating this as an ideal gas.

As usual, the classical treatment is valid if the temperature is not too low, and the gas is dilute enough (we will quantify this more in Chapter 11). We are interested in the probability of the particle being at a position between $\mathbf{r}$ and $\mathbf{r} + d\mathbf{r}$ (in a box of volume $d^3\mathbf{r}$) and with velocity between $\mathbf{v}$ and $\mathbf{v} + d\mathbf{v}$ (in a "box" of "volume" $d^3\mathbf{v}$ in velocity space). We treat our special molecule as being in contact with the other particles as if they formed a heat reservoir at temperature T. Then we know that the probability that we are in this region in an internal state r is given by the canonical distribution, given by

$$\mathcal{P}_r(\mathbf{r}, \mathbf{v}) d^3\mathbf{r} d^3\mathbf{v} \propto e^{-\beta(mv^2/2 + \epsilon_r^{(\text{int})})} d^3\mathbf{r} d^3\mathbf{v}. \tag{9.49}$$

As this is the probability that our molecule is in the given state, $\mathcal{P}_r$ is a **probability density**, which we saw in Section 2.4: In this case, it is the probability per volume per velocity-cubed.

> The expression in Eq. (9.49) comes directly from the canonical distribution, looking at Eq. (8.33) for the partition function for the ideal gas. This is the continuous limit of the sum $\sum_r P_r = \sum_r e^{-\beta E_r}$, so Eq. (8.33) is the integral over the probabilities, which then has the form
>
> $$P_r = \frac{e^{-\beta E_r}}{h_0^{3N}} d^3\mathbf{r}_1 \ldots d^3\mathbf{r}_N d^3\mathbf{p}_1 \ldots d^3\mathbf{p}_N,$$
>
> and thus if we consider only a single molecule, we get Eq. (9.49) (after substituting the velocity for the momentum).

Usually in such a setup we can't determine the specific internal state r—if we are trying to detect this molecule, we just want to know if its position and velocity are what we expect (the internal state is usually unimportant). Thus, in order to allow for any possible internal state, we will sum over them, to obtain

$$\mathcal{P}(\mathbf{r}, \mathbf{v}) d^3\mathbf{r} d^3\mathbf{v} \equiv \sum_r \mathcal{P}_r(\mathbf{r}, \mathbf{v}) d^3\mathbf{r} d^3\mathbf{v} \propto e^{-\beta mv^2/2} d^3\mathbf{r} d^3\mathbf{v}. \tag{9.50}$$

This is nothing new, as it is merely the Boltzmann distribution for an ideal gas, but what is different now is that we will use this expression in this form to study our system.

If we multiply this probability by N, we saw in Exercise 2.6 that it becomes the average number of molecules with position in the range $\mathbf{r}$ and $\mathbf{r} + d\mathbf{r}$, and velocity between $\mathbf{v}$ and $\mathbf{v} + d\mathbf{v}$, which we define as

$$f(\mathbf{r}, \mathbf{v}) d^3\mathbf{r} d^3\mathbf{v} \equiv N\mathcal{P}(\mathbf{r}, \mathbf{v}) d^3\mathbf{r} d^3\mathbf{v} = Ce^{-\beta mv^2/2} d^3\mathbf{r} d^3\mathbf{v}. \tag{9.51}$$

C is just a normalization constant, which we will determine by requiring

$$\int f(\mathbf{r}, \mathbf{v}) d^3\mathbf{r} d^3\mathbf{v} = N. \tag{9.52}$$

9 It is a *velocity* distribution, after all.

Exercise 9.11 Evaluate the integral in Eq. (9.52) to show that

$$C = \frac{N}{V}\left(\frac{\beta m}{2\pi}\right)^{3/2} = n\left(\frac{\beta m}{2\pi}\right)^{3/2}, \tag{9.53}$$

with as usual $n = N/V$.

With the result of Exercise 9.11, we can write

$$f(\mathbf{v})d^3\mathbf{r}d^3\mathbf{v} = n\left(\frac{m}{2\pi k_B T}\right)^{3/2} e^{-mv^2/(2k_BT)}d^3\mathbf{r}d^3\mathbf{v}. \tag{9.54}$$

This is often what is referred to when one mentions the Maxwell velocity distribution. We see that it does not depend on the position $\mathbf{r}$ (which makes sense for an ideal gas), and only depends on the magnitude of the velocity, $v = |\mathbf{v}|$. For a dilute gas, were we to count the number of molecules with a given velocity, this tells us on average how many we should find.

The expression in Eq. (9.54) is often more than we need. For one, perhaps we are only interested in the distribution of velocities in one direction, say the x-direction. Imagine we are trying to determine the likelihood our molecules will deposit on a particular surface in the yz-plane, so the only relevant velocity component is v_x. Then we could integrate Eq. (9.54) over v_y and v_z to obtain the average number density of molecules with an x-component of the velocity between v_x and $v_x + dv_x$,

$$g(v_x)dv_x = \int_{v_y=-\infty}^{\infty}\int_{v_z=-\infty}^{\infty} f(\mathbf{v})d^3\mathbf{v} = n\left(\frac{m}{2\pi k_B T}\right)^{1/2} e^{-mv_x^2/(2k_BT)}dv_x. \tag{9.55}$$

Note this is properly normalized, as $g(v_x)dv_x$ gives the average number density of molecules, we would require $\int g(v_x)dv_x = n$. We can define similar distributions for the other directions, $g(v_y)dv_y$ and $g(v_z)dv_z$.

Exercise 9.12 Show, using Eq. (9.55), that $\overline{v}_x = 0$. Do this first by considering the symmetry of the function you are integrating, and then verify the result by explicitly performing the integral.

Exercise 9.13 Show, using Eq. (9.55), that

$$\overline{v_x^2} = \frac{k_B T}{m}. \tag{9.56}$$

To do this, use the results in Appendix C.

The result in Eq. (9.56) is not surprising, because we are studying a classical system. The equipartition theorem lets us determine immediately that the average kinetic energy is

$$\overline{\frac{1}{2}mv_x^2} = \frac{1}{2}k_B T.$$

The same is true for the y- and z-directions, so we can define the **root-mean-square velocity**,

$$v_{\text{rms}} = \sqrt{\overline{(v_x^2 + v_y^2 + v_z^2)}} = \sqrt{\frac{3k_B T}{m}}, \tag{9.57}$$

which is a useful measure of the speeds we would expect in such a gas. "rms" quantities are used commonly in problems when the average value of the quantity itself vanishes (as $\overline{v}_x$). We can see that v_{rms} increases with temperature which makes sense given the increased energy, while it decreases with the mass (heavier molecules will travel slower). In fact, any speed or velocity must scale like

$$\sqrt{\frac{k_B T}{m}}$$

just from dimensional analysis. The $\sqrt{3}$ in Eq. (9.57) only comes from the explicit calculation.

As Eq. (9.54) is independent of the direction of the velocity, it is often common to use the **Maxwell speed distribution**, denoted as $F(v)dv$: the mean number of molecules per unit volume with a speed $v \equiv |\mathbf{v}|$ in the range v and $v + dv$. We arrive at this by writing $d^3\mathbf{v}$ in spherical coordinates and then integrating $f(\mathbf{v})$ over the angles to get

$$F(v)dv = \int_{\theta=0}^{\pi} \int_{\phi=0}^{2\pi} f(\mathbf{v})v^2 dv \sin\theta d\theta d\phi.$$

With no angular dependence in $f(\mathbf{v})$, the angular integrals simply give a factor of 4π, and so

$$F(v)dv = 4\pi n \left(\frac{m}{2\pi k_B T}\right)^{3/2} e^{-mv^2/2k_B T} v^2 dv. \tag{9.58}$$

This is shown graphically (in two dimensions) in Figure 9.8, where we show the velocity vector $\mathbf{v}$ which can point in any direction (a result of the lack of angular dependence in f). The shaded region shows the allowed values of v (which would be a thin spherical shell in three dimensions). Even though the direction of the velocity doesn't matter, there is still a restriction on v_x and v_y (and v_z if we included a third dimension) because we restrict the speed to be in a range v to $v + dv$.

The average speed is easiest to calculate with the Maxwell speed distribution. Unlike $\bar{v}_x$, it is not zero as v is never negative and thus gives another useful quantity which helps us describe the distribution. This is given by

$$\bar{v} = \int_0^{\infty} vF(v)dv = \sqrt{\frac{8k_B T}{\pi m}}. \tag{9.59}$$

Note that $\bar{v}$ scales as expected, with the explicit integral giving the factor of $\sqrt{8/\pi}$.

Exercise 9.14 Calculate the integral in Eq. (9.59) to verify Eq. (9.59).

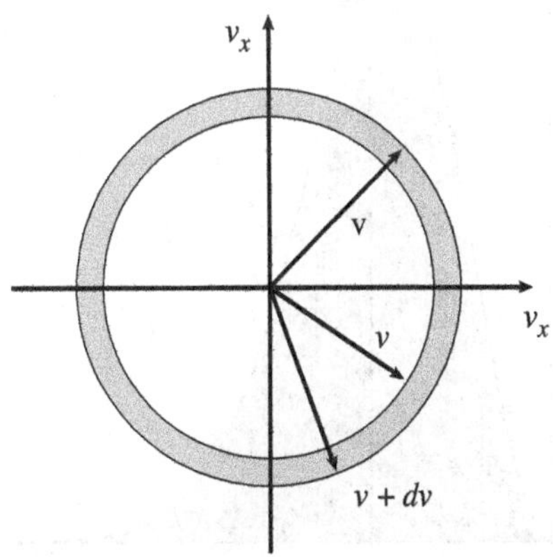

Figure 9.8 The allowed values of v_x and v_y for a classical particle with speed v.

Finally, what is the *most likely* speed that we would measure? This comes from setting the derivative of $F(v)$ to zero, and we would get

$$\tilde{v} = \sqrt{\frac{2k_BT}{m}}. \tag{9.60}$$

Exercise 9.15 Calculate $\tilde{v}$ from $F(v)$ and show that it is indeed given by Eq. (9.60).

Each of these quantities, v_{rms}, $\bar{v}$, and $\tilde{v}$, are all useful in their own ways, and are shown in Figure 9.9.

Computer Exercise 9.1 You can visit the Maxwell Velocity Distribution section of the companion site to simulate a classical ideal gas using a Jupyter notebook, as was used to produce Figure 9.10. The simulation calculates the theoretical value of $\bar{v}$ and computes it from a numerical simulation (these differ by 0.09% in this case, hence the fact they are indistinguishable). You can

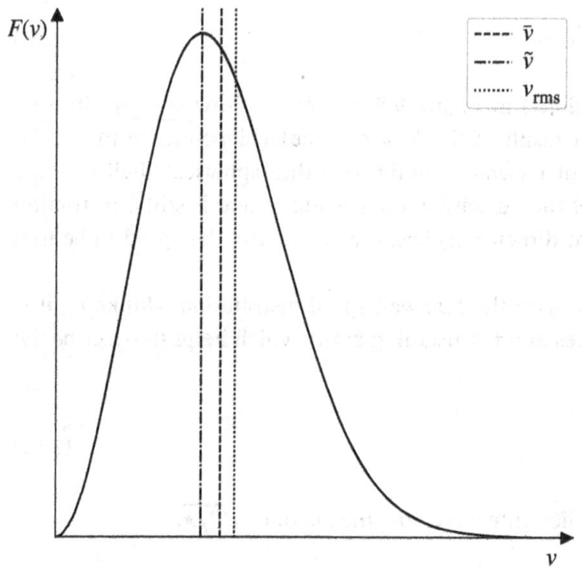

Figure 9.9 The Maxwell speed distribution with the three relevant speeds shown as vertical lines.

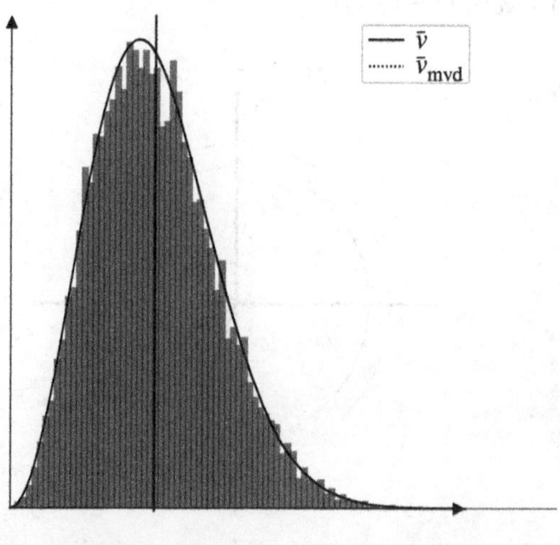

Figure 9.10 Results for $F(v)$ from a simulation of a system of $N = 10^4$ molecules. In this figure we show a histogram counting the number of molecules with given speeds along with Eq. (9.58). Additionally the theoretical average speed and the numerically determined average speed are shown (they differ by 0.09% so cannot be distinguished!).

consider various sizes of systems to see how increasing N gives better agreement between the theory and "experiment."

This distribution has been verified as far back as 1927 [7], and is still used in modern applications. For example, if you have a gas which is passing through a tube, but only want certain velocities of the molecules to continue passing through it, you can use a device known as a **velocity selector** (see the video at Ref. [8]). Such a device has blades that rotate with a particular speed such that only allow molecules with that speed can pass through it. You can also use this to verify the distribution: For each velocity you select, you can count the number of molecules in the gas you detect with that speed to see that it matches the Maxwell velocity distribution.

9.4.2 Molecules Striking a Surface

As mentioned at the beginning of this section, this is our first glimpse at the **kinetic theory of gases**. A detailed study of this field is beyond the scope of this textbook, as much of kinetic theory involves non-equilibrium processes. However, we can still use the tools at hand to look at a process known as **effusion**. This is the process by which molecules move through a very small opening—that is, an opening that is much smaller than the average distance the molecules travel.

To study this, we consider a small area dA on one of the surfaces of the box which contains our gas, as shown in Figure 9.11. We would like to determine how many molecules hit this area dA per unit time, or dN/dt. More commonly, we are interested in the **flux** of particles, or the number of molecules hitting the surface per unit area per unit time, $\Phi_0 = dN/(dtdA)$.

First we'll do a quick estimate of this quantity. A given molecule in our volume will have a velocity with probability given by the Maxwell velocity distribution. It will hit the area dA in a time dt only if it is a distance around $\bar{v}dt$ from the surface, where we will use the average speed as a measure of the velocity of the molecule. However, it also needs to be traveling in the correct direction—it can travel in the $\pm x$-, $\pm y$-, or $\pm z$-direction and we only want those traveling in the positive z-direction (given the coordinate system in Figure 9.11). With $n = N/V$ the number density of molecules in the entire container, then the number of molecules hitting the area dA is given by

$$dN = \frac{1}{6} n \left(\bar{v}dt\right) dA.$$

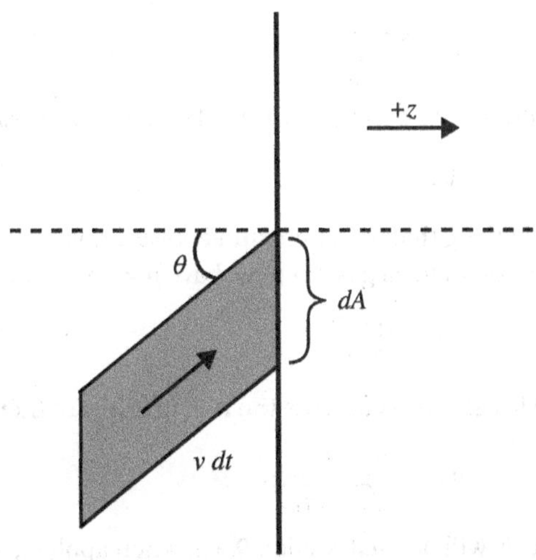

Figure 9.11 The volume of gas that we are considering when determining how many molecules hit a surface of area dA in a time dt.

The factor 1/6 comes from considering only those moving in one of the six possible directions in this rough approximation. In this case, the flux is given by

$$\Phi_0 = \frac{n\bar{v}}{6}. \tag{9.61}$$

A more exact calculation results is this expression with a slight change in the prefactor, from 1/6→1/4. While the exact calculation is not necessary for a qualitative understanding, we will still pursue it so that we can be ready to move beyond the simple approximations used in the Maxwell velocity distribution.

For the exact calculation, we consider the same area element dA of the wall of the container, with the $+z$-axis pointing to the right as in Figure 9.11, so dA lies in the xy-plane. The molecules in our gas have velocities in the range $\mathbf{v}$ and $\mathbf{v} + d\mathbf{v}$, and θ is the angle from the z-axis to the velocity vector which ranges from $-\pi/2 \to \pi/2$ (but we will restrict to the range $0 \to \pi/2$ to consider only those moving in the $+z$-direction), while the azimuthal angle ϕ ranges from 0 to 2π.

During a time dt, molecules will have a displacement of $\mathbf{v}dt$, and those we care about are at most a distance of $|\mathbf{v}dt|$ from the surface (in the figure, the shaded region shows where the particles with this velocity must be to hit the surface in the allowed time). The number of particles that hit the surface is given by the Maxwell velocity distribution,

$$dN = \left[f(\mathbf{v})d^3\mathbf{v}\right]\left(dAv_z dt\right) = \left[f(\mathbf{v})d^3\mathbf{v}\right]\left[dA(v\cos\theta)dt\right],$$

where $v_z = v\cos\theta$ is the component of the velocity in the z-direction. The flux is given by

$$\frac{dN}{dAdt} = f(\mathbf{v})v\cos\theta d^3\mathbf{v}.$$

The total flux of molecules hitting the surface is found by integrating over velocities which have a positive z-component,

$$\Phi_0 = \int_{v_z>0} d^3\mathbf{v} f(\mathbf{v}) v\cos\theta = \pi\int_0^\infty f(v)v^3 dv. \tag{9.62}$$

Exercise 9.16 Fill in the steps between the first and second equality in Eq. (9.62). Recall the restriction on θ to ensure $v_z > 0$.

We could substitute Eq. (9.54) in this equation directly, but let's evaluate this with a little thought. We know that the mean speed is given by

$$\bar{v} = \frac{4\pi}{n}\int_0^\infty f(v)v^3 dv,$$

and so comparing this to Eq. (9.62), we can easily see

$$\Phi_0 = \frac{n\bar{v}}{4}, \tag{9.63}$$

showing that our approximate result is very close to the exact result in Eq. (9.61), as mentioned.

For an ideal gas, then we have $\bar{p} = nk_BT$, and the mean speed is

$$\bar{v} = \sqrt{\frac{8k_BT}{\pi m}}.$$

This allows us to write the flux in terms of thermodynamic quantities,

$$\Phi_0 = \frac{\bar{p}}{\sqrt{2\pi m k_B T}}. \tag{9.64}$$

This will be used Section 9.4.3, when applying the Maxwell velocity distribution to effusion.

9.4.3 Effusion

To study effusion, we imagine we have a small slit in the area upon which we considered the molecules to strike, so that some number of particles will pass through the hole. One can create collimated beams of molecules this way, and it also works as a way to test the Maxwell velocity distribution for an ideal gas.

Using the results from the previous section, we consider a small (but no longer necessarily infinitesimal) hole with area A. The number of molecules passing through the hole per unit time is given by

$$\begin{aligned} A\Phi(\mathbf{v})d^3\mathbf{v} &\propto Af(v)v^3 dt \cos\theta dv\,(\sin\theta d\theta d\phi) \\ &\propto e^{-mv^2/(2k_BT)}v^3 dv\,(\sin\theta d\theta d\phi)\,. \end{aligned} \tag{9.65}$$

Systems which are allowed to interact via effusion have a new equilibrium condition that we have yet to see. Suppose we had a gas divided up into two sections of a container as in Figure 9.12, and there is a small slit between the two sides. On one side, the gas has density n_1, pressure $\overline{p}_1$, and temperature T_1; while on the other side, the gas has a density n_2, pressure $\overline{p}_2$, and temperature T_2. The divider is otherwise thermally insulating.

If the slit is large enough (much larger than the **mean free path**[10] of the molecules), then clearly $T_1 = T_2$ and $\overline{p}_1 = \overline{p}_2$ at equilibrium. In this case, a very large slit would really just be considered "no partition" between the two sides.

However, if the size of the slit is much smaller than the mean free path of the molecules, then the molecules cannot so easily move from one side to the other. Even so, every so often they are able to pass through the slit, and this seems as though it is no longer an equilibrium situation. However, as long as the flux of particles from side 1 to side 2 equals equal the flux of particles from side 2 to side 1, then we have a form of equilibrium. That is, we require

$$\Phi_1 = \Phi_2 \Rightarrow n_1\overline{v}_1 = n_2\overline{v}_2,$$

so using our expressions for the ideal gas, the equilibrium condition becomes

$$\frac{\overline{p}_1}{\sqrt{T_1}} = \frac{\overline{p}_2}{\sqrt{T_2}}. \tag{9.66}$$

The different condition for equilibrium comes from the fact that the system isn't static in the way we often think of this term. However, the system is in a **steady state**, so that at any time the densities on either side of the container remain the same. As an example, imagine a decorative

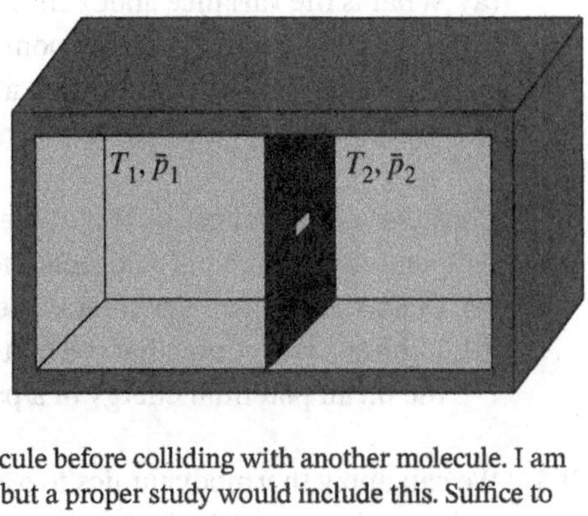

Figure 9.12 A container divided into two sections, with a small slit between the two, allowing gas to effuse from one side to the other until equilibrium is reached.

10 The mean free path is the average distance traveled by a molecule before colliding with another molecule. I am not discussing this in detail because of our ideal gas assumption, but a proper study would include this. Suffice to say, it is a useful distance scale for our problem.

fountain—water is continuously entering the fountain; however, to keep the amount of water constant in the pool, there must be a drain such that the flux of water entering the system must equal the flux of water leaving it.

Effusion commonly occurs in Nature. For example, balloons deflate over time because the gas inside effuses out, and the same can be said of bicycle or automobile tires. This tends to happen over a longer time scale (not seconds but days or even much longer for tires) because the "slits" involved are not actual holes, but merely the spacing of the molecules making up the barrier (balloon or tire).

9.5 Summary

- The canonical ensemble is useful to prove the equipartition theorem, which states that any classical degree of freedom that is separable and appears quadratically in the energy of a state contributes $\frac{1}{2}k_BT$ to the mean internal energy.
- A classical calculation of the specific heat of solids gives a reliable estimate for this quantity when solids are at room temperature, but does not obey the third law of thermodynamics.
- The Einstein model improves upon this by using the quantum mechanical energy levels, but with an approximation which considers all of the natural frequencies of the normal modes to be equal. This model, while it agrees with the classical result in the proper (high temperature) limit and the third law, does not have the proper zero-temperature behavior.
- Even better is the Debye model for specific heats, which treats the normal modes of oscillation as quantized sound waves, or phonons. This model agrees with the classical result and the third law, *and* has a low-temperature behavior which agrees quantitatively with experimental observations.
- Using the canonical ensemble, we can expand our study of paramagnetism to consider systems with arbitrary spins. This results in an expression for the magnetization which agrees extremely well with empirical observations.
- Finally we can study the basics of the kinetic theory of gases via the Maxwell velocity distribution. This still considers the system in a statistical manner but focuses on a single molecule at a time.

Problems

9.1 Do thermal fluctuations affect classical mass-spring systems? That is, consider a mass m connected to a spring with spring constant k at temperature T. Assume the equilibrium position of the mass is at $x = 0$.

(a) What is the variance about equilibrium $\overline{x^2}$ due to thermal fluctuations?

(b) If we consider a system at room temperature, where $k = 10$ N/m, calculate the size of these fluctuations. Would this affect your measurements if the mass oscillates with a maximum displacement of 10 cm?

9.2 Consider a classical gas of N particles with mass m in a cubical box of volume L^3 at a fixed temperature T. Each particle feels a force of gravity mg pointing in the negative z-direction.

(a) Calculate the average kinetic energy of each particle.

(b) If the box is set up so that the bottom of the box is at $z = 0$ and the top at $z = L$, determine the mean potential energy of a particle.

9.3 We can apply thermodynamics to an atom in a harmonic trap with a radial force $\mathbf{F} = -br\hat{r}$ acting on it where b is the "spring constant." A change in the internal energy arises when we change the value of b, so that the first law has the form

$$d\overline{E} = TdS + \overline{\mathcal{A}}db,$$

where $\overline{\mathcal{A}}$ is the generalized force conjugate to the external parameter b, which we can control explicitly. This is discussed nicely in Ref. [9].

(a) Given the definition of the Helmholtz free energy, write down dF. How do you evaluate $\overline{\mathcal{A}}$ in terms of the partial derivative of F? (Be sure to note which variables must be held constant!)

(b) One can show that semi-classically, the single-particle partition function can be approximated by

$$\zeta = \int_0^{\infty} \frac{\epsilon^2}{2(\hbar\omega)^3} e^{-\beta\epsilon} d\epsilon$$

where the factor in front of the exponential is the density of states coming from the degeneracy of the quantum energy levels. Evaluate this integral to find ζ in terms of T, b, and constants.

(c) Using part (b), evaluate the Helmholtz free energy for a system of N identical atoms in such a trap.

(d) Using parts (a) and (c), find the equation of state for this system. That is, evaluate $\overline{\mathcal{A}}$ as a function of b and other thermodynamic variables.

(e) Calculate the energy as a function of T. Does your result agree with the equipartition theorem? Should it?

9.4 One can estimate the Einstein temperature of a solid through a simple process, using the fact that you can relate the effective spring constant of the "springs" in a solid k to the isothermal compressibility κ by

$$k = \frac{a}{\kappa},$$

where a is the average spacing between atoms in the solid (assume a cubic structure). Use this to estimate the Einstein temperature for the solids in the following table (you will need the molar masses, which you can obtain in Appendix B).

Solid	Density (g/cm^3)	Compressibility (Pa^{-1})
Copper	8.96	7.25×10^{-12}
Lead	11.3	2.33×10^{-11}
Silver	10.5	9.93×10^{-12}

9.5 Consider a system with a very large number of degrees of freedom, $f \gg 1$. Assume that the number of states with energy between E and $E + dE$ is proportional to $E^f dE$. Now suppose that the system is in thermal contact with a heat reservoir at temperature $T = 1/(k_B\beta)$.

(a) Explain why the mean energy of the system is given by

$$\overline{E} = \frac{\int E^{f+1} e^{-\beta E} dE}{\int E^f e^{-\beta E} dE}.$$

(b) Define $Z = \int E^f e^{-\beta E} dE$ and show that

$$\overline{E} = -\frac{\partial}{\partial \beta} \ln Z.$$

(c) By changing integration variables in Z, factor out the β dependence of the integral, and then calculate $\overline{E}$ without having to do the integral explicitly.

9.6 Consider the paramagnetic system in Section 9.3 but as a classical system instead of quantum mechanically. Classically, the angular momentum and thus spin can point at any angle; the z-component of the angular momentum is a continuous variable, not discrete as in the quantum case. That means the energy of a spin in a magnetic field in the z-direction $\mathbf{H} = H\hat{z}$ is given by

$$U = -\mu_0 H \cos\theta.$$

Given that our degree of freedom (θ) is continuous, we have to convert our sum over states for the single-particle partition function

$$\zeta = \sum_r e^{-\beta U}$$

to an integral, as well as any other sums that are needed to calculate mean values.

(a) Show that you can write the mean dipole moment of a molecule as

$$\overline{\mu} = \frac{\int_0^{\pi} \mu_0 \cos\theta e^{\beta\mu_0 H\cos\theta} \sin\theta d\theta}{\int_0^{\pi} e^{\beta\mu_0 H\cos\theta} \sin\theta d\theta} \tag{9.67}$$

(b) Evaluate this explicitly, and determine the mean magnetization $\overline{M}_z = N\overline{\mu}/V = n\overline{\mu}$.

(c) Show that the high and low temperature limits of your expression from the previous part to be what one expects. Specifically, in the low temperature case, show that the system obeys Curie's Law. Classically, what is the magnetic susceptibility?

9.7 Let's return to the thermodynamics of a rubber band, but this time consider it to be fastened to a table with a mass M attached to it so it hangs vertically (the acceleration due to gravity is g). Assume the band consists of N linked polymers each with a length a which can be oriented either up or down (as in the figure in Problem 6.2, rotated 90°). Determine the mean length $\overline{\ell}$ of the rubber band as a function of M and the other variables in the problem. (Neglect the kinetic energies or weights of the polymers themselves, or any interaction between the links.)

[The tension in the rubber band comes from the thermal fluctuations in the orientation of the segments. It's one of the amazing consequences of statistical mechanics that a rubber band has a tension even in the absence of any attractive force between the segments.]

9.8 Starting from Eq. (9.24), fill in the steps (using the Debye model assumptions) to calculate Eq. (9.28).

9.9 If you were thinking carefully about the integrals we performed for the Maxwell velocity distribution, you might have thought there was something amiss. We performed the integrals over all values of the speed, but we know that special relativity forbids the velocity of a massive particle to ever reach or exceed the speed of light. Let's consider the expression in Eq. (9.58) and estimate the contribution speeds greater than the speed of light c make to the integral.

(a) First estimate the exponent $-\beta mv^2/2$ for an oxygen molecule with $v = c$ at room temperature.

(b) To calculate the probability that $v \geq c$, we need to evaluate the integral

$$P(v > c) = \frac{1}{n}\int_c^\infty F(v)dv.$$

This cannot be evaluated exactly, but we can estimate it by setting one of the v's in the integrand to c, that is, $v^2 e^{-\beta mv^2/2} \to cve^{-\beta mv^2/2}$. Substitute this into our expression for $F(v)$ and show that you obtain for oxygen gas at room temperature,

$$P(v > c) \sim e^{-10^{12}}.$$

Thus it's safe to include these contributions to the integral, as this probability can safely be considered zero.

9.10 Suppose you have an enclosure containing a gas of atoms which can be described by the Maxwell velocity distribution, each with mass m at absolute temperature T. The atoms emit light which passes (in the z-direction) through a window of the enclosure and can then be observed as a spectral line in a spectroscope. A stationary atom would emit light at a single, well-defined, frequency ν_0, but, because of the Doppler effect, the frequency of the light observed from an atom having a z-component of velocity v_z is shifted to

$$\nu = \nu_0\sqrt{\frac{c + v_z}{c - v_z}},$$

where c is the speed of light. As such, the light at the spectroscope is characterized by some intensity distribution $I(\nu)d\nu$ specifying the fraction of light intensity lying in the frequency range between ν and $\nu + d\nu$.

(a) Show that for $v_z \ll c$, the frequency shift can be approximated as

$$\nu = \nu_0\left(1 + \frac{v_z}{c}\right),$$

This will greatly simplify the following calculations.

(b) Calculate the mean frequency $\overline{\nu}$ of the observed light.

(c) What is the root-mean-square frequency shift, $(\Delta\nu)_{\mathrm{rms}} = \sqrt{\overline{(\nu - \overline{\nu})^2}}$, of the observed light?

(d) Write down the relative intensity distribution $I(\nu)d\nu$ of the light observed in the spectroscope.

9.11 An ideal monatomic gas is in thermal equilibrium at room temperature T so that the velocity distribution is Maxwellian.

(a) If v denotes the speed of a molecule, calculate $\overline{(1/v)}$. Compare this to $1/\overline{v}$.

(b) Find the mean number of molecules per unit volume whose energy lies between ε and $\varepsilon + d\varepsilon$.

9.12 A thin-walled vessel of volume V, kept at constant temperature, contains a gas which slowly leaks out through a small hole of area A. The outside pressure is low enough that leakage back into the vessel is negligible.

(a) Find the time required for the pressure in the vessel to decrease to $1/e$ of its original value. Express your answer in terms of A, V, and the mean molecular speed $\bar{v}$.
(b) A spherical bulb 10 cm in radius is maintained at room temperature (300 K) except for one square centimeter which is kept at liquid nitrogen temperature (77 K). The bulb contains water vapor originally at a pressure of 100 Pa. Assuming that every water molecule striking the cold area condenses and sticks to the surface, estimate the time required for the pressure to decrease to 1 mPa.
Hint: Use part (a), as the condensation is due to the gas striking the small surface.

9.13 Calculate the following mean values for the ideal gas which satisfies the Maxwell velocity distribution. Your answers may contain any of the following variables: m, T, and k_B.
(a) $\overline{v_y}$
(b) $\overline{v_z^2}$
(c) $\overline{v^2 v_y}$
(d) $\overline{v_z^2 v_y}$
(e) $\overline{(v_z + a v_y)^2}$ (where a is a constant)
(f) $\overline{v_x^2 v_z^2}$

9.14 In a molecular beam experiment, the source is a tube containing hydrogen at a pressure $\bar{p}_s = 0.15$ mmHg and at a temperature of $T = 300$ K. In the tube wall is a slit 20 mm by 0.025 mm, opening into a highly evacuated region. Opposite the source slit and one meter away from it is a second detector slit parallel to the first and of the same size. This slit is in the wall of a small enclosure in which the pressure can be measured.
(a) How many hydrogen molecules leave the source slit per second?
(b) How many hydrogen molecules arrive at the detector slit per second?
(c) What is the pressure $\bar{p}_d$ in the detector chamber when a steady state has been reached so that $\bar{p}_d$ is independent of time?

9.15 We can use our study of effusion to estimate the validity of a common science fiction problem. The situation is as follows: A hole is punctured in the hull of a spaceship cabin and there are only minutes before the air runs out. Suppose the cabin is at temperature T, has a volume V, and the hole has an area A.
(a) Estimate how long it takes for half of the gas (treating it as ideal) to escape. Assume only one type of gas of mass m is escaping.
Hint: To do this, assume, unrealistically, that the temperature remains constant.
(b) If the cabin were initially at room temperature with volume 10^3 m^3, estimate the time it would take all of the oxygen gas to escape if the hole had an area of 0.05 m^2.

References

1 J. B. Marion and S. T. Thornton. *Classical Dynamics of Particles and Systems.* Cengage Learning, 5th edition, 2003.
2 J. R. Rumble. *CRC Handbook of Chemistry and Physics.* CRC Handbook of Chemistry and Physics. CRC Press, 2022.

3 E. Lagendijk. A simple determination of the Einstein temperature. *American Journal of Physics*, 68(10):961–962, 2000.

4 A. C. Victor. Heat capacity of diamond at high temperatures. *The Journal of Chemical Physics*, 36(7):1903–1911, 1962.

5 C. Z. Tan. Temperature-dependent refractive index and material dispersion of diamond. *Applied Physics A*, 128(5):412, Apr 2022.

6 W. E. Henry. Spin paramagnetism of Cr^{+++}, Fe^{+++}, and Gd^{+++} at liquid helium temperatures and in strong magnetic fields. *Physical Review*, 88:559–562, Nov 1952.

7 J. A. Eldridge. Experimental test of Maxwell's distribution law. *Physical Review*, 30:931–935, Dec 1927.

8 Institut Laue Langevin. Ill - sans instrument d11 - 1. velocity selector. URL https://www.youtube.com/watch?v=jjPg59qxH3k&t=5s. Accessed: 16 March 2023.

9 M. Ligare. Classical thermodynamics of particles in harmonic traps. *American Journal of Physics*, 78(8):815–819, 2010.

10

Phase Transitions and Chemical Equilibrium

Until now, we have generally discussed systems with only one type of substance and with a fixed number of molecules, and more importantly those that are in a single phase. In this chapter I would like to remove these restrictions, by not only allowing mixtures of different types of molecules, but also by allowing the number of molecules of each type to change. Additionally we have not yet described transitions between different phases, say, from liquid to gas, and we will take that up in this chapter. More importantly we will define more concretely what a phase is so as to understand what we mean when we use this term.

We will begin by discussing the general requirements for a system to be in equilibrium without focusing on the specific details of the microscopic system. This will allow us to keep our conditions general so as to be able to apply them to (pretty much) any thermodynamic system.

After finishing this chapter, you should be able to

- understand phases, the conditions for equilibrium, and the conditions for equilibrium between phases,
- determine how to obtain the properties of the phase transition region from the equation of state, and
- understand chemical equilibrium between general substances and specifically between ideal gases.

10.1 Introduction to Phases

To begin, we should first define what we mean by the **phase** of a substance. If I were to ask you to describe a liquid, solid, or gas, you could probably do so fairly easily, at least qualitatively. For example: You might say that a solid is rigid, while liquids and gases are not, and a gas expands to fill its space, while a liquid does not.[1] These descriptions are perfectly fine initially, but what we really want here is a *quantitative* way to describe these phases.

For one, we would prefer a more concrete description of what we mean by a phase even with these simple forms of matter. First, a rigorous definition of what a phase is will allow us to realize that there are not just the three standard phases of matter (or four, if you consider plasma, or ionized gas, as a separate phase from gas), but there are many other phases of pretty much any substance.

1 And I'm sure you can come up with other ideas.

Statistical Thermodynamics: An Information Theory Approach, First Edition. Christopher Aubin.

Companion website: www.wiley.com/go/Aubin/StatisticalThermodynamics

Usually these other phases are structural in nature (different forms of solid phases, for example), so they may not seem different, but they are. Additionally, the common phases correspond to how substances behave at different temperatures and pressures. Other materials will have different external parameters—such as the magnetic field of a spin system—and as such we need a theoretical way to study phases and the transitions between them more precisely.[2]

To begin to distinguish phases, let us consider how the different forms of a substance, specifically water, behave. The molar volumes of water in its three common phases (solid, liquid, and gas, respectively) are

$$v_s = 19.6\ \text{cm}^3/\text{mol},\ v_\ell = 18\ \text{cm}^3/\text{mol},\ \text{and}\ v_g = 2.24\times 10^4\ \text{cm}^3/\text{mol}. \tag{10.1}$$

Consider what this means: If you have a mole of ice and a mole of liquid water both at 0°C, the volume of the liquid water would be slightly less than that of the ice.[3] More drastically, consider 100°C, the boiling point of water—water vapor takes up over 1200 times the volume of the same amount of liquid water!

Additionally, the molar specific heats at constant pressure of water in its different phases are given at 0°C by

$$c_s = 2.09\ \text{J/g/K},\ c_\ell = 4.2\ \text{J/g/K},\ \text{and}\ c_g = 1.9\ \text{J/g/K}.$$

These are approximately constant for small temperature ranges, but they are notably different for these different phases.[4] More interestingly, the temperature dependences of the specific heats are quite different among the different phases. If we treat the gaseous form as ideal (which we will see is not a good idea for phase transitions), then the specific heat is constant, while the solid form could be treated with the Einstein model of Section 9.2.2 or the Debye model of Section 9.2.4, and either way it is not a constant expression.

This suggests that a phase of a substance is categorized by bulk properties that have not just different values but different dependences on thermodynamic variables. This is not a perfect definition but that won't matter to us, because as we will see, it will not always be clear what phase a substance is in. What we want to ensure is that these differences in bulk properties should come from the equation of state, as that should contain all of the information in the system. I also won't give an "official" definition of what a phase of matter is because we will see that it is not very helpful to understand the concept. Additionally, like hot vs. cold, we will find that in many cases we only care about the phase a substance is in when we compare it to another phase. As such, it will usually suffice to understand how to differentiate two (or more) different phases.

With this in mind, how do we determine when a system will transition from one phase to another? That is, we know that if we have liquid water, and we remove heat at a fixed pressure, then the temperature will decrease until it reaches 0°C, and then a **phase transition** occurs. As we continue to remove heat, the liquid *freezes* to change into the solid phase. How can we see this in the setup of the thermodynamic problem?

The key here is to understand the equilibrium of the two phases—at 0°C, water can coexist in equilibrium in different phases. As such, we should revisit our discussion of equilibrium to

2 Ferromagnets have two different phases as you saw in Problem 6.10: a magnetized phase at low temperatures and an unmagnetized at high temperatures.

3 A fact that we'll discuss in more detail later as this has profound implications for life as we know it on this planet.

4 You might be wondering why I am including the gaseous phase in this comparison at the freezing point of water. Remember that when considering thermodynamic variables we not only are interested in their value at a given temperature but at a given *pressure* as well—at very small (~0.006 atm) pressures, water can be in its gaseous form at 0°C.

then apply those ideas to this problem. We will do this while also discussing how to study small fluctuations a system can make away from equilibrium.

10.2 Equilibrium Conditions

In order to study systems in equilibrium, we will again start with the simplest: an isolated system (even though it is the least useful for our purposes), then move to a system held at constant temperature and pressure. The case of a system held at constant temperature only will be left to the problems, as it's an interesting problem but not a setup that we will use.

10.2.1 Isolated System

For a thermally isolated system in equilibrium, we know that the entropy of the system is an absolute maximum, S_{max}. If the system deviates from equilibrium, which it may due to thermal fluctuations, then the entropy must decrease, or[5]

$$\Delta S = S - S_{max} \leq 0. \tag{10.2}$$

This is shown in Figure 10.1, where I plot the entropy of an ideal two-state system as a function of the energy (although this could easily be the entropy of a system as a function of any parameter). Any fluctuations would necessarily cause the entropy to decrease if the system initially is at the point labeled S_{max}. This maximum in the entropy corresponds to a **stable equilibrium** state—fluctuations of the system cause the entropy to decrease, but then a spontaneous process occurs so as to return the system to equilibrium.

Other forms of equilibrium are allowed, and are shown in Figure 10.2, where I show the entropy as a function of some thermodynamic parameter x for a system. Again, the dot marks our stable

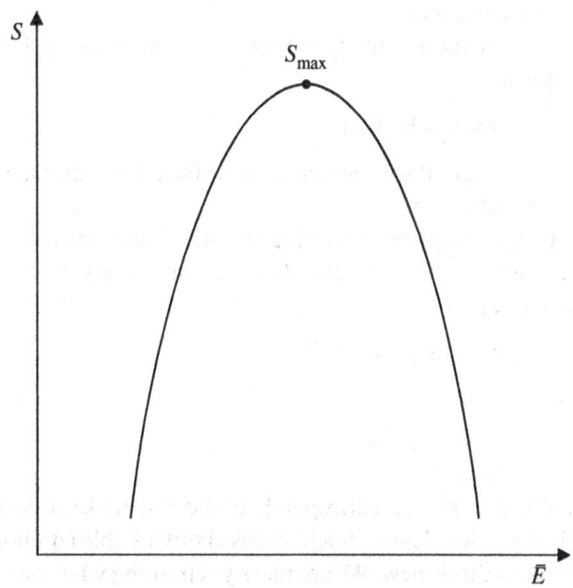

Figure 10.1 A plot of entropy vs. energy for a simple two-state system with the equilibrium state labeled with S_{max}.

5 Don't confuse this ΔS with the entropy change due to a spontaneous process; that must always increase.

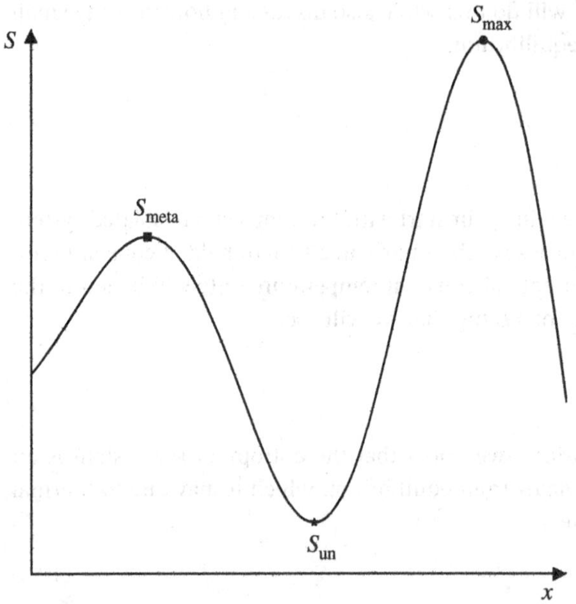

Figure 10.2 The entropy vs. some thermodynamic parameter for a system with a stable equilibrium state (circle), an unstable equilibrium state (star), and a metastable equilibrium state (square).

equilibrium state, but the star marks an **unstable equilibrium** point. At such a point, the entropy is *minimized*, and so $dS = 0$ holds, but $\partial S/\partial x$ is positive, so that any fluctuations away from such a state would move away from equilibrium (much like balancing a marble on the top of a hill: It may be at rest in equilibrium but any small deviation will cause the system to move away from this state).

Finally in this figure, I mark a **metastable equilibrium** point with a square—here the entropy is at a local, but not global, maximum. Small fluctuations would cause the system to return to this state; however, if the fluctuations became large enough, the system would transition to a truly stable equilibrium state.

For any process the heat flow into a thermally isolated system must be zero, and the first law gives us

$$Q = 0 = W + \Delta\bar{E},$$

or $W = -\Delta\bar{E}$. If we keep the volume (or other external parameters) fixed, no work is done, so that $\bar{E} = \text{constant}$.

Considering the underlying statistical state. If $\Omega(x)$ is the number of accessible states for some parameter x of the system, then the entropy is given by $S(x) = k_B \ln \Omega(x)$, and the probability of a given state is

$$P(x) \propto \Omega(x) = e^{S(x)/k_B}, \tag{10.3}$$

or

$$\frac{P(x)}{P_{\text{max}}} = e^{\Delta S/k_B}. \tag{10.4}$$

and recall that ΔS corresponds to the fluctuations away from equilibrium so it is negative. This allows us to calculate fluctuations about a stable equilibrium state for a thermally isolated system. None of this is new: We are merely reframing what we have already worked through with a different point of view. And as we want to study phase transitions, an isolated system is not really that interesting to us here, so from now on we will consider a specific non-isolated system.

10.2.2 A System in Contact with a Heat and Work Reservoir

Within the context of the canonical ensemble, we have considered systems with a fixed temperature (in contact with a heat reservoir), so we could redo the analysis of Section 10.2.1 with this in mind. However, I'll let you work through that case in Problem 10.1, as phase transitions (and many other processes of interest) take place when we have two parameters that we can control (for now temperature *and* pressure).

Consider a system A which is in contact with a heat reservoir A_0 at temperature T_0. Additionally, we will consider A_0 to be a **work reservoir** at pressure p_0. Just as a heat reservoir can gain or lose energy in the form of heat without its temperature changing, a work reservoir can do any amount of work without the pressure of the system changing. Just as we formally defined a heat reservoir by Eq. (6.5), a work reservoir is defined by

$$\left|\frac{\partial p_0}{\partial E_0} W_0\right| \ll p_0, \tag{10.5}$$

with E_0 the energy of the reservoir and W_0 the work done by it. To ensure the number of molecules in A remains unchanged, we will insist the systems are connected via a thermally conducting and movable piston that allows heat to flow between the systems and the volume of the systems to change. The volume of A_0 is so large that it can keep the pressure fixed and the volume is unaffected by this relatively small change. This combined system is shown in Figure 10.3.[6]

Exercise 10.1 The pressure of the reservoir is a function of its energy, $p_0(E_0)$. Derive Eq. (10.5), following a procedure similar to that leading up to Eq. (6.5). Do this by considering the work done by the reservoir W_0 to be much smaller than its energy, E_0.

Many systems in everyday life can be considered as being in contact with such a heat and work reservoir, where the reservoir is the surrounding atmosphere. Clearly temperature and pressure are not *precisely* constant; however, the fluctuations that arise are small enough such that their effects can be considered negligible for many processes. Thus, such formalism can easily be seen to be applicable to phase transitions, which we know occur readily in our lives.

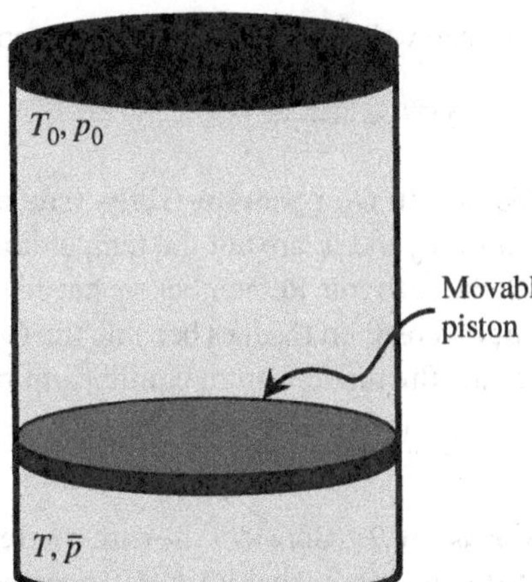

Figure 10.3 An example of a system in contact with a heat and work reservoir (with fixed temperature and pressure).

6 Work reservoirs aren't discussed as often in the subject of thermodynamics although they are as important in many ways as heat reservoirs. Sometimes you might see the term **pressure reservoir**, but I find this misleading. For more details on their role in the subject, I would encourage you to read more about them in Ref. [1].

Just as for the isolated system, the total entropy is maximized when the combined system (the system A plus the reservoir) here is in equilibrium, so

$$S^{(0)} = S + S_0 = \text{maximum},$$

and fluctuations away from equilibrium can be denoted as[7]

$$\Delta S^{(0)} = \Delta S + \Delta S_0 \leq 0.$$

We know for the reservoir that

$$\Delta S_0 = \frac{Q_0}{T_0} = -\frac{Q}{T_0},$$

where we have used $Q_0 = -Q$ because we are interested in writing as many of our quantities as possible in terms of those relating to A and not A_0.

The first law gives us

$$Q = \Delta\bar{E} + W = \Delta\bar{E} + p_0\Delta V + W^*,$$

where W^* is the work done by the system when other external parameters (for example, a magnetic field) change. We see that the work done by changing the volume is just $p_0\Delta V$ because the pressure of the reservoir is constant, and we will set $W^* = 0$ for our analysis here (most systems of interest for us will only involve volume as the external parameter). Putting the first law into $\Delta S^{(0)}$, we get

$$\Delta S^{(0)} = \Delta S - \frac{\Delta\bar{E} + p_0\Delta V}{T_0}, \tag{10.6}$$

and if we multiply by T_0 we obtain,

$$T_0\Delta S^{(0)} = \Delta\left(T_0 S - \bar{E} - p_0 V\right).$$

The quantity in parentheses looks like the Gibbs free energy up to a minus sign, although it's not quite the same. Given this similarity though, we define

$$G_0 \equiv \bar{E} + p_0 V - T_0 S, \tag{10.7}$$

so we can write the change in the entropy of the combined system as

$$\Delta S^{(0)} = -\frac{\Delta G_0}{T_0}. \tag{10.8}$$

I like to call G_0 a **pseudo-Gibbs free energy** to emphasize that it is not the Gibbs free energy because T_0 and p_0 are not the temperature and pressure of the system A unless it is in equilibrium with the reservoir. Remember we haven't yet insisted that A and A_0 are in equilibrium, but of course if they were, then G_0 does become the Gibbs free energy of A. And the main point of this section is that any fluctuations from equilibrium require that $\Delta S^{(0)} \leq 0$, or in terms of G_0,

$$\Delta G_0 \geq 0. \tag{10.9}$$

Exercise 10.2 Consider the case where the non-mechanical work done by the system is non-zero, $W^* \neq 0$, and show that Eq. (10.9) becomes

$$\Delta G_0 \geq -W^*.$$

---

7 Again, recall if $S^{(0)}$ is maximum, then any move away from equilibrium necessarily decreases the entropy.

In this case, $-W^*$ is the maximum amount by which this pseudo-free energy can change. We can use this idea to remark that the Gibbs free energy corresponds to the energy available to the system to do non-mechanical work.

So at equilibrium, G_0 is a *minimum*—and thus this is a constraint that gives us information about how the system A behaves. The probability that a given thermodynamic parameter has a value x is given by

$$\frac{P(x)}{P_{\text{max}}} = e^{\Delta S/k_B}$$
$$= e^{-\Delta G_0(x)/(k_B T_0)}.$$

With $\Delta G_0 = G_0 - G_{0,\text{min}}$, we can write this as

$$P(x) = P_0 e^{-G_0(x)/(k_B T_0)}, \tag{10.10}$$

where we have defined P_0 is a constant which can be determined through normalization. This expression not only will allow us to study phase transitions, but also the conditions for our system to be in a *stable* equilibrium state. Also note that the pseudo-Gibbs free energy is naturally dependent upon temperature and pressure, so it is a natural function to use to study such processes.

10.3 Phase Equilibrium

Let's consider a system with ν moles of a given substance that can exist in two phases. These two phases could be liquid and gas, liquid and solid, etc., but we'll just refer to them as phases 1 and 2, each with ν_1 and ν_2 moles, respectively. I am not worrying about the specifics yet, as I would like to work out a condition for phase equilibrium that will be valid generally.

Of course, we will assume our system is in equilibrium—the two phases both have the same temperature and pressure as the reservoir (so we can say $G_0 = G$), and thus,

$$G(T, \overline{p}) = \bar{E} - TS + \overline{p}V = \text{minimum}.$$

It will be more useful to consider the molar Gibbs free energies of the two phases,

$$g_1(T, \overline{p}) \text{ and } g_2(T, \overline{p}),$$

so that we can write

$$G = \nu_1 g_1 + \nu_2 g_2. \tag{10.11}$$

Given that matter is conserved, if we lose some amount of phase 1, we must gain that same amount of phase 2 and vice versa, so that

$$\nu = \nu_1 + \nu_2 = \text{constant},$$

or

$$d\nu_1 = -d\nu_2.$$

With the requirement that the Gibbs free energy is at a minimum, we can write

$$0 = dG = g_1 d\nu_1 + g_2 d\nu_2 = (g_1 - g_2) d\nu_1. \tag{10.12}$$

This has to be true for an arbitrary change in ν_1, so equilibrium is maintained so long as

$$g_1 = g_2, \tag{10.13}$$

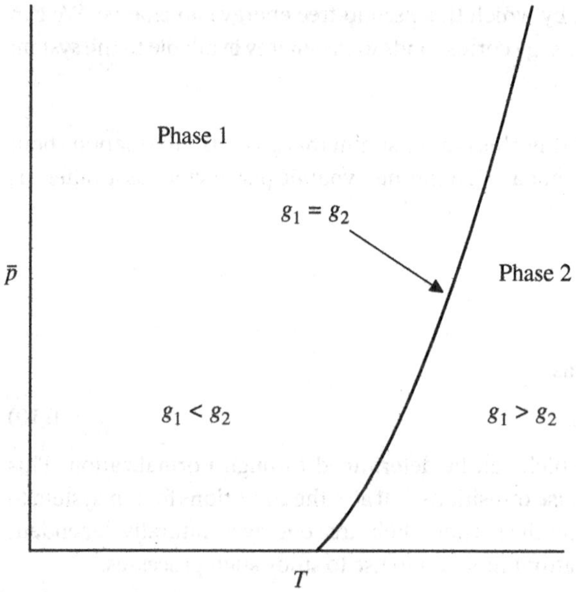

Figure 10.4 A simple example of a phase diagram. The line shows the values of pressure and temperature where the molar Gibbs free energy is the same for both phases.

or their molar Gibbs free energies are equal. If our system is a temperature and pressure such that

$$g_1 < g_2,$$

then the minimum of G would occur if all of our substance is in phase 1 and none in phase 2, or $\nu_1 = \nu$ and $\nu_2 = 0$. Likewise, if the temperature and pressure are such that $g_1 > g_2$, then all of our substance will be in phase 2.

If the temperature and pressure are such that $g_1 = g_2$, then we can't say how much of either phase we have. The system could be in equilibrium with any amount of both phases coexisting. We show this schematically with a **phase diagram** where we plot the line described by $g_1 = g_2$ as a function of $\bar{p}$ and T, and an example of this is shown in Figure 10.4. The line where $g_1 = g_2$ is known as the **phase-equilibrium line**, and on that line you know that both phases can coexist but not how much of either phase there is. On either side of this line (where $g_1 \neq g_2$), the system is entirely in one phase or the other (whichever has a smaller molar Gibbs free energy).

We can determine this phase equilibrium line, or at least the *slope* of it very easily. Consider two points along this line: One is at $(T, \bar{p})$, along this line so that

$$g_1(T, \bar{p}) = g_2(T, \bar{p}),$$

and the other at a slightly different temperature $T + dT$ and pressure $\bar{p} + d\bar{p}$, but the molar Gibbs free energies are still equal,

$$g_1(T + dT, \bar{p} + d\bar{p}) = g_2(T + dT, \bar{p} + d\bar{p}). \tag{10.14}$$

We can Taylor expand g_i for $i = 1, 2$ in the second equation to get

$$g_i(T + dT, \bar{p} + d\bar{p}) = g_i(T, \bar{p}) + \left(\frac{\partial g_i}{\partial T}\right)_p dT + \left(\frac{\partial g_i}{\partial \bar{p}}\right)_T d\bar{p} = g_i(T, \bar{p}) + s_i dT + v_i d\bar{p},$$

where s_i and v_i are the molar entropy and volume, respectively, of phase i. Putting this into Eq. (10.14) and simplifying we have

$$-s_1 dT + v_1 d\bar{p} = -s_2 dT + v_2 d\bar{p},$$

or if we define $\Delta s = s_2 - s_1$ and $\Delta v = v_2 - v_1$ as the differences in the molar entropies and volumes of the two phases,

$$\frac{d\overline{p}}{dT} = \frac{\Delta s}{\Delta v}. \tag{10.15}$$

This equation for the slope of the $\overline{p}$ vs. T equilibrium line is called the **Clausius–Clapeyron equation.**[8] As the left-hand side is an intensive quantity, we could also write this in terms of ΔS and ΔV, the non-molar entropy and volume differences.

In the Clausius–Clapeyron equation, I stress that Δs and Δv are not *changes* in the entropy and volume, but rather they are *differences* in these quantities between the two different phases. We can understand that different phases of a substance have a discontinuity in their entropies and volumes quite easily. For example, consider the molar volumes of different phases of water in Eq. (10.1)—$\Delta v \neq 0$ for each of these cases. Additionally, given that a solid tends to have fewer accessible states (there are more constraints) than a liquid, which in turn has fewer states than a gas, we generally expect

$$s_{\text{solid}} < s_{\text{liquid}} < s_{\text{gas}}, \tag{10.16}$$

so none of them are equal to each other.

During a phase change, heat must flow into or out of the substance, and thus generally there is a change in the entropy of the system. This heat is known as the **latent heat of transformation**, L_{12}: the heat absorbed by a substance in phase 1 when it transforms to phase 2. We saw this briefly during our discussion of heat capacities in Section 6.2.1, where we introduced it in order to study the heat flow between systems while including the possibility of a phase change during an interaction. As stated then, since the process takes place at a constant temperature T, the corresponding entropy change is simply that of a heat reservoir, or

$$\Delta S = S_2 - S_1 = \frac{L_{12}}{T}, \tag{10.17}$$

and thus we can write the Clausius–Clapeyron equation as

$$\frac{dp}{dT} = \frac{L_{12}}{T\Delta V}, \tag{10.18}$$

where keep in mind $\Delta V = V_2 - V_1$. That is, L_{12} is the latent heat when the system changes from phase 1 to phase 2, so the change is the volume would be given by the volume of phase 2 minus the volume of phase 1. If we write the Clausius–Clapeyron in terms of molar quantities, $v_i = V_i/\nu$ and $s_i = S_i/\nu$, then we can write L_{12} in terms of the molar latent heat ℓ_{12}, $L_{12} = \nu\ell_{12}$.

There are several common phase transitions and their associated (molar) latent heats:

Fusion (or freezing): This denotes the liquid to solid transition, with negative latent heat $\ell_f < 0$, as heat must be removed from the system to freeze it.

Melting: This is the reverse of the previous process: When a substance melts, it converts from solid to liquid form, which has a positive latent heat and is related to the latent heat of freezing, $\ell_m = -\ell_f > 0$.

Condensation: This is used for either liquid or gas transitioning to a solid state, with a negative latent heat $\ell_c < 0$.

8 Benoît Clapeyron (1799–1864).

Vaporization (or boiling): This corresponds to the liquid to gas transition, with a positive latent heat which is related to the latent heat of condensation, $\ell_v = -\ell_c > 0$.

Sublimation: Finally, this term is used for a direct solid to gas transition, with a positive latent heat related to the latent heat of condensation, $\ell_s = -\ell_c > 0$.

Obviously there can be different terms or notations than those listed here, and there can sometimes be confusion with the two different latent heats of condensation. However, generally context will usually save us from these complications.

10.3.1 Phase Diagram of Water

Water is a fairly simple substance, and of course is crucial not only to life on Earth but for so many natural processes. I want to look at the phase diagram partially because of the importance of water, but also because it actually has a seemingly benign and yet strange feature. The phase diagram with just the solid, liquid, and vapor phases is shown in Figure 10.5.[9]

Several features of the phase diagram for water are common to many other substances. First, there exists a **triple point**, defined as the temperature and pressure where the liquid, solid, and gas phases can all coexist, and is marked by the circle in the figure. For water, this occurs at $T = 273.16$ K and a pressure of 611.73 Pa (about 0.006 atm), and until May 2019 was the point used to define the kelvin as the SI unit of temperature.

The point on the right where the phase-equilibrium line between the liquid and gas phases ends, marked with a star, is the **critical point**, defined by the critical temperature T_c and critical pressure p_c (first seen along with the critical volume v_c in Problem 7.3). Beyond this point, there is no sharp distinction between the two phases due to the ending of the phase equilibrium line. Suppose our

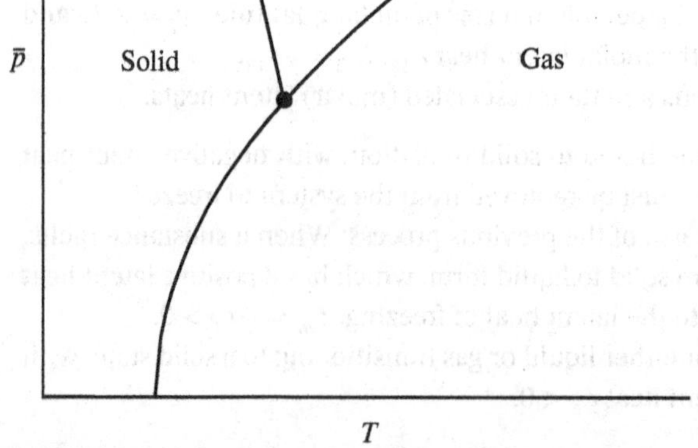

Figure 10.5 A schematic of the phase diagram of water. The star labels the critical point, or the end of a phase equilibrium line. The circle denotes the triple point, where three phases can coexist.

9 The exact details of this diagram should not be taken too seriously, as to show important features it is not to scale. A more accurate and detailed figure can be found online at Ref. [2] for example. In fact, in that figure, you can see other examples of different phases of water.

system is at a pressure below p_c, and a temperature such that it is in the gaseous phase, and we lower the temperature to allow it to transition to liquid (along a horizontal path, say). If, during this process, we cross the phase equilibrium line, we would see the volume of the system decrease sharply, from $v_g \to v_\ell$ in Eq. (10.1). However, if the temperature is changed such that the path we take goes around the critical point in this diagram, thus not actually crossing the line, there is never a point at which we can say that the gas has definitely become a liquid; the transition from one phase to the other is gradual enough so the phase transition is blurred. In Section 10.4 when we discuss how to determine a phase transition from the equation of state, we will define the conditions for a critical point concretely.

The features above are common to many phase diagrams; however, let's move on to what makes this particular diagram special. Generally, as we have discussed before, the entropy of a substance increases as it moves from the solid to the liquid phase, so the latent heat of melting is positive, $\ell_m > 0$ (heat is absorbed in this process). Additionally, for normal substances, the liquid phase tends to be less dense than the solid, and thus $\Delta v = v_\ell - v_s > 0$. As such, the slope of the curve of the solid–liquid equilibrium line, by the Clausius–Clapeyron equation, is ordinarily positive.

Exercise 10.3 Explain why if a liquid is less dense than a solid, then we can say the difference in the molar volumes is positive when transitioning from the solid to the liquid phase, or $\Delta v = v_\ell - v_s > 0$.

However as we can see in Figure 10.5, this is not the case for water, and this is one very important way in which water differs from pretty much every other substance. The slope of the solid–liquid transition line is *negative*, while the entropy still increases when going from the solid phase to the liquid phase. This means that the change in the molar volume must also be negative, or $v_\ell - v_s < 0$ (which we know experimentally to be true): The density of liquid water is greater than that of solid ice! (This is clear from our experience; ice floats on liquid water and thus it must be less dense.) This is an important feature of water, as if it weren't the case, then frozen water would sink to the bottom of the oceans of the world. Because ice freezes and remains on top of the oceans, the lower depths are insulated and are able to maintain life in the warmer waters during the winter months. As life began in the oceans, this simple fact allowed it to evolve during the history of the world. One can argue then that this odd feature of the phase diagram of water is precisely why life exists on this planet as we know it.

10.3.2 Vapor Pressure of an Ideal Gas

We can use the Clausius–Clapeyron equation to determine an expression for the **vapor pressure**: the pressure of the gaseous form of a substance when in equilibrium with another phase (liquid or solid). In terms of the molar latent heat and molar volume change, the Clausius–Clapeyron equation is

$$\frac{dp}{dT} = \frac{\ell}{T\Delta v}.$$

Specifically ℓ is the heat required to go from the "other phase," which here would be liquid or solid, to the gas phase, so $\ell = T\left(s_g - s_o\right) > 0$, with s_o being the entropy of the other phase. Generally, a mole of vapor tends to have a significantly larger volume than a mole of solid or liquid of the same substance, so we can approximate

$$\Delta v = v_g - v_o \approx v_g. \tag{10.19}$$

Exercise 10.4 Consider Eq. (10.1) and determine the percent error in the approximation given in Eq. (10.19) for both the solid–gas and liquid–gas transitions of water.

If we treat the vapor as a classical ideal gas, then $\overline{p}v_g = RT$, we get the equation

$$\frac{d\overline{p}}{dT} = \frac{\ell\overline{p}}{T^2R}. \tag{10.20}$$

If we assume the latent heat is independent of temperature and pressure, we can immediately integrate this to obtain

$$\overline{p} = \overline{p}_0 \exp\left(-\frac{\ell}{RT}\right). \tag{10.21}$$

The assumption that the latent heat is constant is a good approximation for certain ranges of these thermodynamic variables, but of course is a limitation to the utility of this expression (we will determine a more precise expression for this in Section 11.6.2). Within this approximation we see that as the temperature increases then so does the vapor pressure. One note, you will often see Eq. (10.21) referred to as the Clausius–Clapeyron equation; however, this is strictly only true for the two approximations we have made: that $v_g \gg v_o$ and the gas is ideal.

Exercise 10.5 Fill in the steps to obtain Eq. (10.21).

10.4 From the Equation of State to a Phase Transition

We have seen that there is a sharp change in the volume of a system during a phase transition. Additionally, our discussion of phases currently has revolved around what happens at different pressures and temperatures. It is only natural to ask how we can understand a phase transition using the equation of state,

$$\overline{p} = \overline{p}(v, T),$$

for a system with (for simplicity) one species. This should, if it contains *all* the relevant information of our system, allow us to see this discontinuous behavior in v as we change T and $\overline{p}$ when changing from one phase to another. Ordinarily we would begin with the ideal gas equation, it being the simplest equation of state we have; however, in this case I will start right with the Van der Waals equation of state,

$$\left(\overline{p} + \frac{a}{v^2}\right)(v - b) = RT.$$

Once we see how a phase transition arises from this equation, it will be clear why we wouldn't be able to do so from the ideal gas equation.

I plot in Figure 10.6 the pressure as a function of the molar volume for six different temperatures for a Van der Waals gas. In this figure, $T_h > T_\ell$, and the curves are those of increasing temperatures from the bottom up. There are three curves where each volume corresponds to a single value of the pressure (the solid curve as well as the two dotted curves above it). For larger temperature, the cubic nature of the Van der Waals equation is not apparent, so we can write

$$\overline{p} \sim \frac{RT}{v - b},$$

and it looks (roughly) like an ideal gas. There is no difficulty arising here—if I give you a pressure and a temperature, you can unambiguously tell me what the volume of the gas is.

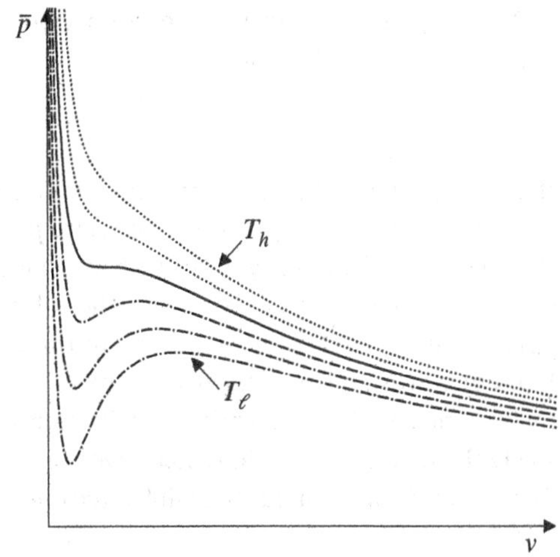

Figure 10.6 The Van der Waals equation of state, $\bar{p}$ vs. v, for different values of the temperature.

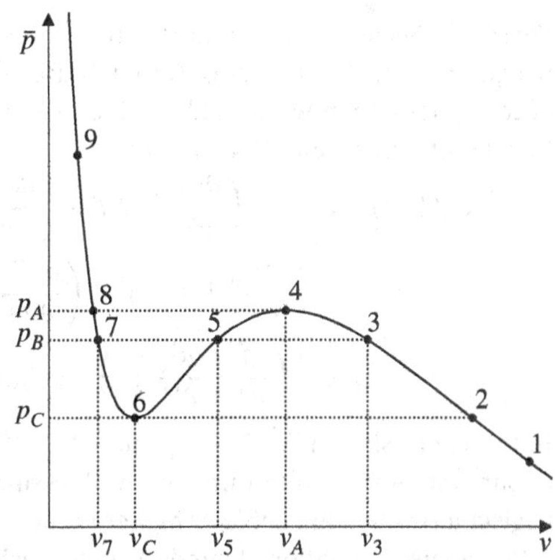

Figure 10.7 The same as Figure 10.6, but for a single temperature where there is a range of pressures that correspond to more than one volume of the substance.

However, for lower temperatures, those shown with the dashed–dotted curves below the solid curve, we see that there are regions where a given pressure and temperature corresponds to *three* possible volumes of our substance. This doesn't make sense, because we expect the equation of state to give us a unique result for the state variables: pressure, volume, and temperature.

Consider one of the curves where this arises, shown in Figure 10.7. I have labeled quite a few points in this figure:

- $p_A \leq \bar{p} \leq p_C$ defines the range of pressures where there is an ambiguity in the volume of the system.
- Volumes v_A and v_C are the volumes corresponding to the pressures p_A and p_C, respectively, *within* the region as shown.
- p_B is some pressure within this range where the volume is ambiguous which we will focus on shortly.
- v_3, v_5, and v_7 are the three possible volumes that correspond to p_B, according to the Van der Waals equation. (The odd numbering that I have chosen will be clearer later.)
- The other unlabeled points will be defined and used shortly.

Of these points, v_A and v_C correspond to volumes when the derivative of the pressure with respect to the volume vanishes, or

$$\left(\frac{\partial \bar{p}}{\partial v}\right)_T = 0. \tag{10.22}$$

If the system has volume v_A or v_C, the pressure is clearly p_A or p_C, respectively. However, the opposite is not true: If the system has pressure p_A, it could have volume v_A or a much smaller volume (less the v_7), and similarly for p_C (it could have a volume v_C or a volume larger than v_3). If the pressure increases above p_A or below p_C, the volume of our system is clear. For those pressures (such as p_B) between p_A and p_C, we have two or three possible volumes according to the equation of state. In other words, in this intermediate region, we do not know what the proper equilibrium state of the system is without revisiting our discussion of equilibrium from earlier in this chapter. It is not solely the equation of state which gives rise to an understanding of this system, but we also need to know something about the Gibbs free energy.

10.4.1 Stable Equilibrium Requirements

We saw in Sections 10.2.2 and 10.3 that equilibrium is reached when G_0 is minimized (and once in equilibrium, G_0 becomes G, the Gibbs free energy). If we change the volume or the temperature of the system, then we should be able to expand $G_0(T, V)$ about the minimum, so long as these fluctuations are small. We can write

$$\begin{aligned} G_0(T,V) = G_{\min} &+ \left(\frac{\partial G_0}{\partial T}\right)_V \Delta T + \left(\frac{\partial G_0}{\partial V}\right)_T \Delta V \\ &+ \left(\frac{\partial^2 G_0}{\partial T^2}\right)_V (\Delta T)^2 + \left(\frac{\partial^2 G_0}{\partial V^2}\right)_T (\Delta V)^2 \\ &+ \frac{1}{2}\left(\frac{\partial}{\partial T}\left(\frac{\partial G_0}{\partial V}\right)_T\right)_V (\Delta T)(\Delta V) + \cdots . \end{aligned} \tag{10.23}$$

In this expression, $\Delta T = T - T_{\text{eq}}$ and $\Delta V = V - V_{\text{eq}}$, where T_{eq} (V_{eq}) is the temperature (volume) of our system in equilibrium, and we'll assume we are close enough to equilibrium that we can neglect terms that are cubic or higher in these differences.

We can use a variation of our derivative crushing method, with the caveat that we are considering the pseudo-Gibbs free energy

$$G_0 = \bar{E} - T_0 S + p_0 V,$$

and not the actual Gibbs free energy. Looking at the first derivative terms, for this function to be an extremum, they must vanish. We then have

$$\left(\frac{\partial G_0}{\partial T}\right)_V \bigg|_{T=T_{\text{eq}}} = 0 = \left(\frac{\partial \bar{E}}{\partial T}\right)_V - T_0\left(\frac{\partial S}{\partial T}\right)_V, \tag{10.24}$$

and since $(\partial \bar{E}/\partial T)_V = T(\partial S/\partial T)_V$, we see that

$$T_{\text{eq}} = T_0 \tag{10.25}$$

in equilibrium. This is not a surprise, merely another way to see what we have determined before: For two systems to be in equilibrium, they must have the same temperature.

Exercise 10.6 Use the requirement that

$$\left(\frac{\partial G_0}{\partial V}\right)_T = 0,$$

and the same method as I did to show that when $V = V_{eq}$ and $T = T_0$, we must have

$$\bar{p} = p_0,$$

as we would expect (the pressures of two systems in equilibrium are equal).

Next we consider the second derivatives, which we have ignored up to now, but are crucial to ensure we are in a *stable* equilibrium. For G_0 to be a minimum, the second derivatives should be positive. We will work through two of the three second derivatives in Eq. (10.23), with the remaining (mixed) derivative relegated to the problems.

We start with the second derivative with respect to temperature. We have

$$\begin{aligned}
\left(\frac{\partial^2 G_0}{\partial T^2}\right)_V &= \left(\frac{\partial}{\partial T}\left(\frac{\partial \bar{E}}{\partial T} - T_0\frac{\partial S}{\partial T}\right)_V\right)_V \\
&= \left(\frac{\partial}{\partial T}\left((T - T_0)\left(\frac{\partial S}{\partial T}\right)_V\right)\right)_V \\
&= \left(\frac{\partial S}{\partial T}\right)_V + (T - T_0)\left(\frac{\partial^2 S}{\partial T^2}\right)_V .
\end{aligned} \tag{10.26}$$

The second term vanishes at equilibrium, because $T_{eq} = T_0$. The first term is related to the heat capacity at constant volume, C_V/T, and if we require that the second derivative is positive for stable equilibrium, then it must be true that

$$C_V > 0. \tag{10.27}$$

We claimed this when we first defined the heat capacity, as it "made sense" that raising the temperature of a system requires the addition of heat. However now we can see this is part of a more general requirement that the heat capacity at constant volume is positive for a system to be in stable equilibrium. This idea, named after Henry Louis Le Châtelier (1850–1936), is known as

Le Châtelier's principle

If a system is in stable equilibrium, then any spontaneous change of its parameters must bring about processes which tend to restore the system to equilibrium.

Next we look at what happens when the volume fluctuates, where we have to include a couple more steps. We have

$$\begin{aligned}
\left(\frac{\partial^2 G_0}{\partial V^2}\right)_T &= \left(\frac{\partial}{\partial V}\left(\frac{\partial E}{\partial V} - T_0\frac{\partial S}{\partial V} + p_0\right)_T\right)_T \\
&= \left(\frac{\partial}{\partial V}\left(T\frac{\partial S}{\partial V} - \bar{p} - T_0\frac{\partial S}{\partial V} + p_0\right)_T\right)_T \\
&= -\left(\frac{\partial \bar{p}}{\partial V}\right)_T .
\end{aligned}$$

We used the first law to go from the first to the second line. From the second to the third line, we set $T = T_0$ (as we are in equilibrium) and that p_0 is a constant so its derivative vanishes. For this second derivative to be positive, we must have

$$\left(\frac{\partial \bar{p}}{\partial V}\right)_T < 0. \tag{10.28}$$

Given the definition of the isothermal compressibility as

$$\kappa = -\frac{1}{V}\left(\frac{\partial V}{\partial \overline{p}}\right)_T,$$

then this requires $\kappa > 0$ for a system to be in stable equilibrium. Again, we put the negative sign in by hand when defining κ, because we wanted it to be positive (and it made sense that $(\partial V/\partial \overline{p})_T < 0$), but now we see it arises from general considerations of stable equilibrium.

In Problem 10.2 you will show that the mixed second derivative,

$$\left(\frac{\partial}{\partial T}\left(\frac{\partial G_0}{\partial V}\right)_T\right)_V = \left(\frac{\partial}{\partial V}\left(\frac{\partial G_0}{\partial T}\right)_V\right)_T$$

vanishes in equilibrium, so it is consistent with our system being in stable equilibrium, and we can drop it from Eq. (10.23) to write

$$\Delta G_0(T, V) = \frac{C_V(\Delta T)^2}{2T_0} + \frac{(\Delta V)^2}{2V_{\text{eq}}\kappa}. \tag{10.29}$$

In this expression we set $T = T_0$ in the first term and $V = V_{\text{eq}}$ in the second as those factors must be evaluated in equilibrium. This result is valid for any system in stable equilibrium, and now we can apply it to phase transitions.

10.4.2 Back to Our Phase Transition

Let's return to our discussion of phase transitions and consider Figure 10.8. This is the same as Figure 10.7, but with the volumes no longer marked. The three special pressures are still labeled, but for this discussion all other points (pressures and volumes) will be referred to by their numbers (e.g., v_1 and p_1). With this figure, we can begin to explain what is happening in the region $p_C < \overline{p} < p_A$. With the requirement of Eq. (10.28), we must exclude any pressure–volume combinations of our system with $v_6 < v < v_4$, because in that region, the slope of our $\overline{p}$ vs. v graph is positive. If our system is at pressure p_B, from the Van der Waal's equation of state, it appears as though either v_3 or v_7 could be the molar volume of the system, but not v_5. But which of these two is the "correct value"?

Actually we will now argue that at this pressure and temperature, our substance will undergo a phase transition. For this temperature, if v is small enough so that $p > p_A$, the slope of this curve

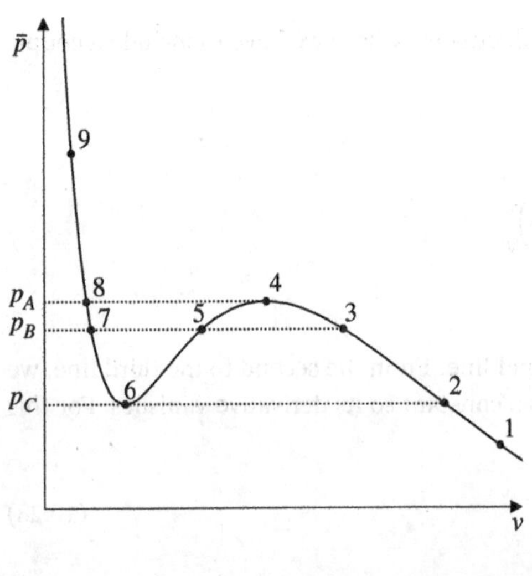

Figure 10.8 The same as Figure 10.7, but with the specific volumes from that figure no longer labeled.

is large and negative, so κ will be very small; the substance is difficult to compress. On the other hand, for $\overline{p} < p_C$, then the slope of this curve is small (and still negative as it should be), so that the compressibility is much larger. This allows us to identify our two phases:

Gas: Gases tend to have large compressibilities, so when $\overline{p} < p_C$ our substance is in a gaseous phase.

Liquid: Liquids tend to be more difficult to compress, so in this case, when $\overline{p} > p_A$, our substance is clearly a liquid.

Suppose our system initially has a pressure less than p_C (in the gaseous phase) and while holding the temperature constant, we do work on the system to decrease the volume. As we do so, the pressure will increase according to the figure, but eventually we enter this region where it is unclear, from the Van der Waals equation, when this transition from gas to liquid occurs. We know from experience that as we continue to compress the gas, instead of the pressure of the system increasing, the gas will condense into a liquid, but this is not apparent in our equation of state.

To determine how this happens, we need to look at the Gibbs free energy of the system as we move along this curve. The molar Gibbs free energy is given by

$$dg = -sdT + vd\overline{p} = vd\overline{p}, \tag{10.30}$$

where in the second equality we used the fact that we are holding the temperature constant. We can determine the molar free energy at some point i on the graph relative to a point, (v_1, p_1), by integration

$$g_i - g_1 = \int_{p_1}^{p_i} vd\overline{p}. \tag{10.31}$$

The right-hand side of this expression corresponds to the area between the curve and the vertical axis (to the *left* of the curve, not below the curve as we are accustomed). As we perform this integral, we see that as we go from p_1 to p_i with $i \leq 4$, the value of g_i increases as we are adding to the area to the left. From $p_4 \rightarrow p_6$, we are now moving down and subtracting from the area, so we get a negative contribution to the integral, and g_i will decrease, and finally from $p_6 \rightarrow p_9$ (and beyond), g_i will increase. Each stage as described here is shown in Figure 10.9. The resulting plot for $g_i - g_1$ as a function of the pressure is shown in Figure 10.10, where I have labeled the three pressures, p_A, p_B, and p_C. Additionally in this figure, the circles mark the points g_2, g_4, g_6, and g_8, while the star denotes the point where we have $g_3 = g_5 = g_7$.

Looking at this figure, we see that in the region where $p_C \leq \overline{p} \leq p_A$, there are two or three possible values for g for each pressure. As we require that g must be minimized for our system to be in equilibrium, then at each pressure, the proper, *physical* value of the Gibbs free energy is the smallest of the possible values. As we increase the pressure, we would move along the curve as usual from g_1 to the starred point. At that point, as we continue to compress the gas (decreasing the volume), the pressure would remain the same (p_B), as would the Gibbs free energy, so we would remain there. During this time, the phase transition occurs as the substance converts to a liquid. Once the phase transition is complete, the pressure again begins to increase while we compress the system, and we move along the curve from the star to g_8.

Considering the volume of the system during this process, from $p_C \rightarrow p_B$, the system will be in the gaseous phase, and the volume of the gas can be obtained from the Van der Waals equation. At p_B and v_3, however, as we try to decrease the volume, the pressure will remain constant and the volume *does* decrease, but because the gas is changing to a liquid, and we know the molar volume of

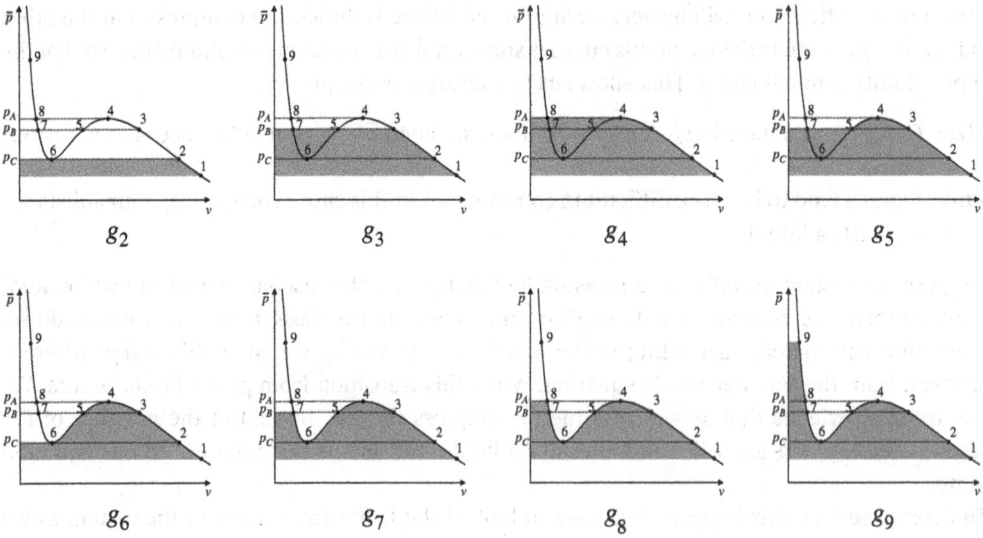

Figure 10.9 The area to the left of the Van der Waals curve for a given temperature, at different choices for the upper limit on the integral for $g_i - g_1$.

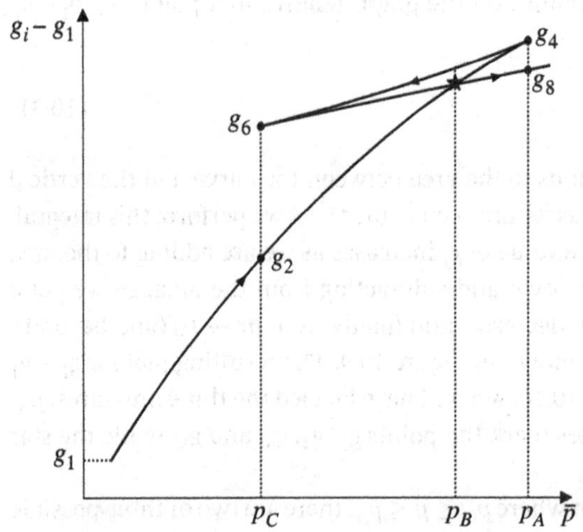

Figure 10.10 The molar Gibbs free energy vs. the pressure as a result of the integral in Eq. (10.31). The star denotes the points $g_3 = g_5 = g_7$, while the circles are labeled accordingly.

the liquid phase is distinctly different from that of the gaseous phase (this corresponds to our Gibbs free energy remaining constant at the starred point in Figure 10.10). Once the system reaches v_7, then the system has completely transitioned to the liquid phase, and the pressure begins to increase again, with the volume determined by the Van der Waals equation again.

I show the proper dependence that pressure has on volume in Figure 10.11. Between $v_1 \equiv v_\ell$ and $v_7 \equiv v_g$, the substance is neither liquid nor gas, but a combination of the two phases, with the former being the volume of the liquid phase and the latter being that of the gaseous phase. The molar volume of the entire system in this region

$$v = xv_\ell + (1-x)v_g, \tag{10.32}$$

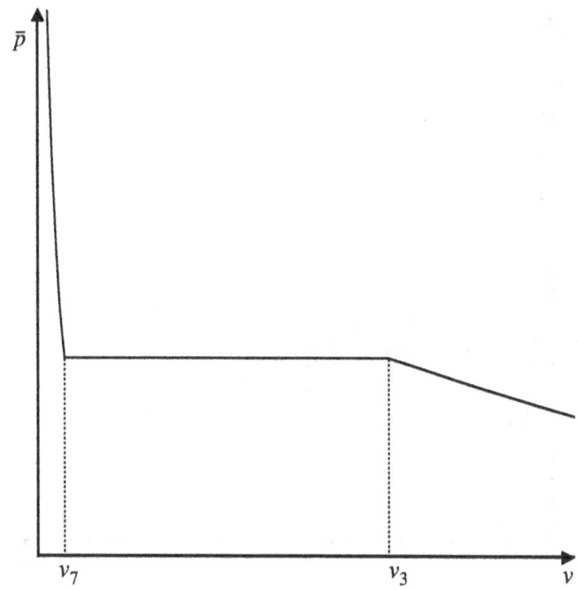

Figure 10.11 The proper dependence pressure has on volume for a given temperature for a Van der Waals gas. The horizontal line indicates the phase transition region.

with x the fraction of the substance that is in the liquid phase, and this is denoted by the horizontal line in the figure. At this point we see that the molar Gibbs free energies of each phase are equal, as shown in the figure, or

$$g_\ell = g_g,$$

and this value of g corresponds to the star in Figure 10.10. This is not something new: It is the requirement we saw earlier in Eq. (10.13), that the two phases must have the same molar Gibbs free energy to coexist.

We can determine the pressure p_B where this occurs using the **Maxwell construction**. This uses the fact that $g_\ell = g_g$ at all points during the phase transition, or $g_\ell - g_g = 0$. This difference, $g_\ell - g_g$, merely corresponds to the shaded region shown in Figure 10.12, as we could see from our discussion above. In this case instead of integrating starting at g_1 as we did before, we integrate from the point (v_3,p_B) to (v_5,p_B) to obtain that (positive) area A_1. We then integrate from (v_5,p_B) to (v_7,p_B), getting a negative area A_2. Moving the line defined by p_B up or down will change the two areas shown, on the left and the right. In order to find the proper pressure, we insist that p_B be chosen so that $A_1 + A_2 = 0$, or the magnitudes of these areas must be equal.

As we increase the temperature, we see that this ambiguity eventually disappears, and above that temperature (the solid line in Figure 10.6), this clear phase transition disappears as well. The point where this occurs is the **critical point** discussed earlier, and it is found to occur at the point where both the first and second derivatives of the equation of state vanish,

$$\left(\frac{\partial \bar{p}}{\partial v}\right)_T = 0, \text{ and } \left(\frac{\partial^2 \bar{p}}{\partial v^2}\right)_T = 0. \tag{10.33}$$

Using these equations, you can determine the **critical temperature** T_c and the **critical volume** v_c, and with these and the equation of state, the **critical pressure** p_c. In Problem 7.3, you found that for the Van der Waals equation of state,

$$RT_c = \frac{8a}{27b}, \; v_c = 3b, \text{ and } p_c = \frac{a}{27b^2}. \tag{10.34}$$

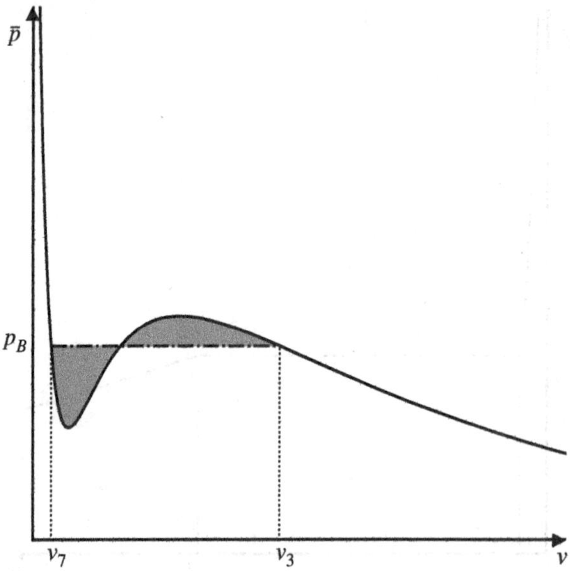

Figure 10.12 The Maxwell construction for the phase transition point: Vary the pressure p_B until the two shaded regions have the same area.

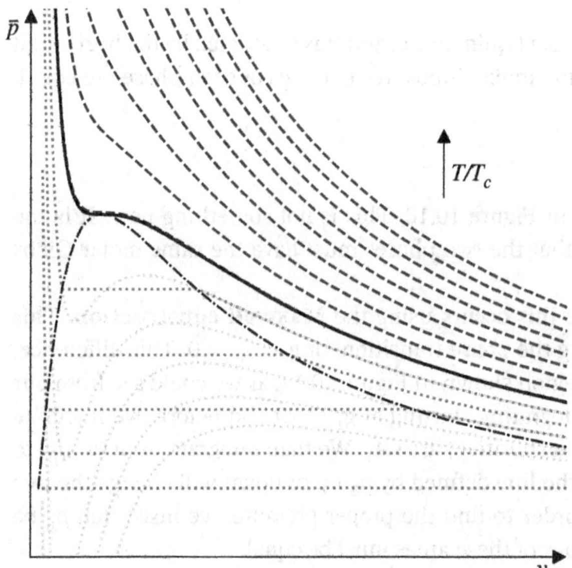

Figure 10.13 The proper $\overline{p}$ vs. v diagram for various temperatures (as opposed to Figure 10.6) after performing the Maxwell construction.

and if you write this equation in terms of dimensionless parameters, $p' = \overline{p}/p_c$ (and similarly for T and v) it takes a more universal form,

$$p' = \frac{8T'}{3v' - 1} - \frac{3}{v'^2}. \tag{10.35}$$

This is the form I used to produce the figures in this chapter (so I didn't have worry about which exact gas I was considering).

In Figure 10.13, I update Figure 10.6, showing what we would obtain for $\overline{p}$ vs. v after applying the Maxwell construction to various pressures in the phase transition region. The solid curve is the curve at the critical temperature, the dotted curves are below this temperature (so a phase transition occurs at the appropriate pressure, and the original shape of the Van der Waals curves is shown slightly faded), and the dashed curves are above the critical temperature, so no phase transition is evident. Additionally, the dotted-dashed curve denotes the points where $(\partial^2 \overline{p}/\partial v)_T = 0$.

Computer Exercise 10.1 You can visit the Phase Transition section of the companion site to work with code (modified from Ref. [3]) that can perform the required numerical integrations to the Maxwell construction for this gas. Try modifying this to incorporate a different equation of state (such as the equations of state in Problem 10.11 or 10.12).

Usually we are familiar with phase transitions occurring as the temperature of a system changes at fixed pressure, while we have developed these ideas for a fixed temperature as the pressure changes. The only reason to consider the pressure is that our expression for the Gibbs free energy is an integral of the pressure as a function of the volume. However, one could solve the Van der Waals equation for the temperature, and repeat the entire analysis, obtaining the same results.

Exercise 10.7 Solve Eq. (10.35) for T' and plot (using Mathematica or other software) T' vs. v' for different values of the pressure above and below the critical point. Argue from the figures you see that if $p' < 1$, then you see evidence of a phase transition by comparing the plots to what we saw previously.

The phase transitions we are discussing are called **first-order phase transitions**. These phase transitions, using Ehrenfest's original classification system,[10] correspond to those where the Gibbs free energy is continuous across the phase boundary, but its first derivatives are *discontinuous*, such as

$$\left(\frac{\partial g}{\partial T}\right)_p = -s, \text{ and } \left(\frac{\partial g}{\partial \overline{p}}\right)_T = v.$$

This is clear from the Clausius–Clapeyron equation: Both Δs and Δv are non-zero, and the right-hand side is a ratio of these two. At the same time, the heat capacity (or specific heat) is infinite,

$$c_V = T\left(\frac{\partial s}{\partial T}\right)_V \to \infty,$$

which can be seen in the discontinuity of the entropy; the derivative is ill-defined. This was our original motivation for defining the latent heat of transformation, so this does not seem too far fetched. Additionally, the other first derivatives (α and κ) are infinite, as they involve the change in the (discontinuous) volumes as we cross the phase boundary.

There are other types of phase transitions, and again if we consider Ehrenfest's original classification scheme, we could study **second-order phase transitions**. In this case, the entropy and volume (the "first derivatives" of g) are continuous during the phase transition, while the heat capacity and other second derivatives are *discontinuous* and not infinite. An example of a second-order transition is the ferromagnetic phase transition in materials such as iron. Here the magnetization, which is the first derivative of the free energy with respect to the applied magnetic field strength, increases continuously from zero as the temperature is lowered below the **Curie temperature** (this is defined by the temperature below which the magnetization of the iron turns on, which was briefly discussed in Problem 6.10). The magnetic susceptibility (the second derivative of the free energy with the field) is discontinuous across the phase transition.

10 Paul Ehrenfest (1880–1933). For a nice historical discussion of this, see Ref. [4].

The difficulty with this exact classification scheme is that there are some phase transitions that do not fall into such simple categories. I won't worry about this here, as for our purposes this scheme will be acceptable, but it's good to know that there is a lot more to phase transitions than we are covering here. The behavior of systems during phase transitions is quite a rich field of study in general, especially around the critical point, but beyond what we can get to here. I would encourage you to look into this in more detail, such as what can be found in chapter 10 of Ref. [5] or Ref. [6].

10.4.3 Density Fluctuations

We can use our expression for G_0 above to study our system for small deviations from equilibrium. Recall that the probability density behaves like $\mathcal{P} \propto \exp\left[-G_0/(k_B T_0)\right]$, and if we fix the temperature of our system to T_0, then we can examine how this behaves for small volume fluctuations. We can write

$$\mathcal{P}(V)dV = P_0 e^{-G_0(V)/(k_B T)} dV,$$

for small dV, with P_0 the normalization constant. We have already expanded G_0 about the minimum, where we obtained

$$G_0(V) = G_{\text{min}} + \frac{(\Delta V)^2}{2V_{\text{eq}}\kappa}.$$

As before, $\Delta V = V - V_{\text{eq}}$ and V_{eq} is the volume of the system in equilibrium (when $\overline{p} = p_0$). Thus we have

$$\mathcal{P}(V)dV = P_0' \exp\left[-\frac{(V - V_{\text{eq}})^2}{2k_B T_0 V_{\text{eq}}\kappa}\right] dV, \tag{10.36}$$

which is just the Gaussian distribution with variance

$$\overline{(\Delta V)^2} = k_B T_0 V_{\text{eq}}\kappa. \tag{10.37}$$

For a fixed number of molecules N in our system, any changes in the volume cause a change in the density $n = N/V$. These fluctuations can be found by differentiating n,

$$\Delta n = -\frac{n_{\text{eq}}}{V_{\text{eq}}}\Delta V, \tag{10.38}$$

with n_{eq} the number density of molecules in equilibrium. The origin of the minus sign is clear: For a fixed number of molecules, if the volume increases, then the number density must decrease.

Exercise 10.8 Derive Eq. (10.38) by considering $n = N/V$ and evaluating dn. In the end, set $dn = \Delta n$ and $dV = \Delta V$.

Exercise 10.9 Show that the variance in the density follows from above immediately and that it is

$$\overline{(\Delta n)^2} = \frac{n_{\text{eq}}^2 k_B T_0 \kappa}{V_{\text{eq}}}. \tag{10.39}$$

Let's consider this in the context of our discussion of a system near the phase transition line. At temperatures below the critical point, during the phase transition, we have $\left(\partial\overline{p}/\partial v\right)_T = 0$ (this corresponds to the horizontal line in Figure 10.13, found using the Maxwell construction). Thus during this transition, $\kappa \to \infty$, or rather, the compressibility is undefined. We can understand this because the compressibility is different for different phases of a substance, so it is unclear which we should consider during the transition. Putting this into Eq. (10.39), this implies that $\overline{(\Delta n)^2} \to \infty$ during this transition, which again makes physical sense. The molar volumes of gas and liquid are very different, so as the system undergoes this transition, there will be a wide variation in what we would measure for the density. Once the transition completes, then both the compressibility and the density are well-defined again.

What we can say from this is that with a proper equation of state, combined with our conditions for stable equilibrium, we can determine the properties of a system in two different phases. While a bit more difficult, it is possible to come up with an equation of state which would properly allow for three phases of matter (see Ref. [7] for example). Being able to fully describe our system with the equation of state (and other thermodynamic principles) is an important goal here. As you can imagine, the more realistic the equation of state, the more difficult this process would be.

10.5 Different Phases as Different Substances

In our previous discussion of phase transitions, we considered a substance in two possible phases, where the phases were distinguished by different bulk properties of the system. However, what if we instead just considered the two phases as different materials?

To do this, we look back at the entropy of the system of molecules of type 1 and type 2. The entropy is a function of the energy, volume, and number of molecules of each type, or

$$S(\bar{E}_1, V_1, N_1, \bar{E}_2, V_2, N_2),$$

and this is a maximum when the system is in equilibrium. Because the entropy is extensive, we can write this as

$$S(\bar{E}_1, V_1, N_1, \bar{E}_2, V_2, N_2) = S_1(\bar{E}_1, V_1, N_1) + S_2(\bar{E}_2, V_2, N_2),$$

and looking at a small change in the entropy, we get

$$dS = dS_1 + dS_2,$$

with (for $i = 1$ or 2)

$$\begin{aligned} dS_i &= \left(\frac{\partial S}{\partial \bar{E}_i}\right)_{V_i,N_i} d\bar{E}_i + \left(\frac{\partial S}{\partial V_i}\right)_{E_i,N_i} dV_i + \left(\frac{\partial S}{\partial N_i}\right)_{V_i,E_i} dN_i \\ &= \frac{1}{T_i} d\bar{E}_i + \frac{\bar{p}_i}{T_i} dV_i + \left(\frac{\partial S}{\partial N_i}\right)_{V_i,E_i} dN_i. \end{aligned} \tag{10.40}$$

The last term is new for us, and we will define it as

$$\mu_i \equiv -T_i \left(\frac{\partial S}{\partial N_i}\right)_{V_i,E_i}, \tag{10.41}$$

the **chemical potential** for the ith substance in our system.[11] I have already introduced this when we covered derivative crushing in Section 6.4, even though at that time I said nothing about it. I will discuss what it means physically shortly, but first let's just use it in our discussion of equilibrium between our two substances.

11 Not to be confused with the magnetic moment of a spin system; generally this won't be an issue for us.

As usual, when the entropy is maximized, we have $dS = 0$, and any change in the entropy of one system must be opposite the change in the other, or $dS_1 = -dS_2$, so

$$\frac{1}{T_1}d\bar{E}_1 + \frac{\overline{p}_1}{T_1}dV_1 - \frac{\mu_1}{T_1}dN_1 = -\left(\frac{1}{T_2}d\bar{E}_2 + \frac{\overline{p}_2}{T_2}dV_2 - \frac{\mu_2}{T_2}dN_2\right). \tag{10.42}$$

However, as usual we will assume our combined system is thermally isolated, in a rigid box, and now we also require the *total* number of molecules is fixed, or

$$\bar{E}^{(0)} = \bar{E}_1 + \bar{E}_2 = \text{constant},$$
$$V^{(0)} = V_1 + V_2 = \text{constant, and}$$
$$N^{(0)} = N_1 + N_2 = \text{constant.}$$

With these requirements, Eq. (10.42) becomes

$$\frac{1}{T_1}d\bar{E}_1 + \frac{\overline{p}_1}{T_1}dV_1 - \frac{\mu_1}{T_1}dN_1 = \frac{1}{T_2}d\bar{E}_1 + \frac{\overline{p}_2}{T_2}dV_1 - \frac{\mu_2}{T_2}dN_1, \tag{10.43}$$

and as before we want this to be true regardless of the independent changes in the energy, volume, or number of molecules, so we get two of our standard conditions for two systems being in equilibrium: $T_1 = T_2$ and $\overline{p}_1 = \overline{p}_2$. These conditions are what we have seen several times when considering thermal and mechanical equilibrium, respectively. The new condition is for a new type of equilibrium, **chemical equilibrium**,[12]

$$\mu_1 = \mu_2. \tag{10.44}$$

Exercise 10.10 Is the chemical potential an extensive quantity or an intensive quantity?

This last equilibrium condition seems new, but we actually have already seen it, and to understand where, let's look more into the chemical potential. For a single species, we can write a more general version of the first law of thermodynamics by solving Eq. (10.40) for $d\bar{E}$ (ignoring the subscripts for now),

$$d\bar{E} = TdS - \overline{p}dV + \mu dN. \tag{10.45}$$

In this form, we see that we can also write the chemical potential as

$$\mu = \left(\frac{\partial \bar{E}}{\partial N}\right)_{V,S}, \tag{10.46}$$

which allows us to understand μ a bit better: It is the amount of energy added to the system when we add a particle to it (or it is the amount of energy required in order to add a particle to the system). Additionally, returning to our Legendre transforms of Section 6.3.2, the μdN term just "comes along for the ride," so we have

$$dF = -SdT - \overline{p}dV + \mu dN, \tag{10.47}$$

$$dG = -SdT + Vd\overline{p} + \mu dN, \tag{10.48}$$

$$dH = TdS - \overline{p}dV + \mu dN, \tag{10.49}$$

12 I could call this *phase* equilibrium given the context here, but we will shortly generalize this to other sorts of systems, so I will use this more general term.

from which we can say

$$\mu = \left(\frac{\partial F}{\partial N}\right)_{T,V}, \tag{10.50}$$

$$\mu = \left(\frac{\partial G}{\partial N}\right)_{T,p}, \tag{10.51}$$

$$\mu = \left(\frac{\partial H}{\partial N}\right)_{S,p}. \tag{10.52}$$

Exercise 10.11 Derive Eqs. (10.47)–(10.52).

Exercise 10.12 Evaluate the chemical potential for a classical monatomic ideal gas using the (fudged) partition function we calculated before,

$$\ln Z_f = N\left[\ln\left(\frac{V}{N}\right) + \frac{3}{2}\ln T + \frac{3}{2}\ln\left(\frac{2mk_B\pi}{h_0^2}\right) + 1\right]. \tag{10.53}$$

Calculating μ from the Helmholtz free energy is the simplest method, and you should obtain

$$\mu = -k_B T\left[\ln\left(\frac{V}{N}\right) + \frac{3}{2}\ln T + \frac{3}{2}\ln\left(\frac{2mk_B\pi}{h_0^2}\right)\right]. \tag{10.54}$$

Notice that if we did *not* use fudged classical statistics, this expression would be incorrect!

Returning to the condition that $\mu_1 = \mu_2$ in equilibrium, we can write the Gibbs free energy for our system (recall Eq. (10.11)) in terms of the Gibbs free energies per molecule,[13]

$$G(T, \overline{p}, N_1, N_2) = N_1 g_1(T, \overline{p}) + N_2 g_2(T, \overline{p}), \tag{10.55}$$

which leads us, using Eq. (10.51) and ensuring we differentiate with respect to the proper number of molecules, to

$$\mu_1 = g_1 \text{ and } \mu_2 = g_2,$$

and so the Gibbs free energy per molecule for each substance must be equal (and thus so must the molar Gibbs free energies). This was just our condition from Eq. (10.13).

10.5.1 Systems with Many Components

Generalizing our results above is quite simple, if we (i) want to have many different types of molecules, and (ii) do not want to focus solely on phase transitions. We merely write the entropy of the entire system as

$$S = S(\bar{E}, V, N_1, N_2, \ldots, N_m),$$

where we have m types of molecules and N_i is the number of molecules of the ith type. The differential of the entropy becomes

$$dS = \left(\frac{\partial S}{\partial \bar{E}}\right)_{V,N} d\bar{E} + \left(\frac{\partial S}{\partial V}\right)_{E,N} dV + \sum_{i=1}^{m}\left(\frac{\partial S}{\partial N_i}\right)_{E,V,N} dN_i,$$

13 There is a slight difference between this and Eq. (10.11): Now we are defining g_1 and g_2 as the Gibbs free energies *per molecule*, while in Eq. (10.11) they were the molar free energies, but those two quantities only differ by a factor of Avogadro's constant.

and we are going to be a little sloppy with the notation with so many variables: N in the subscript refers to all N_j's for the first two terms and all that are not N_i in each term in the final term. The chemical potentials for the ith molecule are then

$$\mu_i \equiv -T\left(\frac{\partial S}{\partial N_i}\right)_{E,T,N},$$

and the first law of thermodynamics, Eq. (10.45), and all of the equations for the thermodynamic potentials, Eqs. (10.47)–(10.49), hold with $\mu dN \to \sum_{i=1}^{m} \mu_i dN_i$.

We can also come up with new Legendre transforms to eliminate the N_i's as dependent variables for μ_i, but these are not used as much as those we have seen before (see Problem 10.17 for an example). From those, there is a whole host of new Maxwell relations from the new mixed partials, such as

$$\left(\frac{\partial \mu_i}{\partial V}\right)_{S,N} = -\left(\frac{\partial \bar{p}}{\partial N_i}\right)_{S,V,N},$$

which can be necessary for some problems (see Problem 10.15). Usually those additional relations will only be introduced as needed, so I will let you consider them on your own. All of our original Maxwell relations still hold, so long as we remember that the numbers of particles are held constant along with the other constant variables in all of those previous partial derivatives, that is,

$$\left(\frac{\partial \bar{p}}{\partial T}\right)_V \to \left(\frac{\partial \bar{p}}{\partial T}\right)_{V,N}.$$

10.5.2 Gibbs–Duhem Relation

We are finally in a position to be able to discuss the missing ingredients in step 2 of our derivative crushing procedure from Section 6.4, by studying the extensive nature of thermodynamic quantities. If we consider the energy of a system, treating it as a function of its natural variables, S, V, and N,[14]

$$\bar{E} = \bar{E}(S, V, N). \tag{10.56}$$

All of the variables in this expression are extensive, so we know that if we scale the system by some amount z, then we have

$$S \to zS, \qquad V \to zV, \qquad N \to zN, \tag{10.57}$$

and we must also have $\bar{E} \to z\bar{E}$. Specifically,

$$\bar{E}(zS, zV, zN) = z\bar{E}(S, V, N). \tag{10.58}$$

Let's set $z = 1 + \delta z$, with $|\delta z| \ll 1$ corresponding to an infinitesimal amount. We can expand $\bar{E}(zS, zV, zN)$ about the point $\delta z = 0$ to get

$$\bar{E}(zS, zV, zN) = \bar{E}(S, V, N) + \left(\frac{\partial \bar{E}}{\partial S}\right)_{V,N} dS + \left(\frac{\partial \bar{E}}{\partial V}\right)_{S,N} dV + \left(\frac{\partial \bar{E}}{\partial N}\right)_{V,S} dN + \cdots. \tag{10.59}$$

14 For now I will just focus on systems with a single type of molecule, but this can readily be generalized to any number of species.

We also know the change in the entropy is given by

$$dS = zS - S = (z - 1)S = (\delta z)S,$$

and similarly for V and N, so that (dropping the $\cdots$),

$$\begin{aligned}\bar{E}(zS, zV, zN) &= \bar{E}(S, V, N) + \delta z\left[\left(\frac{\partial \bar{E}}{\partial S}\right)_{V,N} S + \left(\frac{\partial \bar{E}}{\partial V}\right)_{S,N} V + \left(\frac{\partial \bar{E}}{\partial N}\right)_{V,S} N\right] \\ &= \bar{E}(S, V, N) + \delta z\left(TS - \overline{p}V + \mu N\right).\end{aligned} \tag{10.60}$$

As we also know that $\bar{E}(zS, zV, zN) = z\bar{E} = (1 + \delta z)\bar{E}$, then we end up with

$$\bar{E} = TS - \overline{p}V + \mu N, \tag{10.61}$$

which is known as **Euler's theorem for homogeneous functions**, derived originally by Leonhard Euler (1707–1783). Differentiating this, we get

$$d\bar{E} = TdS + SdT - \overline{p}dV - Vd\overline{p} + \mu dN + Nd\mu, \tag{10.62}$$

and we can see the first, third, and fifth terms are nothing more than $d\bar{E}$ from the first law of thermodynamics. Eliminating this and solving for $Nd\mu$, we get

$$Nd\mu = -SdT + Vd\overline{p}. \tag{10.63}$$

This is known as the **Gibbs–Duhem relation**, which we can also write in terms of the entropy and volume per molecule,

$$d\mu = -sdT + vd\overline{p}. \tag{10.64}$$

This relation[15] is the last ingredient needed for crushing derivatives: It is required to eliminate any derivative with μ in it after bringing the chemical potential to the numerator.

Example 10.1 Consider the adiabatic expansion of a gas, where we were interested in the pressure change as we changed the volume at constant entropy. We showed in Eq. (7.5)

$$d\overline{p} = \left(\frac{\partial \overline{p}}{\partial V}\right)_{S,N} dV = -\frac{C_p}{C_V}\frac{1}{V\kappa}dV.$$

Here we have made it clear that N is also being held constant. We can now determine how the chemical potential of the system changes during this adiabatic process, by crushing

$$d\mu = \left(\frac{\partial \mu}{\partial V}\right)_{S,N} dV. \tag{10.65}$$

The Gibbs–Duhem relation allows us to write

$$\left(\frac{\partial \mu}{\partial V}\right)_{S,N} = \frac{1}{N}\left[-S\left(\frac{\partial T}{\partial V}\right)_{S,N} + V\left(\frac{\partial \overline{p}}{\partial V}\right)_{S,N}\right].$$

It's easy then to continue the rest of the steps of the crushing procedure to get

$$d\mu = \frac{1}{NC_V\kappa}\left(ST\alpha - C_p\right)dV. \tag{10.66}$$

While it does depend on the entropy, any *derivatives* that enter are the heat capacities, α, and κ as before.

15 Named after Gibbs and Pierre Duhem (1861–1916).

Exercise 10.13 Using Eq. (10.54) for the monatomic ideal gas, as well as our expression for the entropy in Eq. (8.47), verify Euler's theorem, Eq. (10.61), explicitly in that case.

Exercise 10.14 Suppose you have m different species, with $N_1, N_2, \ldots, N_m$ molecules of each type. Show that the more general Gibbs–Duhem relation is given by

$$\sum_{i=1}^{m} N_i d\mu_i = -SdT + Vd\overline{p}.$$

10.6 Chemical Equilibrium

In Section 10.5 we saw that moving from systems with a single type of molecule to those with many types is not too difficult, and this leads us quite naturally to being able to study chemical processes. We still focus only on systems in (or near) equilibrium, so we will stick with somewhat simple chemical processes—it would take more machinery than we have currently to consider very complex processes.

An example of such a process is the **dissociation** of water vapor. If you have a system with water vapor (with chemical formula H_2O), some of the hydrogen and oxygen naturally dissociates, or separates, from each other. The system will have some amount of water as well as some oxygen (O_2) and hydrogen (H_2) molecules, and they will coexist in equilibrium. We write this as

$$2H_2 + O_2 \rightleftharpoons 2H_2O, \tag{10.67}$$

where we have balanced the chemical equation (we assume that mass is conserved in our system).

> I won't go into balancing chemical equations in great detail, as it is introduced in an introductory chemistry course, but whenever you are given such a chemical equation, it's best to check that this has been done. Be sure you have the same number of *atoms* of each type on both sides of the equation, by multiplying the coefficient in front of a molecule by the number of atoms, which is given by the subscript (so on the right-hand side, we have $2 \times 2 = 4$ H atoms, and $2 \times 1 = 2$ O atoms, and we can see the same is true on the left-hand side). Some equations may be more complicated than this, of course, but the basic procedure will still work. This will be the example equation I will work with during this section, while also generalizing our previous work to any number of different types of molecules.

This system can be thought of as a system with N_{H_2} hydrogen molecules, N_{O_2} oxygen molecules, and N_{H_2O} water molecules. The double arrow in our equation lets us know that this isn't a *reaction* per se, but either side is a viable state for a set of hydrogen and oxygen atoms to be in; this is an equilibrium state. In more generality, we will eventually allow for the possibility k different molecules in our system ($k = 3$ for the dissociation of water), and we will denote them as X_i, with $i = 1, \ldots, k$. In our example, we could denote $X_1 = H_2, X_2 = O_2$, and $X_3 = H_2O$,[16]

$$2X_1 + X_2 \rightleftharpoons 2X_3. \tag{10.68}$$

This allows us to discuss this equation in a more abstract way. In order to turn this into a mathematical equation, we need to arbitrarily decide on a direction for the process. In our example, we will choose

$$2H_2O \rightarrow 2H_2 + O_2, \tag{10.69}$$

16 Of course, in this form, checking if the equation is balanced is not possible.

and any molecules on the left-hand side are considered the **reactants** and those on the right-hand side are called the **products** of the process. We then write the chemical equation in a **standard form** by moving the reactants to the same side of the equation as the products, and replace the arrow with an equality, to get

$$2H_2 + O_2 - 2H_2O = 0. \tag{10.70}$$

Going back to a general system, we would write

$$\sum_{i=1}^{k} x_i X_i = 0, \tag{10.71}$$

where the x_i's are known as the **stoichiometric coefficients**, which as we saw, are determined by properly balancing our chemical equation. For our example, we have $x_1 = x_{H_2} = +2, x_2 = x_{O_2} = +1$, and $x_3 = x_{H_2O} = -2$. The overall signs of the coefficient are arbitrary: All of the products will have the same sign as each other and the opposite sign to all of the reactants; only the *relative* signs matter. (And we arbitrarily define which are the products and which are the reactants when we "choose a direction" in our equilibrium reaction equation.) These coefficients tell us the relative relationships among the *changes* of each species, and could in principle be fractions, but generally it is common to use integers.

At any given point we have N_i molecules (or ν_i moles) of the ith species. While in equilibrium these could change, with the restriction that if the number of one species changes, the others must change as well according to the ratios of the stoichiometric coefficients. So if in our example of the dissociation of water, the system lost two water molecules, it *must* gain two hydrogen molecules and one oxygen molecule. This means that the changes in the numbers dN_i are all proportional to these stoichiometric coefficients, or

$$dN_i = \lambda x_i, \tag{10.72}$$

where λ is the same for all molecules and would be defined by the specific change being discussed. In our example where we lost two water molecules, $\lambda = +1$.

Exercise 10.15 Given the relationship between the number of molecules N_i and the number of moles ν_i, show that Eq. (10.72) can also be written as

$$d\nu_i = \lambda' x_i, \tag{10.73}$$

with λ' still a constant which is the same for all species. How is λ' related to λ?

Exercise 10.16 The chemical composition of propane is C_3H_8, and this combines with hydrogen gas H_2 to produce methane CH_4. Write a balanced chemical equation for this and determine the stoichiometric coefficients, by treating methane as the product.

Taking the system to be in stable equilibrium, we know that the Gibbs free energy is minimized and can write

$$dG = -SdT + Vd\overline{p} + \sum_i \mu_i dN_i = 0,$$

where we now have to use the more generalized version discussed before, including a chemical potential for each substance. We will consider simple chemical reactions which are performed at constant temperature and pressure, so we can say

$$\sum_i \mu_i dN_i = 0. \tag{10.74}$$

With our relationship between dN_i and the coefficients x_i, we can write this as

$$\sum_i \mu_i x_i = 0. \tag{10.75}$$

This constraint will allow us to relate the chemical potentials to each other. Additionally, once we have determined the entropy (or any of the thermodynamic potentials) for a system, the chemical potentials follow, and we can determine the relationships between the numbers of molecules (or moles) of each substance in the system. We'll do this explicitly for ideal gases in Section 10.7. For the dissociation of water, we see that the chemical potentials are related by

$$+2\mu_{H_2} + \mu_{O_2} - 2\mu_{H_2O} = 0. \tag{10.76}$$

Exercise 10.17 Using the result from Exercise 10.16, determine the equation which relates the chemical potentials of propane, hydrogen, and methane.

Notice that Eq. (10.75) is really just a statement of conservation of energy: Energy that is "lost" when we lose some water is just gained in the form of hydrogen and oxygen.

10.7 Chemical Equilibrium Between Ideal Gases

As shown in Problem 10.1, if we consider a system only in contact with a heat reservoir, we find that the Helmholtz free energy F is also minimized when in stable equilibrium, which means we can write (assuming constant temperature and no work being done, valid for most chemical reactions of interest here),

$$dF = \sum_i \mu_i dN_i = \sum_i \mu_i x_i = 0.$$

The resulting relationship is the same among the chemical potentials, but it is nice to frame it with F instead of G, because we know that the Helmholtz free energy is easier to determine from the partition function via $F = -k_B T \ln Z$. And this way we can quite readily determine the chemical potentials from F with

$$\mu_i = \left(\frac{\partial F}{\partial N_i}\right)_{T,V,N}.$$

In general, however, determining the chemical potential from this is not simple, given that it is only for very simple systems that we can easily calculate the partition function and thus F.

However, for a classical ideal gas, this is simple, because we consider the molecules all to be (essentially) independent of each other. The energy of the system in a state r is given by

$$E_r = \varepsilon_1(r_1) + \varepsilon_2(r_2) + \cdots + \varepsilon_N(r_N),$$

where $r = \{r_1, r_2, \ldots, r_N\}$ and r_i describes the state of one of the N molecules in the system. We also allow for different species in our system, perhaps k different species, so that $N = \sum_{j=1}^{k} N_j$ is the total number of molecules, and N_j is the number of molecules of the jth type. For a monatomic gas, ε_i is just the kinetic energy and the state will be defined solely by the momentum of the molecule.

In general, for a non-monatomic gas, the gas will additionally have internal degrees of freedom, and for now we will allow for this to be true. This means we could write

$$\varepsilon_i(r_i) = \frac{\mathbf{p}^2}{2m_i} + \varepsilon_i^{(\text{int})},$$

where $\varepsilon_i^{(\text{int})}$ will depend on relative positions and momenta of the atoms composing the molecule. It will *not* depend upon degrees of freedom of other molecules in the system. The partition function for distinguishable molecules is given by

$$Z = \sum_{r_1, r_2, \ldots} e^{-\beta[\varepsilon_1(r_1)+\varepsilon_2(r_2)+\cdots]},$$

and because the energies only depend on degrees of freedom for an individual molecule, we can factor the sum,

$$Z = \left[\sum_{r_1} e^{-\beta\varepsilon_1(r_1)}\right]\left[\sum_{r_2} e^{-\beta\varepsilon_2(r_2)}\right]\cdots = \prod_{i=1}^{N}\left[\sum_{r_i} e^{-\beta\varepsilon_i(r_i)}\right].$$

For each of these N factors, we have N_j factors that are identical for the molecule of type j, and we again define the single-particle partition function as

$$\zeta_j = \sum_r e^{-\beta\varepsilon_j(r)}$$

so the partition function becomes

$$Z = \zeta_1^{N_1}\zeta_2^{N_2}\ldots\zeta_k^{N_k}.$$

In this case, because the chemical potential is calculated from a derivative with respect to N_j, we must use fudged classical statistics to ensure we properly account for the indistinguishability of identical particles. Instead of Z, we will use Z_f, with

$$Z_f = \frac{\zeta_1^{N_1}\zeta_2^{N_2}\ldots\zeta_k^{N_k}}{N_1!N_2!\ldots N_k!} = Z_{f1}Z_{f2}\ldots Z_{fk}, \tag{10.77}$$

where Z_{fj} is the (fudged) partition function for the jth type of molecule,

$$Z_{fj} = \frac{1}{N_j!}\zeta^{N_j} \tag{10.78}$$

That the partition functions for each species multiply, and all physical quantities arise from the logarithm of Z_f, then we simply have, from now on dropping the limits on the sums for brevity,[17]

$$\begin{aligned}
\ln Z_f &= \sum_j \ln Z_{fj},\\
\bar{E}(T) &= \sum_j \bar{E}_j(T),\\
\bar{p} &= \sum_j \bar{p}_j,\\
S(T,V) &= \sum_j S_j(T,V), \text{ and}\\
F(T,V) &= \sum_j F_j(T,V).
\end{aligned}$$

We have seen $\bar{p}_j$ before: This is the **partial pressure** of the gas of type j, which is the pressure exerted by this gas if it were the only gas in the volume V. It is straightforward to derive the equation of state (with $n_j = N_j/V$ as usual)

$$\bar{p}_j = n_j k_B T,\ \bar{p} = \sum_j \bar{p}_j = nk_BT,$$

17 Note that we know from before that the mean energy for a classical ideal gas is independent of the volume, so we omit the volume dependence here.

which you derived in Problem 8.11.

For the jth gas, we find

$$\ln Z_{fj} = N_j \ln \zeta_j - \ln N_j! = N_j \left(\ln \zeta_j - \ln N_j + 1\right), \tag{10.79}$$

having used Stirling's formula in the second equality. The Helmholtz free energy is then

$$\begin{aligned} F &= -k_B T \ln Z_f \\ &= -k_B T \sum_j N_j \left(\ln \zeta_j - \ln N_j + 1\right) \\ &= -k_B T \sum_j N_j \left[\ln\left(\frac{\zeta_j}{N_j}\right) + 1\right]. \end{aligned} \tag{10.80}$$

The chemical potential for the jth gas follows immediately from this,

$$\begin{aligned} \mu_i &= \left(\frac{\partial F}{\partial N_i}\right)_{T,V,N} \\ &= -k_B T \ln(\zeta_j/N_j). \end{aligned} \tag{10.81}$$

Exercise 10.18 Fill in the steps to obtain Eq. (10.81).

Exercise 10.19 How would Eq. (10.81) differ if we did not use fudged classical statistics? Determine what μ_j would be in this case.

Using the fact that $\sum_j \mu_j x_j = 0$, we can derive the **law of mass action**. We start from[18]

$$\begin{aligned} \Delta F &= \sum_j \mu_j x_j \\ &= -k_B T \sum_j \ln(\zeta_j/N_j) x_j \\ &= -k_B T \sum_j \ln \zeta_j x_j + k_B T \sum_j \ln N_j x_j \end{aligned}$$

and we define the **standard free energy change of reaction** as

$$\Delta F^0 \equiv -k_B T \sum_j x_j \ln \zeta_j, \tag{10.82}$$

so

$$\Delta F = \Delta F^0 + k_B T \sum_j x_j \ln N_j.$$

ΔF^0 depends on the temperature and volume of the system, but not on the number of molecules. This can be thought of as the energy required to construct a system at temperature T and volume V, with no molecules yet, but expecting a particular set of species in the system (say H_2, O_2, and H_2O in our dissociation of water example from earlier). Then, in addition to that energy, the term $k_B T \sum_j x_j \ln N_j$ is the energy connected with creating the actual N_j molecules of each type themselves.

In equilibrium, $\Delta F = 0$, so we can derive

$$N_1^{x_1} N_2^{x_2} \ldots N_k^{x_k} = \exp\left(-\frac{\Delta F^0}{k_B T}\right) \equiv K_N(T, V), \tag{10.83}$$

18 We use ΔF and not dF because the right-hand side of this equation no longer has any differentials.

where we have introduced the **equilibrium constant**,

$$K_N(T,V) \equiv \exp\left(-\frac{\Delta F^0}{k_B T}\right) = \zeta_1^{x_1} \dots \zeta_k^{x_k}. \tag{10.84}$$

While K_N is called a constant, this is somewhat of a misnomer, as it does depend on T and V, through the individual partition functions. It does *not* depend on the number of molecules, although it is implicitly dependent on the number of *types* of molecules, k. The subscript N on K_N reminds us that this is related to the product of the numbers N_j of each type of molecule.

Exercise 10.20 Derive Eq. (10.83).

It is easy to determine this constant if we know how much of each type of species we have. Returning to our example involving the dissociation of water, we would get

$$K_N(T,V) = \frac{N_{H_2}^2 N_{O_2}}{N_{H_2O}^2}. \tag{10.85}$$

Because K_N depends on T and V, we see that the relative concentrations of each type of molecule will vary if we change these thermodynamic variables. If we can determine $K_N(T,V)$ from other means (which we will do shortly), then we can determine precisely what fraction of molecules are in what state as we change the temperature and volume. We will do a more explicit example after we rewrite $K_N(T,V)$ in a different form.

Exercise 10.21 The numbers of molecules N_j in our system will generally be very large, so as we have done before, it is common to consider the numbers of moles ν_j of each species. By simply using the relationship between ν_j and N_j, show that you can define a new equilibrium constant

$$K_\nu(T,V) = \nu_1^{x_1} \nu_2^{x_2} \dots \nu_k^{x_k},$$

and relate K_ν to K_N.

For a classical ideal gas, the integrals over the positions for each molecule merely give a factor of the volume for each single-particle partition function, so

$$\zeta_j(T,V) = V\zeta_j'(T).$$

$\zeta_i'(T)$ comes from integrating the momentum variables as well as possible internal degrees of freedom (so we could extend this to non-monatomic ideal gases). We can rewrite K_N as

$$\begin{aligned} K_N(T,V) &= \left(V^{x_1} \dots V^{x_k}\right) \zeta_1'^{x_1} \zeta_2'^{x_2} \dots \zeta_k'^{x_k} \\ &= V^x \zeta_1'^{x_1} \zeta_2'^{x_2} \dots \zeta_k'^{x_k} \end{aligned}$$

with $x = \sum_j x_j$. Then we have

$$N_1^{x_1} N_2^{x_2} \dots N_k^{x_k} = V^x \zeta_1'^{x_1} \zeta_2'^{x_2} \dots \zeta_k'^{x_k},$$

or

$$n_1^{x_1} n_2^{x_2} \dots n_k^{x_k} = \zeta_1'^{x_1} \zeta_2'^{x_2} \dots \zeta_k'^{x_k} \equiv K_n(T), \tag{10.86}$$

where this new equilibrium constant is independent of V because each ζ_j' depends only on the temperature. This form is more useful, and the n_j's are more accurately referred to as **concentrations**, as they are the number densities of each substance.

Example 10.2 Consider the reaction

$$H_2 + CO_2 \rightleftharpoons H_2O + CO,$$

such that we set up this system with $n^0_{H_2}$ and $n^0_{CO_2}$ molecules (per unit volume) of hydrogen and carbon dioxide gas (respectively) at some temperature T. What is the number density of water as a function of the equilibrium constant, $K_n(T)$?

We define the left-hand side as the reactants (as that is what we start with) so $H_2O + CO - H_2 - CO_2 = 0$, which determines the stoichiometric coefficients

$$x_{H_2O} = x_{CO} = -x_{H_2} = -x_{CO_2} = 1.$$

From this, the law of mass action states

$$K_n(T) = \frac{n_{H_2O} n_{CO}}{n_{H_2} n_{CO_2}},$$

which allows us to determine what the concentrations would be at different temperatures, assuming we know $K_n(T)$. We define the number density of water (which is the same as that of the carbon monoxide) as y, so

$$n_{H_2O} = n_{CO} = y.$$

The number densities of the other molecules are not independent of y. Using $dN_j = \lambda x_j$ from above, we can write

$$n_{H_2} = n^0_{H_2} - y, \qquad n_{CO_2} = n^0_{CO_2} - y.$$

We have for the law of mass action

$$\frac{y^2}{(n^0_{H_2} - y)(n^0_{CO_2} - y)} = K_n,$$

which is a simple quadratic equation we can solve to obtain

$$y = n^0 K_n \frac{1 \pm \sqrt{(1 - 2\xi^0)^2 + 4(1 - \xi^0)\xi^0 / K_n}}{2(K_n - 1)}, \tag{10.87}$$

where I have defined the initial fraction of hydrogen gas, $\xi^0 = n^0_{H_2}/n^0$ and the total number of reactants initially, $n^0 = n^0_{CO_2} + n^0_{H_2}$. From this point, if we knew K_n from the partition function (non-trivial for non-monatomic gases), then we could determine how much water we have at any given point as a function of the initial amount of hydrogen we had.

Exercise 10.22 Fill in the steps to obtain Eq. (10.87).

Example 10.3 While we would need to know $K_n(T)$ explicitly to determine anything quantitative, we can use the results from the previous example to determine the amount of hydrogen that gives a maximum yield of water at a given temperature. That is, what value of ξ^0 maximizes y, or rather, when is

$$\frac{\partial y}{\partial \xi^0} = 0?$$

Evaluating this derivative we get

$$\frac{\partial y}{\partial \xi^0} = \mp n^0 \frac{1 - 2\xi^0}{\sqrt{1 + 4(1 - K_n)\xi^0 (1 - \xi^0 / K_n)}}. \tag{10.88}$$

Setting this equal to zero we obtain

$$\xi^0 = 1/2,$$

so we need to start with equal parts of hydrogen and carbon dioxide to get the most amount of water at any temperature (regardless of $K_n(T)$). This could have been guessed: As the stoichiometric coefficients are all equal, the maximum yield would come from having equal amounts of each substance.

Exercise 10.23 Fill in the steps to obtain Eq. (10.88).

Note we didn't choose a sign for y in Eq. (10.87), because we don't have an expression for $K_n(T)$. To determine the correct sign, we would have to evaluate this expression to see if any unphysical results appeared (for example, if $y < 0$, this would not make sense). There are several examples in the problems which allow you to study the law of mass action further.

10.8 Summary

- For a system to be in stable equilibrium when in contact with a heat and work reservoir, the (pseudo) Gibbs free energy must be a minimum, which requires the heat capacity at constant volume and the isothermal compressibility to be positive (Le Châtelier's principle).
- The entropy and volume are discontinuous during a (first-order) phase transition and the slope of the phase equilibrium line can be determined with the Clausius–Clapeyron equation.
- The requirements for stable equilibrium allow us to observe phase transitions in a substance while studying the equation of state (if it permits such transitions). Using the Maxwell construction, we can determine the proper pressure vs. volume plot for a substance at all temperatures, including the phase transition region.
- When considering systems with multiple components (and changing numbers of components), we introduce the chemical potential to describe the energy required to add molecules to our system.
- Again using the requirements for stable equilibrium, we can understand the chemical equilibrium of many systems using the law of mass action. This allows us to determine the relative concentrations of molecules in a reaction if we know how the equilibrium constant depends upon temperature.

Problems

10.1 Consider a system only in contact with heat reservoir and not a work reservoir as we discussed in Section 10.2.2. Work through the steps in that section but instead of defining a pseudo-Gibbs free energy, define a **pseudo-Helmholtz free energy**,

$$F_0 = \bar{E} - T_0 S.$$

Show that $-\Delta F_0 \geq W$, so if no work is done $W = 0$, then

$$\Delta F_0 \leq 0.$$

From this show that the probability of the system being in a state with a thermodynamic parameter x is given by

$$P(x) \propto e^{-F_0(x)/(k_B T_0)}.$$

10.2 Consider a system in contact with a work and heat reservoir as in Section 10.2.2 and allow for fluctuations in *both* the temperature and pressure. In this case, determine the *normalized* probability $\mathcal{P}(V,T)dVdT$ that the volume of this portion lies between V and $V+dV$ and that its temperature lies between T and $T+dT$. We worked through much of this already, so you simply need to show the mixed second derivative in Eq. (10.23) vanishes and normalize it.
Hint: In normalizing the probability, you can to take the limits on T and V to be $-\infty$ to $+\infty$ instead of 0 to $-\infty$. Why?

10.3 Consider the pseudo-Gibbs free energy $G_0 = E - T_0 S + p_0 V$, but now treat it as a function of T and $\overline{p}$ instead of T and V. Calculate the fluctuations in $G_0(T,\overline{p})$ about the equilibrium point and show that Le Châtelier's principle holds. That is, show that our requirement that G_0 is at a minimum in equilibrium requires

$$C_p \geq 0, \text{ and } \kappa > 0.$$

The second of this we saw in Section 10.2.2, and the first of these we had assumed when we introduced heat capacity.

10.4 Suppose that a certain liquid boils at 400 K at around one atmosphere and that a 5% rise in the pressure raises the boiling point by 1 K. Estimate the latent heat of vaporization in joules per mole, assuming you can treat the vapor as an ideal gas.

10.5 The vapor pressures p (in mmHg) of solid and liquid chlorine are given by

$$\ln p_{\text{solid}} = 24.32 - \frac{3777}{T}, \text{ and}$$

$$\ln p_{\text{liquid}} = 17.89 - \frac{2669}{T}.$$

(a) What is the temperature of the triple point?
(b) What are the latent heats of sublimation and vaporization at the triple point?
(c) What is the latent heat of melting at the triple point?
Hint: The forms of these equations imply the assumptions of Section 10.3.2 hold here.

10.6 Repeat Problem 10.5 for ammonia, where the equations given are instead

$$\ln p_{\text{solid}} = 23.03 - \frac{3754}{T}, \text{ and}$$

$$\ln p_{\text{liquid}} = 19.49 - \frac{3063}{T}.$$

10.7 People can skate on ice because a thin layer of water forms when we stand (and then skate) on ice. Often, a simple explanation is **regelation**, where the pressure of one standing on ice causes the freezing point of the water to decrease below 0°C, thus melting the ice. One can use the Clausius–Clapeyron equation to estimate how much the freezing point changes when a person on skates stands on ice, to see if this in fact a viable explanation.
Do this by approximating $dp/dT \approx \Delta p/\Delta T$, and estimate ΔT when the pressure on ice changes by a person of mass 75 kg standing on a pair of skates with area 0.5 mm × 20 cm. The density of water is 1 g/mL, the density of ice is 0.92 g/mL, and the latent heat of fusion for ice is 334 J/g.

Hint: You should find this temperature change to be very small, too small in fact for this to be the cause of the layer of water that appears when skating, so clearly there are other factors that give us this layer of water which allows us to skate.

10.8 Determine how the molar latent heat of transformation depends upon temperature, $d\ell/dT$. ℓ is the molar latent heat when a system goes from phase 1 to phase 2 at the temperature T and pressure $\overline{p}$. While this is a total derivative, you will need to crush some partial derivatives. Note that some parameters (such as the final first derivatives that may appear) will be different for each phase, while others such as $\overline{p}$ and T are not.
Hint: Start by writing $\ell = T\Delta s = T(s_2 - s_1)$, and think of each of the s_i's as functions of T and $\overline{p}$.

10.9 For the system in Problem 10.8, calculate the pressure dependence of the molar latent heat of transformation instead.

10.10 In a second-order phase transition, there is no difference in entropy or volume between the two phases and therefore the Clausius–Clapeyron equation cannot be applied. Two **Ehrenfest equations** can, however, be derived. They follow from the equality in entropy or volume across the phase transition, and they relate $d\overline{p}/dT$ to differences in some or all of α, C_p, and/or κ between the two phases.

(a) To get the first Ehrenfest equation, set $dS_1 = dS_2$, find $d\overline{p}/dT$ in terms of the differences in α and C_p between the two phases.
(b) For the second Ehrenfest equation, set $dV_1 = dV_2$ and find $d\overline{p}/dT$ in terms of the differences in α and κ between the two phases.

10.11 Suppose we have the **Dieterici equation of state,**

$$\overline{p}(v - b) = RT \exp\left(-\frac{a}{RTv}\right).$$

(a) Evaluate the critical constants p_c, T_c, and v_c in terms of a and b.
(b) Rewrite the equation of state in terms of the reduced variables $p' = \overline{p}/p_c$ (and similarly for v' and T').
(c) Plot p' vs. v' for several values of T' above and below the critical point (use Mathematica or similar software for this, putting them all on the same plot). Does this equation of state allow for a phase transition? Explain using your plots.

10.12 Repeat Problem 10.11 for the **Berthelot equation of state,**

$$\overline{p} = \frac{RT}{v - b} - \frac{a}{Tv^2}.$$

10.13 Consider a classical monatomic ideal gas of electrons in thermal equilibrium at temperature T in a container of volume V in the presence of a uniform electric field with magnitude $\mathcal{E}$ and pointed in the $+x$-direction. The energy of a state for one electron is $\varepsilon = \mathbf{p}^2/2m - e\mathcal{E}x$ at a given x.

(a) Calculate $\zeta(x)$, the partition function per molecule in a small volume ΔV at distance x. You can treat the volume ΔV to be small enough that $x \approx$ constant.
(b) From $\zeta(x)$, determine F, the Helmholtz free energy.

(c) Calculate the chemical potential μ of an element of volume of such a gas as a function of $\overline{p}$, T, and x.
(d) Use the requirement that μ is constant to determine the dependence that the pressure p has on T and x.

10.14 Consider the differentials of the potentials given by Eqs. (10.45), (10.47), (10.48), and (10.49). Given the extra terms in these expressions, each of these give two additional Maxwell relations. Determine all eight of these new Maxwell relations by equating the mixed second derivatives.

10.15 How does the chemical potential change in the following processes?
(a) In the free expansion of a gas, and
(b) in the Joule–Thomson process.
Hint: Determine the relevant derivative and crush it for each case.

10.16 With the introduction of the chemical potential, we can study the process of heating a room. When doing so, we will find that the entropy of the room actually *decreases* in the process, which sounds counterintuitive! For this we will assume the air in the room can be described by a classical ideal gas.
(a) First consider entropy to be a function of temperature, volume, and energy, and argue that if the volume and energy remain constant, we have

$$dS = \left(\frac{\partial S}{\partial T}\right)_{E,V} dT.$$

(b) Use the chain rule for the partial derivative above to write

$$dS = \left(\frac{\partial N}{\partial T}\right)_{E,V} \left(\frac{\partial S}{\partial N}\right)_{E,V} dT,$$

and use the equation of state for an ideal gas and the definition of the chemical potential to write

$$dS = \frac{\mu}{k_B T} \frac{\overline{p}V}{T^2} dT.$$

(c) Argue from our determination of the chemical potential of an ideal gas in Eq. (10.54) that $\mu < 0$ and thus $dS < 0$.
(d) Find the finite change in the entropy and the heat flow into the system if the room increases from some initial temperature T_i to a final temperature T_f.
See Ref. [8] for a more general calculation of this process as well as a nice discussion of it.

10.17 Calculate the chemical potential for a Van der Waals gas, using Eqs. (8.53) and (8.58). Write your final result in terms of the parameters a and b used in the Van der Waal's equation of state. To simplify your result, you can assume that $n = N/V$ is small so you can Taylor expand in an appropriate place.

10.18 The **grand potential** for a system with one type of molecule is a Legendre transform defined by

$$\Omega \equiv \overline{E} - TS - \mu N. \tag{10.89}$$

(a) Determine $d\Omega$, simplifying your expression using the first law.
(b) What are the natural variables that Ω depends upon?
(c) Determine all of the Maxwell relations that arise from this potential.

10.19 Consider the following chemical reaction between ideal gases:

$$\sum_{j=1}^{k} x_j X_j = 0.$$

Let the temperature be T, the total pressure be $\overline{p}$. Denote the partial pressure of the jth species by $\overline{p}_j$. Show that the law of mass action can be put into the form

$$\overline{p}_1^{x_1} \overline{p}_2^{x_2} \dots \overline{p}_k^{x_k} = K_p(T),$$

where the constant $K_p(T)$ depends only on T. What is K_p in terms of K_n?

10.20 Recall Example 10.2, where at a fixed temperature T, carbon dioxide and hydrogen gas interact to form carbon monoxide and water vapor. If we maintain the temperature T but increase the volume, what happens to the relative concentration of CO_2 compared to the other gases? That is, does it increase, decrease, or stay constant?

10.21 Consider the example of the dissociation of water vapor. Suppose initially there are ν_0 moles of water vapor at a low enough temperature in a volume V so that it is entirely undissociated (there is no hydrogen or oxygen gas). As we raise the temperature, dissociation will occur as we discussed. If ξ is the fraction of water molecules which are dissociated at any temperature T, determine how ξ relates to the total pressure of the gas $\overline{p}$ and the equilibrium constant $K_p(T)$ from Problem 10.19.

10.22 Show that K_n increases as T is increased, if the change in the energy of a system is positive, in agreement with Le Châtelier's principle. In other words, show that

$$\frac{d \ln K_n}{dT} = \frac{\Delta E}{k_B T^2}.$$

10.23 A simple model for the partition function for a liquid treats the liquid as if the molecules formed a gas of molecules moving independently, with the following assumptions:

i. Each molecule has an additional binding energy in the form of a constant potential energy $-u_0$, arising from the average interaction with the other molecules.
ii. The volume each liquid molecule can travel through is given by $N_\ell v_\ell$, where v_ℓ is the constant volume available per molecule in the liquid phase.

Otherwise, our expression for the partition function of an ideal gas, Eq. (10.78), for example, holds (with the same single-particle partition function).

(a) Within this model, what is the partition function for a liquid with N_ℓ molecules?
(b) What is the chemical potential μ_ℓ for N_ℓ molecules of liquid at the temperature T?
(c) Write down μ_g, the chemical potential of an ideal gas in terms of the volume per molecule of the gas $v_g = V_g/N_g$ and ζ.
(d) If the two phases are in equilibrium, $\mu_g = \mu_\ell$. Use this to determine the vapor pressure of the liquid at a temperature T.

(e) What is $\Delta s = s_g - s_\ell$, the molar entropy difference between the gas and liquid in equilibrium at this T and $\overline{p}$?
(f) From Δs, determine the molar heat of vaporization ℓ. Show that $\ell = N_A u_0$ if $u_0 \gg k_B T$. [Do not assume $v_g \gg v_\ell$!]
(g) The **boiling point** T_b is defined as the temperature where the vapor pressure equals 1 atm. What is ℓ / RT_b in terms of the variables in the problem?
(h) Show that the order of magnitude of ℓ / RT_b is a number of order 10 for ordinary liquids. (This result is called **Trouton's rule**, named after Frederick Trouton (1863–1922).)
(i) Compare this simple theory with experiment by looking up the densities and molecular weights of some liquids online, computing ℓ / T_b, and comparing with the experimental ratio of ℓ / T_b.

10.24 Consider the reaction

$$N_2 + O_2 \rightleftharpoons 2NO$$

or

$$2NO - N_2 - O_2 = 0.$$

Suppose that at some very low temperature, we start with a density n_0 of pure NO molecules in some fixed volume. The temperature is then raised to temperature T, at which the reaction above may occur. Let $K_n(T)$ be the equilibrium constant for this reaction, and let ξ be the fraction of NO molecules that remain at any given temperature ($\xi = 1$ initially).

(a) Solve for ξ as a function of K_n. Note you'll have two solutions for your quadratic equation—be sure to explain *physically* why you chose the solution you did (recalling that $0 \leq \xi \leq 1$).
(b) Show that if $K_n \gg 1$, almost all of the NO molecules remain in the system.
(c) Show that if $K_n \ll 1$, about $\sqrt{K_n}/2$ of the NO molecules remains in the system.
(d) Show that if $K_n = 1$, one-third of the NO molecules remains in the system.
(e) Treat these molecules, as classical "monatomic" ideal gases,[19] but let them each have a constant negative potential energy: $-E_{NO}$, $-E_{N_2}$, and $-E_{O_2}$ (these are binding energies that can represent what the internal degrees of freedom will contribute to the energy). Calculate $K_n(T)$ in terms of the binding energies, the masses of N and O, T, and h_0. You can assume the masses of the diatomic molecules are just the sum of the atoms that make up the molecules.
(f) Is there any ambiguity in your result for K_n? That is, remember that h_0 is an arbitrary parameter; does your result depend on this parameter?

10.25 Consider the recombination of hydrogen ions (and the ionization of hydrogen atoms)

$$H^+ + e^- \rightleftharpoons H.$$

Suppose that at some very low temperature, we start with a density n_0 of pure hydrogen atoms in some fixed volume. The temperature is then raised to temperature T, at which

19 Yes, a terrible idea because these are all diatomic molecules! But this allows us to find a concrete result for now. We will be able to correct this in Chapter 11 when we calculate the partition function for a diatomic molecule.

some ionization may occur. Let $K_n(T)$ be the equilibrium constant for this reaction, and define $x \equiv n_0 K_n(T)$.

(a) Show that if $x \gg 1$, only one in every $\sqrt{x}$ of the H atoms is ionized.
(b) Show that if $x \ll 1$, the gas is almost entirely ionized.
(c) Show that if $x = 1$, ~62% of the H atoms are ionized.
(d) Treat the H atoms, and the H^+ and e^- ions, as classical monatomic ideal gases, but let the H atoms have a constant negative potential energy $-E_0$ (the binding energy). Calculate $K_n(T)$ in terms of E_0; the masses of H, H^+, and e^-; T; and h_0.
(e) Explain why your result for K_n cannot be valid here.
(f) Ignore this fact and plug in numerical values to get $K_n(T)$ at $T = 10^4$ K, $T = 10^5$ K, and $T = 10^6$ K. Use $E_0 = 13.6$ eV and $h_0 = h$ (Planck's constant).[20]
(g) Calculate the fraction of ionized H atoms at $T = 10^4$ K, $T = 10^5$ K, and $T = 10^6$K. Use $n_0 = 10^{23}/\text{L}$.

10.26 Suppose you have 1/2 mol of H_2S, 3/4 mol of H_2O, 2 mol of H_2, and 1 mol of SO_2 which are allowed to react in a vessel maintained at a temperature of 300 K and a pressure of 10^4 Pa.

(a) Write down a balanced chemical reaction, with H_2S and H_2O as the reactants.
(b) Write down the condition for equilibrium in terms of the chemical potentials.
(c) Show that the number of H_2 moles is given by

$$n_{H_2} = 2 - 3\lambda,$$

where λ is the proportionality constant defined in Eq. (10.72), and write similar equations for the other three components. For each component, find the value of λ where n_i vanishes.
(d) Show that $\lambda_{max} = 2/3$ and $\lambda_{min} = -3/8$. Which components are depleted in these cases?
(e) Assume the nominal solution of the equilibrium condition gives $\lambda = 1/4$. What fraction of each component is in the mixture?
(f) Suppose that we increase the pressure, and this changes the nominal value of λ to 4/5. What fraction of each component is in the mixture now?

References

1 J. Anacleto. Work reservoirs in thermodynamics. *European Journal of Physics*, 31(3):617, Apr 2010.

2 Wikimedia Commons. URL https://commons.wikimedia.org/wiki/File:Phase_diagram_of_water.svg. Accessed: 16 March 2023.

3 C. Hill. URL https://scipython.com/blog/the-maxwell-construction/. Accessed: 17 March 2023.

4 G. Jaeger. The Ehrenfest classification of phase transitions: Introduction and evolution. *Archive for History of Exact Sciences*, 53(1):51–81, May 1998.

5 H. B. Callen. *Thermodynamics and An Introduction to Thermostatistics*. Wiley, New York, NY, 2nd edition, 1985.

20 Max Planck (1858–1947).

6 L. P. Kadanoff, W. Götze, D. Hamblen, R. Hecht, E. A. S. Lewis, V. V. Palciauskas, M. Rayl, J. Swift, D. Aspnes, and J. Kane. Static phenomena near critical points: Theory and experiment. *Reviews of Modern Physics*, 39:395–431, Apr 1967.

7 C. Mo, G. Zhang, Z. Zhang, D. Yan, and S. Yang. A modified solid–liquid–gas phase equation of state. *ACS Omega*, 7(11):9322–9332, Mar 2022.

8 H. J. Kreuzer and S. H. Payne. Thermodynamics of heating a room. *American Journal of Physics*, 79(1):74–77, 2011.

11

Quantum Statistics

Up until now, we have only studied statistical mechanics with classical systems. Of course, we have used quantum energy levels in some cases (such as the Einstein model for the specific heat of a solid), but we still treated the thermodynamic problem classically. In fact, this is precisely what caused the Gibbs paradox (from the overcounting of states) of Section 8.5, where we had to introduce a fudge factor to account for the fact that we were treating our system with distinguishable particles when we shouldn't have been.

Now we will formulate the problem from a quantum mechanical point of view (after introducing a new method). Not only will this allow us to treat quantum mechanical systems more correctly, but we will be able to show that fudged classical statistics is the proper classical limit of quantum statistics.

After finishing this chapter, you should be able to

- formulate a system in contact with a heat reservoir and a particle reservoir,
- restate the statistical problem to calculate the occupation number for classical particles and two classes of quantum particles (fermions and bosons),
- show the classical limit of the quantum problem is indeed fudged classical statistics, and
- work through a few simple applications of these new techniques.

11.1 Grand Canonical Ensemble

In Chapter 10 we considered processes where the number of particles in the system could change, primarily in the context of phase transitions or chemical reactions. We can formalize this further from the beginning by considering the **grand canonical ensemble**, which is, in a sense, an extension of the canonical ensemble.

11.1.1 A System in Contact with a Particle Reservoir

Now we will consider a system A in contact with a system A' that acts as both a heat reservoir at constant temperature T and a **particle reservoir**. The particle reservoir can be considered an

Statistical Thermodynamics: An Information Theory Approach, First Edition. Christopher Aubin.

Companion website: www.wiley.com/go/Aubin/StatisticalThermodynamics

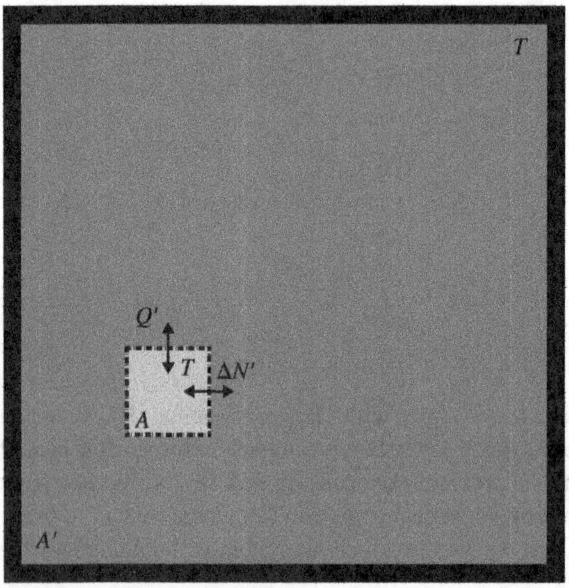

Figure 11.1 A system A (with rigid boundaries) in thermal contact with a heat and particle reservoir A' at temperature T. The combined system is thermally isolated from the outside world.

infinite source or sink of particles for our system A, much like electric ground works as such an infinite source of electrons, or the ocean is a particle reservoir for water.[1] This is shown in Figure 11.1, which is very similar to Figure 6.1 but now we allow particles to pass through the rigid, but thermally conducting, barrier between A and A'.

As in the case of the canonical ensemble, we would like to determine the probability that the system A will be in a microstate r with energy E_r and number of particles N_r. Later we will consider possibly changing the volume or other external parameters, but for now we'll keep those fixed. (By now, from experience, you should feel OK doing this, as adding in these additional changes merely amounts to adding a variable that is held constant when partial derivatives enter the picture.)

As in previous cases, the combined system $A^{(0)} = A + A'$ is isolated from the surroundings, so the total number of particles as well as the total energy are fixed, or

$$N_r + N' = N^{(0)} = \text{constant}, \tag{11.1}$$

$$E_r + E' = E^{(0)} = \text{constant}. \tag{11.2}$$

Additionally, the reservoir is much larger than the system A, so we require

$$N_r \ll N' \quad \text{and} \quad E_r \ll E', \tag{11.3}$$

where the latter requirement we have used before. If we wanted to be rigorous with this statement, remember that the requirement $E_r \ll E'$ comes from Eq. (6.5), where the change in the temperature of the heat reservoir when we change its energy is negligible. We can (and will) consider a similarly rigorous statement for the former requirement shortly—in this case it turns out to be the chemical potential of the reservoir which will not change (significantly) when we change the number of particles. For now we will use the simpler statement in Eq. (11.3).

The probability of A being in a given state is proportional to the number of states in the reservoir (see the discussion before Eq. (8.3) as well as Exercise 8.1), so that

$$P_r \propto \Omega'(E^{(0)} - E_r, N^{(0)} - N_r). \tag{11.4}$$

1 These are closer analogies to a particle reservoir than they were for the heat or work reservoirs!

Exercise 11.1 Using the arguments of Section 8.1.2, verify Eq. (11.4).

Following the same steps we used for the canonical ensemble, let us expand the logarithm of the probability about $E_r = 0$ and $N_r = 0$, to obtain

$$\begin{aligned} \ln P_r &= \ln \Omega' + (\text{const}) \\ &\approx \ln \Omega'(E^{(0)}, N^{(0)}) - \left(\frac{\partial \ln \Omega'}{\partial E'}\right)_{N'} E_r - \left(\frac{\partial \ln \Omega'}{\partial N'}\right)_{E'} N_r + (\text{const}). \end{aligned} \tag{11.5}$$

The first term is merely β, and the second is related to the chemical potential,

$$\left(\frac{\partial \ln \Omega'}{\partial N'}\right)_{E'} = -\beta\mu, \tag{11.6}$$

because we know that

$$\left(\frac{\partial S}{\partial N}\right)_E = -\frac{\mu}{T},$$

and $S = k_B \ln \Omega$. Just as in the case of β, we omit the prime on μ, because they are equal for A and A' when in equilibrium.

Exercise 11.2 Instead of stating that $N_r \ll N'$ is our condition for A' to be a particle reservoir, we instead require that

$$\left|\frac{\partial \mu'}{\partial \bar{E}'} N'\right| \ll \mu'. \tag{11.7}$$

Derive this expression using the same method as that used to derive Eq. (6.5). In this case you should consider μ' to be a function of the energy $\bar{E}'$.

Thus we find the probability that the system is in the state labeled by r is

$$P_r = \frac{1}{\mathcal{Z}} e^{-\beta(E_r - \mu N_r)}, \tag{11.8}$$

where

$$\mathcal{Z} = \sum_r e^{-\beta(E_r - \mu N_r)} \tag{11.9}$$

is the **grand partition function**, which is required to properly normalize our probability in Eq. (11.8), and this is known as the **grand canonical distribution**.

Just as with the canonical ensemble (or any system with a known probability distribution), we can determine the mean energies, mean numbers of particles, and mean values of other quantities with

$$\bar{E} = \frac{1}{\mathcal{Z}} \sum_r E_r e^{-\beta(E_r - \mu N_r)} = -\frac{\partial \ln \mathcal{Z}}{\partial \beta}, \tag{11.10}$$

$$\overline{N} = \frac{1}{\mathcal{Z}} \sum_r N_r e^{-\beta(E_r - \mu N_r)} = \frac{1}{\beta}\frac{\partial \ln \mathcal{Z}}{\partial \mu}, \text{ and} \tag{11.11}$$

$$\bar{y} = \frac{1}{\mathcal{Z}} \sum_r y_r e^{-\beta(E_r - \mu N_r)}. \tag{11.12}$$

In this last line, y can be any quantity (E^2, N^2, p, etc.) of interest. In the first two lines, we wrote the mean values in terms of appropriate derivatives of the logarithm of the grand partition function as we did before. For other quantities, as we saw for the canonical ensemble, we will try to find appropriate derivatives to simplify calculations as needed. Also, a similar argument to that in Section 8.7 can be made to show that this distribution can also describe a system with fixed mean energy and fixed $\overline{N}$, and from those one could determine β and μ from the above equations as in Problem 8.4. (You can use the constraint that $\overline{N}$ is fixed for an alternate derivation of the grand canonical distribution in Problem 11.1.)

Exercise 11.3 Show that the variance in the number of particles in our system can be written as

$$\overline{(\Delta N)^2} = \overline{N^2} - \overline{N}^2 = \frac{1}{\beta^2}\frac{\partial^2 \ln \mathcal{Z}}{\partial \mu^2} = \frac{1}{\beta^2}\left(\frac{\partial \overline{N}}{\partial \mu}\right)_{T,V}. \tag{11.13}$$

In the second equality I made it explicit that T and V are held constant.

11.1.2 Connecting $\mathcal{Z}$ to Thermodynamics

In order to connect this to thermodynamics, we would like to relate the grand partition function to the entropy, which then allows us to relate it to the ordinary partition function.

Exercise 11.4 It is quite simple to find this relationship by starting with the entropy in its most general form. Show that substituting Eq. (11.8) into $S = -k_B \sum_r P_r \ln P_r$, you obtain

$$S = k_B\left(\ln \mathcal{Z} + \beta \bar{E} - \beta\mu\overline{N}\right). \tag{11.14}$$

From Exercise 11.4, and because we know from the canonical ensemble that $S = k_B\left(\ln Z + \beta\bar{E}\right)$, we can relate Z to $\mathcal{Z}$ by

$$\ln Z = \ln \mathcal{Z} - \beta\mu\overline{N}. \tag{11.15}$$

Recall that when we connected the microcanonical and canonical ensembles, we argued that for large systems, the energy in the former ensemble is the equal to the mean energy of the latter.[2] The same is true here, as for large systems, $\overline{N}$ (for the grand canonical ensemble) is equal to N for the canonical ensemble. Again this requires that the variance as related to $\overline{N}$ is small, or in the grand canonical ensemble,

$$\frac{\overline{(\Delta N)^2}}{\overline{N}^2} \ll 1. \tag{11.16}$$

Thus, the probability P_r of being in a given state must drop rapidly as the number of molecules in the system moves away from $\overline{N}$ such that we can safely say that $P_r = 0$ to an excellent approximation.

Example 11.1 For the classical ideal gas, we can calculate the grand partition function simply. Our sum over states in r includes a sum over the number of particles, so we can write (including a fudge factor of $N!$ to avoid the Gibbs paradox)

$$\begin{aligned}\mathcal{Z} &= \sum_r e^{-\beta(E_r-\mu N_r)} \\ &= \sum_{N=0}^{\infty}\frac{1}{N!}\sum_{r'} e^{-\beta(E_{r'}-\mu N)}.\end{aligned}$$

2 Really this is an approximate statement, but the difference for macroscopic systems is negligibly small, so as usual I will just say "equal."

In this expression, r' now corresponds to the states allowed for a given number of particles, and as such we can write this as

$$\mathcal{Z} = \sum_{N=0}^{\infty} \frac{1}{N!} e^{\beta\mu N} \zeta^N,$$

where ζ is the single-particle partition function. This sum is nothing more than the expansion of the exponential (note the importance of the fudge factor!), so

$$\mathcal{Z} = \exp\left(e^{\beta\mu}\zeta\right).$$

Remember that ζ is only a function of T and V. Thus we have

$$\overline{N} = e^{\beta\mu}\zeta,$$

and

$$\overline{(\Delta N)^2} = e^{\beta\mu}\zeta,$$

so we can see that

$$\frac{\overline{(\Delta N)^2}}{\overline{N}^2} = \frac{1}{e^{\beta\mu}\zeta} = \frac{1}{\overline{N}}.$$

Thus, as we saw with the canonical ensemble when considering the energy, this ratio of ΔN to $\overline{N}$ falls off like $1/\sqrt{\overline{N}}$, so we can see that the two ensembles here can be used equivalently for large systems.

Exercise 11.5 Use the result for $\mathcal{Z}$ in Example 11.1 and our expressions for the fudged partition function,

$$\ln Z_f = N \ln \zeta - N \ln N + N,$$

and the chemical potential,

$$\mu = -k_B T \ln\left(\frac{\zeta}{N}\right),$$

to verify Eq. (11.15), assuming $\overline{N} = N$.

We can see the connection between the canonical and grand canonical ensembles using another method to derive the grand partition function. This relies on the last statement in Section 11.1.1, that we can say that $P_r = 0$ when N is not very close to $\overline{N}$. As we saw in Example 11.1, the partition function Z is dependent on the number of particles, so we write $Z(N)$.

Exercise 11.6 For the monatomic ideal gas, using fudged classical statistics, we had

$$\ln Z_f(N) = N\left[\ln\left(\frac{V}{N}\right) + \frac{3}{2}\ln T + \frac{3}{2}\ln\left(\frac{2mk_B\pi}{h_0^2}\right) + 1\right]$$

Using Planck's constant for h_0, show that this is a rapidly increasing function of N. Specifically, assume the system is that of argon gas at room temperature, show that

$$Z(N) \sim \frac{e^{KN}}{N^N},$$

where K is a number of order 100.

(Continued)

(Continued)

Suppose we have the partition function for some number of particles, N', but are interested in Z at a different value, N. To start, we first realize that because $Z(N')$ is a rapidly increasing function of N', we can multiply it by a rapidly *decreasing* function, such as

$$Z(N')e^{-\xi N'},$$

where $\xi > 0$ is chosen so that this product is peaked at $N' = N$. ξ can be found by differentiation, and you'd find

$$\xi = \left(\frac{1}{Z}\frac{\partial Z}{\partial N'}\right)\Big|_{N'=N} = \left(\frac{\partial \ln Z}{\partial N'}\right)\Big|_{N'=N'}$$

which is merely the chemical potential up to a factor of $-\beta$.

An expansion of $Z(N')e^{-\xi N'}$ about $N' = N$ gives (see Problem 11.2)

$$Z(N')e^{-\xi N'} = Z(N)e^{-\xi N}\left[1 + \frac{1}{2}\left(\frac{\partial^2 \ln Z}{\partial N^2}\right)(N - N')^2 \cdots\right]. \tag{11.17}$$

As we expect Z to scale exponentially with N, such as our example above for argon, then

$$\left(\frac{\partial^2 \ln Z}{\partial N^2}\right) \sim \frac{1}{N},$$

and so even if N' is as much as $\sim \pm\sqrt{N}$ away from N (so $N' = N \pm \sqrt{N}$), this second term is still extremely small compared with unity. For systems with $N \sim 10^{24}$ particles, this is quite a large range.

We can then sum over all values of N' to select only those terms near $N' = N$, or

$$\sum_{N'} Z(N')e^{-\xi N'} = Z(N)e^{-\xi N}\Delta N'. \tag{11.18}$$

$\Delta N' \ll N$ is the width of the maximum and as N increases, $\Delta N'$ rapidly decreases. The left-hand side is just the grand partition function, so

$$\mathcal{Z}(N) = \sum_{N'} Z(N')e^{-\xi N'},$$

and taking the logarithm of both sides gives us

$$\ln Z = \xi N + \ln \mathcal{Z} + \cdots$$

where we have neglected $\ln \Delta N'$ (which we expect to be of order one for large N). Note that this argument relies on the fact that N is very large so that fixing N and fixing $\overline{N}$ would give the same results, just as we did with the energy of the system when comparing the canonical and microcanonical ensembles.

11.2 Classical vs. Quantum Statistics

Before considering statistical mechanics from the quantum point of view, we first should consider what makes the microscopic world different when we consider a system classically or quantum mechanically. The first issue we have already seen when discussing the Gibbs paradox: For a truly classical system, the particles are *distinguishable*, so that we can always label each particle and follow its trajectory. Quantum mechanically this is not possible, not even theoretically. The second issue for a quantum system, which is connected to the indistinguishability of the particles, is that there are two classes of particles which have very different behavior. First we will discuss the symmetry requirements of a quantum system that give rise to these two different classes that are not necessary in the classical case, and then we will to restate the problem at hand. That is, we want to treat the problem differently in this case due to the odd nature of quantum mechanics.

11.2.1 Symmetry Requirements

When describing a classical system, we need to be able to know the positions and momenta of every particle in the system as a function of time. Thus a state R of a three-dimensional system of N particles could be described as

$$R = \{(\mathbf{r}_1, \mathbf{p}_1), (\mathbf{r}_2, \mathbf{p}_2), \dots, (\mathbf{r}_N, \mathbf{p}_N)\}.$$

The way this is ordered is such that particle 1 has position $\mathbf{r}_1$ and momentum $\mathbf{p}_1$, particle 2 has position $\mathbf{r}_2$ and momentum $\mathbf{p}_2$, and so forth. However, we could also have a state

$$R' = \{(\mathbf{r}_2, \mathbf{p}_2), (\mathbf{r}_1, \mathbf{p}_1), \dots, (\mathbf{r}_N, \mathbf{p}_N)\},$$

where particle 2 has position $\mathbf{r}_1$ and momentum $\mathbf{p}_1$, particle 1 has position $\mathbf{r}_2$ and momentum $\mathbf{p}_2$, and the other positions and momenta are the same as in R. Classically, these are two distinct states, as would be any other combinations, but these are not *practically* distinct, hence the Gibbs paradox.

Quantum mechanically these would be the same state, although we do not describe a state in the same way. A quantum mechanical system is described by the **wavefunction**,

$$\psi\left(\alpha_1, \dots, \alpha_N\right),$$

where the α_i's are the important variables (or quantum numbers) which are required to properly define the system. As above, our convention is that in this case, particle 1 is in the state defined by α_1, particle 2 is in the state α_2, and so forth. In the simplest case, this might just be the position (or momentum, but not both) of the particle (and this is the most commonly known variable for an undergraduate course), but it could also include the spin of the particle. In more advanced treatments, the state of the system would be described by a state vector, and the α_i's could refer to such quantum numbers such as integers required to denote the energy of the particle, its angular momentum, and so forth.

Example 11.2 As a couple of examples of relevant quantum numbers, we revisit several quantum systems already discussed.

1. The one-dimensional simple harmonic oscillator, where the energy of a state is given by

$$\varepsilon_n = \left(n + \frac{1}{2}\right)\hbar\omega.$$

Here the state α would be described by $n = 0, 1, 2, \dots$, the relevant quantum number.

2. A system with total spin $\mathbf{J}$ has two relevant quantum numbers. The state α is determined by the total spin (squared) $J^2 = j(j+1)\hbar^2$, with $j = 0, 1/2, 1, \dots$, and for a given j, the z-projection of the spin, $J_z = m_j\hbar$, with m_j ranging from $-j$ to j in integer steps. j and m_j are the two quantum numbers, and either (or both) might be relevant for the energy of a system. For example, for a rigid rotator with moment of inertia I, we have

$$E_j = \frac{j(j+1)\hbar^2}{2I},$$

while for a spin in a magnetic field H in the z-direction,

$$U_{m_j} = -m_j \mu_0 H.$$

For a quantum system, unlike a classical system, R and R' correspond to the same state. That is, interchanging the properties of any two particles does not lead to a distinct state. In the quantum system, we compare the wavefunctions (for now let's set $N = 2$ for simplicity) $\psi\left(\alpha_1, \alpha_2\right)$ and $\psi\left(\alpha_2, \alpha_1\right)$. These should correspond to the same physical situation because we cannot distinguish the two particles. Physical observables come from the wavefunction through its modulus squared, $|\psi|^2$, so for these two wavefunctions to correspond to the same state, we require

$$\left|\psi(\alpha_1, \alpha_2)\right|^2 = \left|\psi(\alpha_2, \alpha_1)\right|^2.$$

For systems in equilibrium, we can omit the time dependence which is implicit in the wavefunction, and as such we can write ψ as a real function, so we require

$$\Psi(\alpha_1, \alpha_2) = \pm\Psi(\alpha_2, \alpha_1). \tag{11.19}$$

Thus we can classify any quantum particle as either

Bosons: Which are described by wavefunctions that are **symmetric** under the interchange of any two (and take the positive sign in Eq. (11.19)), or

Fermions: Whose wavefunctions take the negative sign and are thus **antisymmetric** under this interchange.

Bosons are named after Satyendra Nath Bose (1894–1974), while fermions are named after Enrico Fermi (1901–1954). For our two-particle example, then, ψ would not be an appropriate wavefunction to describe the state, but instead we define

$$\Psi_\pm(\alpha_1, \alpha_2) \equiv \frac{1}{\sqrt{2}}\left[\psi(\alpha_1, \alpha_2) \pm \psi(\alpha_2, \alpha_1)\right] = \pm\Psi_\pm(\alpha_2, \alpha_1). \tag{11.20}$$

The factor in front is there to normalize the state (although we will not need to worry about that here). Bosons are described by Ψ_+ while fermions are described by Ψ_-, and in each case we see that Eq. (11.19) holds. This symmetry requirement gives an important distinction between bosons and fermions: The former can be in the same state, because if $\alpha_1 = \alpha_2$, then $\Psi_+(\alpha_1, \alpha_1) \neq 0$, while the same is not true for fermions because $\Psi_-(\alpha_1, \alpha_1) = 0$. This is just a formal statement of the **Pauli exclusion principle.**[3]

Returning to a wavefunction for an N-particle state, where $\Psi(\alpha_1, \alpha_2, \ldots, \alpha_N)$ will be used to describe the complete (properly symmetrized) wavefunction for the state of our system. In general this is a difficult function to determine, and luckily we will not need to. My purpose here was to get some insight into how quantum particles interact to construct our statistical theory. We can consider three types of statistics:

Maxwell–Boltzmann (MB) statistics: These are *truly classical* systems such that if we were to use a wavefunction to describe the state, there would be no restrictions on it. The wavefunction of the system Ψ is the same as ψ, and we consider $\psi(\alpha_1, \alpha_2, \ldots, \alpha_N)$ and $\psi(\alpha_2, \alpha_1, \ldots, \alpha_N)$ (for example) to be a distinct state. Any number of particles can be in any of the states denoted by the α_i's. However, it is important to stress that this is not the same as fudged classical statistics!

3 Wolfgang Pauli (1900–1958).

Bose–Einstein (BE) statistics: These are quantum systems whose wavefunctions are symmetric, so that upon interchange of any of the N particles, the state Ψ would be the same, or

$$\Psi(\alpha_1, \dots, \alpha_j, \dots, \alpha_k, \dots, \alpha_N) = \Psi(\alpha_1, \dots, \alpha_k, \dots, \alpha_j, \dots, \alpha_N),$$

for all $1 \leq j, k \leq N$. Ψ will have $N!$ terms when we write this in terms of the individual ψ's from above. In this case, any of the α_i's can be the same, so that any state can have any number of bosons in it.

Fermi–Dirac (FD) statistics: These are quantum systems[4] whose wavefunctions are antisymmetric, so that upon interchange of any of the N particles, the state Ψ changes sign, or

$$\Psi(\alpha_1, \dots, \alpha_j, \dots, \alpha_k, \dots, \alpha_N) = -\Psi(\alpha_1, \dots, \alpha_k, \dots, \alpha_j, \dots, \alpha_N),$$

for all $1 \leq j, k \leq N$. Like the Bose–Einstein case, Ψ will have $N!$ terms when we write this in terms of the individual ψ's from above, but now with various terms with minus signs in them. In this case, if any of the α_i's are the same, then Ψ vanishes, so that each state can only have at most a single fermion in it.

Exercise 11.7 Consider a three-particle system, where a given microstate could be described as $\psi(\alpha_1, \alpha_2, \alpha_3)$. Write down the fully symmetric wavefunction Ψ_+ and the fully antisymmetric wavefunction Ψ_-. Note there are six terms in both, and any interchange of two α_i's should give the same result in the former case but an overall minus sign in the latter case. Show that for the fermionic case, if *any* two α_i's are the same, it vanishes (you don't need to for the bosonic case, all the terms are positive so it trivially doesn't vanish upon interchange).

The next step is to apply this to the statistical problem. While α_i can correspond to any quantum number, we are most interested in the quantum number(s) that denote the energy of a particle. Along with this, to know the energy of the complete macrostate, we need to know the number of particles with each allowed energy. Most quantum systems of interest are bound states, such that the energy levels accessible to the system are discrete. In many cases, there are an infinite number of possible energy states, but many times there is a finite number of states, or the temperature is such that only a finite number of states are sufficiently occupied.

Example 11.3 Suppose we have a three-state system, with single-particle energy levels given by ε_1, ε_2, and ε_3. In this system, there are two identical particles, and each could be in one of these possible states (and they don't have any significant interaction with each other, so this is a **quantum ideal gas**). What possible states are allowed, if these are treated as Maxwell–Boltzmann (MB) particles, Bose–Einstein (BE) particles, or Fermi–Dirac (FD) particles?

In the MB case, we can label them, and we'll denote them A and B. For the BE and FD cases, they cannot be distinguished, so no such labels are included. We will use the canonical ensemble (for now we will keep the number of particles fixed) to study this system, so we also want the total energy of each state. The possible ways to fill these two particles in these three energy levels are shown in Figure 11.2 for the MB case, Figure 11.3 for the BE case, and Figure 11.4 for the FD case. Additionally, I have labeled the different states (R) arbitrarily, where we can list the possible energy E_R of each as

$$E_1 = 2\varepsilon_1,\ E_2 = 2\varepsilon_2,\ E_3 = 2\varepsilon_3,$$

4 Named after not only Enrico Fermi, but Paul Dirac (1902–1984).

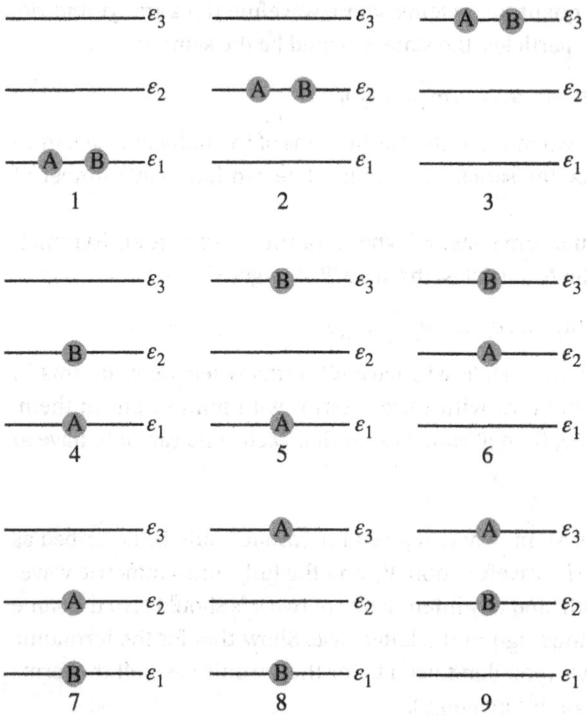

Figure 11.2 Allowed states for the system in Example 11.3 if the particles obey Maxwell–Boltzmann statistics.

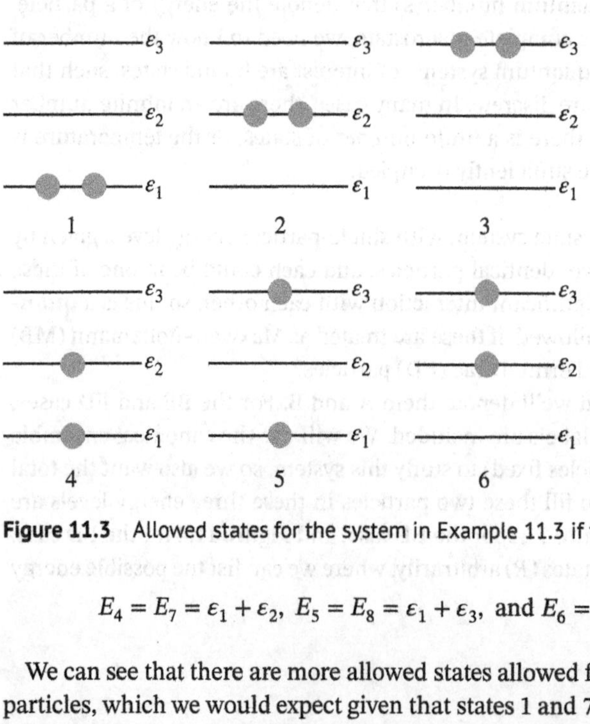

Figure 11.3 Allowed states for the system in Example 11.3 if the particles obey Bose–Einstein statistics.

$$E_4 = E_7 = \varepsilon_1 + \varepsilon_2,\ E_5 = E_8 = \varepsilon_1 + \varepsilon_3,\ \text{and}\ E_6 = E_9 = \varepsilon_2 + \varepsilon_3.$$

We can see that there are more allowed states allowed for MB particles than for either BE or FD particles, which we would expect given that states 1 and 7, for example, are considered distinct for MB, but are the same for both quantum systems. Additionally there are more possible states for BE

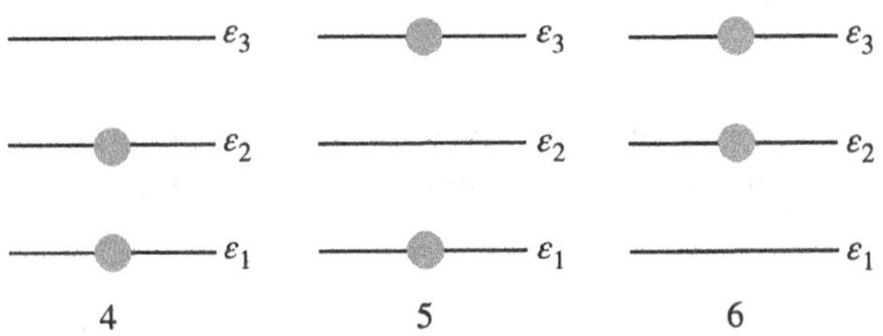

Figure 11.4 Allowed states for the system in Example 11.3 if the particles obey Fermi–Dirac statistics.

systems than FD, again because in the latter case we can never have more than one particle in a single state.

Exercise 11.8 Consider the different systems in Example 11.3. Apply the canonical ensemble to this problem, by calculating the partition function for the MB case, the BE case, and the FD case. Show that $Z_{MB} = Z_{FD} + Z_{BE}$.

Exercise 11.9 For the partition functions in Exercise 11.8, suppose $\varepsilon_1 = 0$, $\varepsilon_2 = \varepsilon$, and $\varepsilon_3 = 2\varepsilon$, and the temperature is chosen such that $e^{-\beta\varepsilon} = 1/2$. Calculate the mean energy of each system. You should find $\bar{E}_{FD} > \bar{E}_{MB} > \bar{E}_{BE}$. Explain why this should be the case.

The restrictions for each different type of system (MB, BE, and FD) are all we need to work out the partition function and thus determine everything we need for a system (as you did in Exercises 11.8 and 11.9). However, for reasons that we'll discuss later, we are going to reformulate the problem in terms of the average number of particles in each state. As we do this, as mentioned previously, we focus on the **quantum ideal gas** of N particles in a volume V and neglect interactions between them. The total energy of a given state R is

$$E_R = n_1\varepsilon_1 + n_2\varepsilon_2 + \cdots = \sum_r n_r\varepsilon_r, \tag{11.21}$$

where n_r is the number of particles in the quantum state r with energy ε_r.[5] These energies come from solving the eigenvalue equation for the Hamiltonian (often called the time-independent Schrödinger equation), $\hat{H}\psi = E\psi$ for a single particle, but that is a problem for a quantum mechanics course—we will merely use the results from such calculations.[6] For now we will require that the total number of particles remains fixed, so that

$$N = \sum_r n_r. \tag{11.22}$$

The partition function is simple to calculate (in principle), and in this notation is given by

$$Z = \sum_R e^{-\beta(n_1\varepsilon_1 + n_2\varepsilon_2 + \cdots)}, \tag{11.23}$$

where the sum R refers to the sum over all possible states subject to the condition in Eq. (11.22). We used this above to calculate the mean energy, or any other thermodynamic quantity we have already studied once we have the partition function.

5 This is precisely how we determined the E_R's in Example 11.3.
6 For a multi-particle system this equation becomes exponentially more difficult, and as we have seen, we cannot solve very difficult problems in physics! That's not entirely true, but it requires techniques beyond the level of this book.

11.3 The Occupation Number

However, let's consider a different quantity: the mean number of particles in a given state s, the **occupation number**, or as it is sometimes known, the **quantum distribution function**. This is given by

$$\overline{n}_s = \frac{1}{Z}\sum_R n_s e^{-\beta E_R} = \frac{\sum_R n_s e^{-\beta(n_1\varepsilon_1+n_2\varepsilon_2+\cdots)}}{\sum_R e^{-\beta(n_1\varepsilon_1+n_2\varepsilon_2+\cdots)}}. \tag{11.24}$$

We can use our derivative trick again to simplify this calculation, where

$$\overline{n}_s = -\frac{1}{\beta}\frac{\partial \ln Z}{\partial \varepsilon_s}. \tag{11.25}$$

Exercise 11.10 Show that you can write Eq. (11.24) as Eq. (11.25).

When evaluating the sum over R for the partition function, we have two possible restrictions: one for all types of particles and one only for FD statistics. The first restriction, as we discussed above, is that the total number of particles is fixed, as in Eq. (11.22).[7] For MB particles, the particles are distinguishable, so determining the occupation number will be slightly different than for the two truly quantum systems. For FD statistics, there is an additional restriction, that the number of particles in any given state r can only take on one of two values,

$$n_r = 0 \text{ or } 1,$$

since we can never have two particles in the same quantum state.

Before calculating the occupation number for each case, let's pause to discuss why this is a useful quantity. Consider the mean energy as an example, where

$$\bar{E} = \sum_R E_R P_R = \sum_R \left(n_1\varepsilon_1 + n_2\varepsilon_2 + \cdots\right) P_R,$$

with P_R the probability of being in a state R. We can rewrite this as

$$\bar{E} = \varepsilon_1 \sum_R n_1 P_R + \varepsilon_2 \sum_R n_2 P_R + \cdots = \overline{n}_1\varepsilon_1 + \overline{n}_2\varepsilon_2 + \cdots,$$

or

$$\bar{E} = \sum_r \overline{n}_r \varepsilon_r. \tag{11.26}$$

Thus knowing the occupation number for each state, we can calculate the mean energy (and in fact any other mean value) of our system. For a general observable y, which has a value y_r when the system is in a state r, we would have

$$\bar{y} = \sum_r \overline{n}_r y_r. \tag{11.27}$$

If we can calculate the occupation number as a function of the energies (and other parameters) from the partition function in more generality, it can be more easily applied to essentially any problem. Additionally, this will allow for a starting point for more complex problems in condensed matter physics and other advanced topics (by expanding upon our calculations of $\overline{n}_r$ for MB, FD, or BE statistics). This will be discussed in a little more detail in Chapter 12. In Sections 11.3.1–11.3.4, we will derive $\overline{n}_r$ for each of the three different types of particles.

---

7 There is one special case of BE statistics, when we discuss photons, where this is not the case.

11.3.1 Maxwell–Boltzmann Distribution Function

We start with the Maxwell–Boltzmann (MB) distribution function so that we can connect the current approach with our classical results from before. The partition function for N MB particles, with single-particle energies ε_r, is

$$Z = \sum_R e^{-\beta(n_1\varepsilon_1+n_2\varepsilon_2+\cdots)}.$$

This is trickier than other partition function calculations we have done, because the sum implicitly contains a restriction on the number of particles. Additionally, for a given energy E_R, we have to count how many different ways we can assign the particles to different energy levels while maintaining this energy. While doing so we have to take into account the distinguishability of the particles.

For example, consider the case in Exercise 11.8, while looking at the MB case in Example 11.3. Each of the middle three states have the same energy as each of the last three states, and so if we want our sum to only include $e^{-\beta E_R}$ terms which are distinct, we would include factors of two in front of these particular terms when evaluating our sum for Z. This is obvious when we evaluate the partition function in this simple case, but once we wish to extend this to more particles (even five or ten particles), it becomes difficult to count the degeneracies in this manner.

When considering a large number (or rather a variable number) of particles, we refer to our discussion of combinatorics (see the box on page 9). For each possible state R, we have n_1, $n_2, \ldots$ particles in their respective single-particle states, and the N total particles combine to form a state with energy E_R. To obtain this proper energy, we can choose any one of our N particles at a time to fill up these energies (starting with filling ε_1, then once n_1 are in that energy level, continue with ε_2 and so forth). Because we can choose any of the particles, there are $N!$ ways this can be done. Of those n_1 particles with energy ε_1, the order in which we chose them is irrelevant, so we need to divide $N!$ by $n_1!$ to account for this overcounting. We continue for each of the numbers of particles in each level to get

$$\frac{N!}{n_1!n_2!\cdots} \tag{11.28}$$

combinations for a given energy. Luckily because $0! = 1$, this will work even when particular levels have zero particles in them.

Exercise 11.11 Consider a case of $N = 5$ particles with energy $E_R = 2\varepsilon_1 + 2\varepsilon_2 + \varepsilon_4$. Using the analogy above Eq. (11.28), show that it does correspond to the number of ways to obtain this energy.

Equation (11.28) allows us to rewrite the sum over R as an explicit sum over the different n_r's, so we have

$$Z = \sum_{n_1,n_2,\cdots} \frac{N!}{n_1!n_2!\cdots} e^{-\beta(n_1\varepsilon_1+n_2\varepsilon_2+\cdots)}.$$

The sums over each n_r range from 0 to N in each case, and we require

$$\sum_r n_r = N.$$

This is not difficult, because we can write

$$Z = \sum_{n_1,n_2,\cdots} \frac{N!}{n_1!n_2!\cdots} e^{-\beta n_1\varepsilon_1} e^{-\beta n_2\varepsilon_2}\cdots$$

$$= \sum_{n_1,n_2,\ldots} \frac{N!}{n_1!n_2!\cdots} \left(e^{-\beta\varepsilon_1}\right)^{n_1} \left(e^{-\beta\varepsilon_2}\right)^{n_2} \ldots ,$$

which, by the multinomial theorem, is simply[8]

$$Z_{\rm MB} = \left(e^{-\beta\varepsilon_1} + e^{-\beta\varepsilon_2} + \cdots\right)^N. \tag{11.29}$$

From Eq. (11.25),

$$\overline{n}_s^{(\rm MB)} = N \frac{e^{-\beta\varepsilon_s}}{e^{-\beta\varepsilon_1} + e^{-\beta\varepsilon_2} + \cdots},$$

which we can write in terms of the single-particle partition function

$$\zeta = \sum_r e^{-\beta\varepsilon_r},$$

as

$$\overline{n}_s^{(\rm MB)} = N \frac{e^{-\beta\varepsilon_s}}{\zeta} = NP_s. \tag{11.30}$$

This is just the Boltzmann (or canonical) distribution that we had before, and thus it is just the classical result. In the quantum mechanical context, we call it the **Maxwell–Boltzmann distribution function**. The only restriction on $\overline{n}_s^{(\rm MB)}$ is that it cannot be greater than N, but other than this, we can have any number of particles in any given state. The constraint $\sum_r \overline{n}_r = N$ holds as well, which is easy to see.

Exercise 11.12 Fill in the steps to obtain Eq. (11.30).

Exercise 11.13 Given that many quantum systems have an infinite number of energy levels, it might seem bizarre that many times we look at systems with a finite number of energy levels. However, often it makes sense to make this approximation. Consider an electron trapped in a one-dimensional box of length 5 nm, so that the energy levels are $E_n = n^2\hbar^2\pi^2/(2mL)$. Show that the ratios of probabilities between the ground state and the first three excited states are $\sim 5.7, 103, 5995$, respectively. Thus this system could be treated as a three-state system to an excellent approximation.

Example 11.4 For the MB case, the results for a discrete set of states can be determined from simply calculating the partition function as we did in the past. We can use Example 11.3 to calculate $\overline{n}_s$ for each state. Using the results of this table, we get for $\overline{n}_1$, the mean number of particles in state 1, to be

$$\begin{aligned}
\overline{n}_1 &= \frac{1}{Z_{\rm MB}} \left[(1)e^{-\beta(\varepsilon_1+\varepsilon_2)} + (0)e^{-\beta(\varepsilon_2+\varepsilon_3)} + (1)e^{-\beta(\varepsilon_1+\varepsilon_3)} + (2)e^{-2\beta\varepsilon_1} + (0)e^{-2\beta\varepsilon_2}\right. \\
&\quad \left. + (0)e^{-2\beta\varepsilon_3} + (1)e^{-\beta(\varepsilon_1+\varepsilon_2)} + (0)e^{-\beta(\varepsilon_2+\varepsilon_3)} + (1)e^{-\beta(\varepsilon_1+\varepsilon_3)}\right] \\
&= \frac{2\left[e^{-\beta(\varepsilon_1+\varepsilon_2)} + e^{-\beta(\varepsilon_1+\varepsilon_3)} + e^{-2\beta\varepsilon_1}\right]}{2e^{-\beta(\varepsilon_1+\varepsilon_2)} + 2e^{-\beta(\varepsilon_1+\varepsilon_3)} + 2e^{-\beta(\varepsilon_2+\varepsilon_3)} + e^{-2\beta\varepsilon_1} + e^{-2\beta\varepsilon_2} + e^{-2\beta\varepsilon_3}} \\
&= \frac{2e^{-\beta\varepsilon_1}\left(e^{-\beta\varepsilon_1} + e^{-\beta\varepsilon_2} + e^{-\beta\varepsilon_3}\right)}{\left(e^{-\beta\varepsilon_1} + e^{-\beta\varepsilon_2} + e^{-\beta\varepsilon_3}\right)^2} \\
&= \frac{2e^{-\beta\varepsilon_1}}{e^{-\beta\varepsilon_1} + e^{-\beta\varepsilon_2} + e^{-\beta\varepsilon_3}}.
\end{aligned} \tag{11.31}$$

8 At this point we'll label Z appropriately as the MB partition function.

We can see this is exactly what one would obtain using Eq. (11.30). In the MB case, both approaches lead to identical results.

Exercise 11.14 Evaluate $\overline{n}_2$ and $\overline{n}_3$ for the system in Example 11.3, and show that the results are also the same as one would obtain using Eq. (11.30).

11.3.2 Photon Distribution Function

The first quantum system we will look at will be a special case of bosons: **photons**, or quantized electromagnetic radiation. It may seem strange to consider a "box of light," as it were, but it is the simplest case to study because we actually do not need to impose the restriction that the number of photons in a system remains fixed: Photons can readily be absorbed and emitted from the walls of the container. Additionally, while it is an idealization, it will be a sneak peak for our discussion of blackbody radiation in Chapter 12.

As before we wish to calculate the partition function

$$Z = \sum_R e^{-\beta(n_1\epsilon_1+n_2\epsilon_2+\cdots)},$$

but unlike the MB case, we now treat the photons as indistinguishable, and as stated already, we do not require $\sum_r n_r = N$. Additionally, photons do not interact at all (they are the perfect ideal gas!), so we can split the sum over the states R as

$$Z = \left(\sum_{n_1=0}^{\infty} e^{-\beta n_1\epsilon_1}\right)\left(\sum_{n_2=0}^{\infty} e^{-\beta n_2\epsilon_2}\right)\cdots,$$

where there is no upper limit to the number of photons in a given state. Each of these sums is identical to each other, and they are each just a geometric series, so we have

$$\sum_{n=0}^{\infty} e^{-\beta n\epsilon} = \frac{1}{1-e^{-\beta\epsilon}}.$$

Our partition function is then

$$Z = \left(\frac{1}{1-e^{-\beta\epsilon_1}}\right)\left(\frac{1}{1-e^{-\beta\epsilon_2}}\right)\cdots, \tag{11.32}$$

or

$$\ln Z = -\sum_r \ln(1-e^{-\beta\epsilon_r}), \tag{11.33}$$

so that using Eq. (11.25), we immediately obtain

$$\overline{n}_s = \frac{1}{e^{\beta\epsilon_s}-1}. \tag{11.34}$$

This is known as the **Planck distribution function**, as Planck first postulated an expression of this form to solve the ultraviolet catastrophe, which we will discuss later. We can see that there is no restriction on this expression: $0 \le \overline{n}_s < \infty$. This distribution function is shown in Figure 11.5, and we see it blows up as $\epsilon \to 0$.

Exercise 11.15 Fill in the steps to obtain Eq. (11.34).

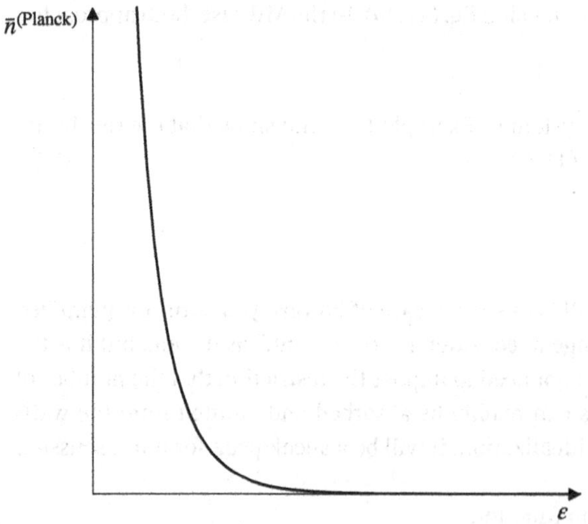

Figure 11.5 The Planck distribution function as a function of the single-particle energy ε.

11.3.3 Bose–Einstein Statistics

Now let's consider bosons more generally. As mentioned before, photons are a special case of bosons; photons are massless bosons. It costs no energy to create a photon; the only energy requirement arises when you want to have the photon in a particular energy state. For *massive* bosons, you need some energy just to create the particle and *then* energy to put it in some state. Thus, the problem is more complicated because we must again require $\sum_r n_r = N$. We will continue to consider the system as an ideal gas, but we cannot split up the summations as in the photon case due to this restriction on the number of particles.

For small systems where we can enumerate all of the possible states, such as that shown in Example 11.3, this restriction is not an issue. In this case, we can simply use the probability of a given state as determined by the Boltzmann distribution, just as we did for the MB case.

Exercise 11.16 Show, using Example 11.3, that the mean number of particles in each state is given by

$$\bar{n}_1 = \frac{e^{-\beta\varepsilon_1}\left(2e^{-\beta\varepsilon_1} + e^{-\beta\varepsilon_2} + e^{-\beta\varepsilon_3}\right)}{Z_{\text{BE}}},$$

$$\bar{n}_2 = \frac{e^{-\beta\varepsilon_2}\left(e^{-\beta\varepsilon_1} + 2e^{-\beta\varepsilon_2} + e^{-\beta\varepsilon_3}\right)}{Z_{\text{BE}}}, \quad \text{and}$$

$$\bar{n}_3 = \frac{e^{-\beta\varepsilon_3}\left(e^{-\beta\varepsilon_1} + e^{-\beta\varepsilon_2} + 2e^{-\beta\varepsilon_3}\right)}{Z_{\text{BE}}}.$$

This method becomes tedious, as we would expect, when applied to larger systems. However, as long as the number of particles is large enough, we can use the equivalency of the canonical and grand canonical ensembles. Therefore, instead of requiring N to be fixed, we require $\overline{N}$, the *mean* number of particles, to be fixed.

To this end, let's instead calculate the grand partition function, $\mathcal{Z}$, which is given by

$$\mathcal{Z} = \sum_R e^{-\beta(E_R - \mu N)} = \sum_R e^{-\beta[(n_1\varepsilon_1 + n_2\varepsilon_2 + \cdots) - (n_1\mu + n_2\mu + \cdots)]} \tag{11.35}$$

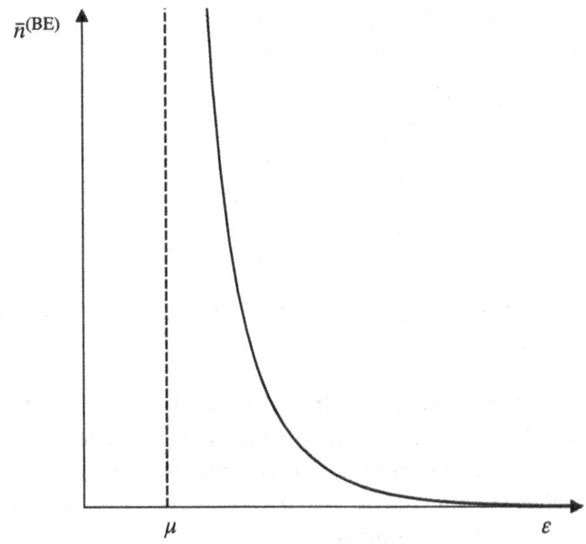

Figure 11.6 The Bose–Einstein distribution function shown as a function of the single-particle energy ε. The vertical dashed line corresponds to $\varepsilon = \mu$.

The sum is unrestricted, and now we can break up the sums as in the photon case,

$$\mathcal{Z} = \left(\sum_{n_1=0}^{\infty} e^{-\beta(\varepsilon_1-\mu)n_1}\right)\left(\sum_{n_2=0}^{\infty} e^{-\beta(\varepsilon_2-\mu)n_2}\right)\cdots \tag{11.36}$$

This is the same as the expression for the photon partition function with $\varepsilon_r \to \varepsilon_r - \mu$, just a product of independent geometric sums

$$\mathcal{Z} = \left(1 - e^{-\beta(\varepsilon_1-\mu)}\right)^{-1}\left(1 - e^{-\beta(\varepsilon_2-\mu)}\right)^{-1}\cdots, \tag{11.37}$$

and so using Eq. (11.25), we obtain

$$\overline{n}_s^{(\mathrm{BE})} = \frac{e^{-\beta(\varepsilon_s-\mu)}}{1 - e^{-\beta(\varepsilon_s-\mu)}} = \frac{1}{e^{\beta(\varepsilon_s-\mu)} - 1}. \tag{11.38}$$

This is known as the **Bose–Einstein distribution function**, and we show it in Figure 11.6. Note that the result here is quite different than those of Exercise 11.16. The results of that exercise hold only if N is fixed, while this expression is valid when $\overline{N}$ is fixed.

For a Bose–Einstein gas, μ is no longer referred to as the chemical potential, and in principle, μ is known from the mean number of particles, because recall $\sum_s \overline{n}_s = \overline{N}$ (a fact you will use in the problems). We see that $\overline{n}_s^{(\mathrm{BE})}$ blows up at $\varepsilon = \mu$, and more specifically we can never have the chemical potential be greater than the ground state energy. This is not an issue, because for such a gas, μ is always negative (in some cases very close to zero, but negative for massive bosons). You will see in Problem 11.4 that when μ is positive, unphysical results for $\overline{n}_s^{(\mathrm{BE})}$ will arise. One last note regarding μ, we can see from Eq. (11.38) that $\mu = 0$ for a photon (or any other massless particle).

11.3.4 Fermi–Dirac Statistics

As in the case of BE particles, we first consider the case of a small number of FD particles, where we can use Example 11.3 to determine the mean number in each state. For fermions, there are only three allowed states, so it is not as complicated as the case for bosons or MB particles.

Exercise 11.17 Show that for the system in Example 11.3 that the mean number of particles in each state is given by

$$\overline{n}_1 = \frac{e^{\beta\varepsilon_2} + e^{\beta\varepsilon_3}}{Z_{\text{FD}}} = 1 - \frac{e^{\beta\varepsilon_1}}{Z_{\text{FD}}},$$
$$\overline{n}_2 = \frac{e^{\beta\varepsilon_1} + e^{\beta\varepsilon_3}}{Z_{\text{FD}}} = 1 - \frac{e^{\beta\varepsilon_2}}{Z_{\text{FD}}}, \text{ and}$$
$$\overline{n}_3 = \frac{e^{\beta\varepsilon_1} + e^{\beta\varepsilon_2}}{Z_{\text{FD}}} = 1 - \frac{e^{\beta\varepsilon_3}}{Z_{\text{FD}}}.$$

After determining $\overline{n}_s^{(\text{FD})}$ later, we will see that the results again are quite different, just as in the BE case.

For large systems, again we will use the grand canonical ensemble to determine the occupation number for fermions. We treat the mean number of particles $\overline{N}$ to be fixed, and now we have the additional constraint that n_r can only be zero or one for any of the states r. The steps we have to follow are the same as the BE case, but our sums are no longer infinite.

We have, just as before,

$$\mathcal{Z} = \sum_R e^{-\beta(E_R - \mu N)} = \sum_R e^{-\beta[(n_1\varepsilon_1 + n_2\varepsilon_2 + \cdots) - (n_1\mu + n_2\mu + \cdots)]}, \tag{11.39}$$

and again the sum is unrestricted. We can break up the sums as

$$\mathcal{Z} = \left[\sum_{n_1=0}^{1} e^{-\beta(\varepsilon_1 - \mu)n_1}\right]\left[\sum_{n_2=0}^{1} e^{-\beta(\varepsilon_2 - \mu)n_2}\right]\cdots. \tag{11.40}$$

Each of these is not a geometric sum as in the BE case, but rather is even simpler as they are just finite sums with only two terms. We get

$$\mathcal{Z} = \left[1 + e^{-\beta(\varepsilon_1 - \mu)}\right]\left[1 + e^{-\beta(\varepsilon_2 - \mu)}\right]\cdots. \tag{11.41}$$

Using Eq. (11.25), we readily obtain

$$\overline{n}_s^{(\text{FD})} = \frac{1}{e^{\beta(\varepsilon_s - \mu)} + 1}. \tag{11.42}$$

This is the **Fermi–Dirac distribution function**, and is shown in Figure 11.7.

Exercise 11.18 Use Eq. (11.25) to get Eq. (11.42) from Eq. (11.41).

This distribution has some interesting features that distinguishes it from the Bose–Einstein distribution. First of all, the denominator is positive and always greater than one, so that we are guaranteed to have

$$\overline{n}_s^{(\text{FD})} \leq 1,$$

as required by the Pauli Exclusion Principle.

Here, $\mu \equiv \mu_F$ is called the **Fermi energy**,[9] and it has a different interpretation than in the bosonic case. To see this, let us consider the zero temperature limit. As $T \to 0$, $\beta \to \infty$, and the exponential in the denominator of Eq. (11.42) has one of two behaviors,

$$e^{\beta(\varepsilon_s - \mu_F)} \to \begin{cases} \infty & \text{if } \varepsilon_s > \mu_F \\ 0 & \text{if } \varepsilon_s < \mu_F \end{cases}. \tag{11.43}$$

9 Many sources tend to use ϵ_F for the Fermi energy, while I'm sticking with μ_F just to remain consistent with our previous work.

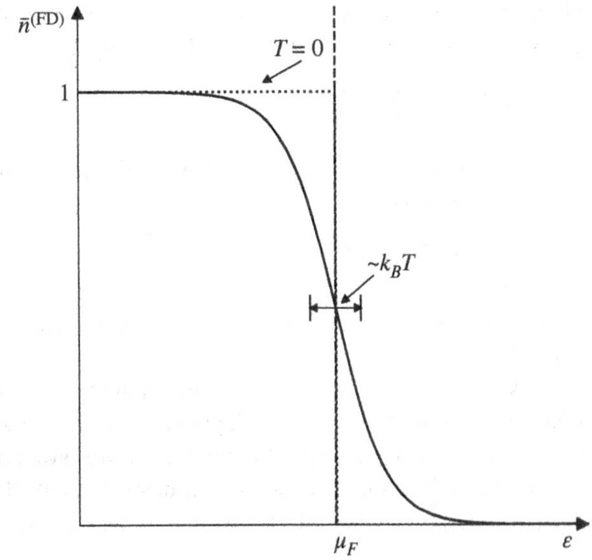

Figure 11.7 The Fermi–Dirac distribution function as a function of the single-particle energy ε. The dotted curve shows the $T = 0$ expression, along with Fermi energy μ_F, and the width $\sim k_BT$ of the "smearing" at non-zero temperature.

This implies

$$\overline{n}_s^{(FD)} \rightarrow \begin{cases} 0 \ \text{if}\, \varepsilon_s > \mu_F \\ 1 \ \text{if}\, \varepsilon_s < \mu_F \end{cases}. \tag{11.44}$$

At zero temperature, all of the energy levels below the Fermi energy are filled, while all of those above this energy are empty. For a given number of particles, we find the Fermi energy by filling all of the lowest energies, and we define μ_F as the energy of the last-filled state when we run out of particles.

For non-zero T, as we can see in Figure 11.7 (which also shows the zero-temperature case with the dotted line), not all of the levels below μ_F are filled, but instead μ_F marks the energy below which there is a higher probability that the states are filled. There is a "smearing" of the filled energy levels with a width of order k_BT. This is a basic foundation for many condensed matter applications, which we will touch on briefly later when we discuss a realistic Fermi gas in Chapter 12.

One final note on these derivations. You can show that you get the same result if you use the canonical ensemble and require a fixed number of particles (see pages 340-342 in Ref. [1] for example), with a little extra effort.

11.4 Classical Limit

Our results for the occupation numbers for BE and FD systems, Eqs. (11.38) and (11.42), respectively, are quite different from each other, and completely different from our MB case, Eq. (11.30). The latter of these is a classical system, so we now would like to determine what happens in the classical limit to our quantum occupation numbers $\overline{n}_s^{(FD)}$ and $\overline{n}_s^{(BE)}$.

To begin, we look at the partition functions for both FD and BE statistics, starting from the grand partition function,

$$\ln \mathcal{Z} = \pm \sum_r \ln(1 \pm e^{-\beta(\varepsilon_r - \mu)}), \tag{11.45}$$

where the upper sign here and later will be for FD statistics and the lower for BE statistics. From this we get the ordinary partition function using Eq. (11.15),[10]

$$\ln Z = -\beta\mu N \pm \sum_r \ln(1 \pm e^{-\beta(\varepsilon_r - \mu)}). \tag{11.46}$$

We require that the number of particles is fixed, so when summing $\overline{n}_r$ over all states for either bosons or fermions,

$$\sum_r \overline{n}_r = \sum_r \frac{1}{e^{\beta(\varepsilon_r - \mu)} \pm 1} = N. \tag{11.47}$$

As discussed when we introduced the grand canonical ensemble, this requirement is how we will determine μ.

To take the classical limit of a quantum system, we have often assumed the energy of the system is large enough so that $k_B T$ is much greater than the relevant (quantum) energy scale of the theory. As such, we are considering a "high temperature" system, specifically that T is large enough such that the concentration of particles is sufficiently small. That is, for a fixed volume, the number of particles is relatively *small* (but still large in the statistical sense).

Example 11.5 As usual, it seems tricky to accomplish the goal of having a value of N that is very large while n is very small at the same time. However, this is only because these systems are less familiar to us in our everyday experience. Consider oxygen gas at room temperature, which has a density of 1.43 kg/m^3. With a molar mass of 32 g/mol, this corresponds to a number density around $n = 3 \times 10^{25}$ molecules/m^3, which, in a cubic region that is 10 cm on a side corresponds to $N = 3 \times 10^{22}$ molecules. Both N and n seem quite large, but at this density, if the molecules are uniformly distributed over the volume, each molecule has a volume of around $v = 3 \times 10^{-26}$ m^3 to itself (on average). Molecules tend to have a length scale of the order of 1 Å = 10^{-10} m, so the volume each molecule takes up is on the order of 10^{-30} m^3. This is significantly smaller than v, so clearly n is a small enough density for the classical limit, while still allowing for N to be statistically large. (At lower temperatures, the density will eventually increase to make this no longer valid.)

For N to be small, then Eq. (11.47) must satisfy

$$\overline{n}_r = \frac{1}{e^{\beta(\varepsilon_r - \mu)} \pm 1} \ll 1,$$

and thus,

$$e^{\beta(\varepsilon_r - \mu)} \gg 1. \tag{11.48}$$

This requirement is the same for both BE and FD particles. Additionally, we can see this condition arise from a different point of view. We have often stated that a system can be considered classical when the occupied levels have energies that are large compared with $k_B T$ (which is greater than the relevant quantum scale, e.g., $\hbar\omega$ for the simple harmonic oscillator). In this case, we would consider the energy relative to μ, so that $\varepsilon_r - \mu \gg k_B T$, or $\beta(\varepsilon_r - \mu) \gg 1$, which is consistent with Eq. (11.48).

With Eq. (11.48), our expression for $\overline{n}_r$ becomes (for both bosons and fermions),

$$\overline{n}_r \approx e^{-\beta(\varepsilon_r - \mu)}. \tag{11.49}$$

10 I won't distinguish between N and $\overline{N}$ here, as we assume that $N \gg 1$ and thus $\overline{N} \approx N$.

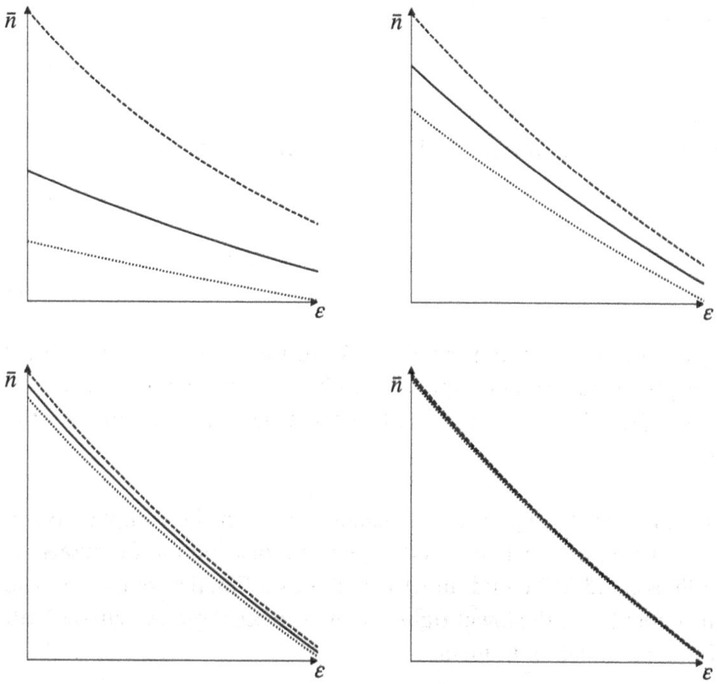

Figure 11.8 The quantum distribution functions for FD (dotted lines), BE (dashed lines), and MB (solid lines) statistics as a function of energy as we increase the energy of the state (the upper left is the lowest energy range and the lower right is the highest range).

This allows us to easily determine μ from the expression

$$N = \sum_r e^{-\beta(\varepsilon_r-\mu)}$$
$$\Rightarrow e^{\beta\mu} = \frac{N}{\sum_r e^{-\beta\varepsilon_r}} = \frac{N}{\zeta}, \tag{11.50}$$

where we have written this in terms of the single-particle partition function ζ we defined for MB statistics. We can then write the quantum occupation number in the classical limit as

$$\bar{n}_r = N\frac{e^{-\beta\varepsilon_r}}{\zeta},$$

which is just the MB expression for the occupation number. At this point, we might want to argue that Maxwell–Boltzmann statistics are the proper classical limit of both Bose–Einstein and Fermi–Dirac statistics (and for $\bar{n}$, it is true, which can be seen in Figure 11.8).

However, as we shall see, this is not the complete story. We also need to look at the partition function,

$$\ln Z = -\beta\mu N \pm \sum_r \ln\left[1 \pm e^{-\beta(\varepsilon_r-\mu)}\right], \tag{11.51}$$

where again the upper (lower) sign refers to FD (BE) statistics. Because $\beta(\varepsilon_r - \mu) \gg 1$, we have $e^{-\beta(\varepsilon_r-\mu)} \ll 1$ and we can expand the logarithm, where $\ln(1 + x) \approx x$ for small x, to get

$$\ln Z \approx -\beta\mu N + \sum_r e^{-\beta(\varepsilon_r-\mu)}. \tag{11.52}$$

Exercise 11.19 Substitute our expression for μ in Eq. (11.50) to show that

$$\ln Z = N \ln \zeta - N \ln N + N. \tag{11.53}$$

For large N, let's apply Stirling's formula in reverse, so Eq. (11.53) becomes

$$\ln Z = N \ln \zeta - \ln N!, \tag{11.54}$$

or

$$Z = \frac{\zeta^N}{N!}. \tag{11.55}$$

Note that this is *not* the MB partition function, but instead it is precisely what we obtained for *fudged classical statistics.* This explicitly justifies our approach to fix the Gibbs paradox. It wasn't merely a trick that just "gets the right answer," but it actually is the proper classical limit for the quantum mechanical problem.

Computer Exercise 11.1 You can visit the Quantum Statistics section of the companion site to visualize what is shown in Figure 11.8, which shows the quantum distribution functions for FD (dotted lines), BE (dashed lines), and MB (solid lines) statistics as a function of energy. The energy range increases from the upper left to the lower right, and we can see these two curves both approach the MB expression for the occupation number.

11.4.1 From Quantum States to Classical Phase Space

We have seen that the quantum mechanical partition functions reduce to the fudged classical partition function in the appropriate limit; however, we also need to make another connection between two seemingly disparate ideas. For a classical system, due to the continuous nature of the degrees of freedom, we had to divide up phase space into small regions such that

$$\Delta x_i \Delta p_{xi} = h_0,$$

and the same for y and z, where i ranges from $1, \dots, N$ for each of our particles. In the limit that Δx_i and Δp_{xi} become infinitesimally small, we obtained integrals that allowed us to evaluate the number of states (microcanonical ensemble), the partition function (canonical ensemble), or even the grand partition function (grand canonical ensemble). Similar methods were used in quantum systems, to make the summations doable; however, now we would like to formalize this, while also connecting the quantum and classical enumeration of states.

The disconnect seems significant, as a classical particle is described by its position $\mathbf{r}(t)$ and momentum $\mathbf{p}(t)$ and a quantum particle is described by its wavefunction, $\psi(\mathbf{r}, t)$. While the complete (properly symmetrized) wavefunction of the system is technically required for this discussion, we focus for now on the ideal gas, so just the single-particle wavefunction is required.

For a non-relativistic quantum particle of mass m, the energy is related to the **wave vector** $\mathbf{k} = \mathbf{p}/\hbar$ via

$$\varepsilon = \frac{\hbar^2 \mathbf{k}^2}{2m} + U(\mathbf{r}),$$

and of course for an ideal gas, $U(\mathbf{r}) \approx 0$. In such a case, we treat this as a **free particle,** so it is described by a plane wave,

$$\psi(\mathbf{r}, t) = A e^{i(\mathbf{k}\cdot\mathbf{r} - \omega t)} = \psi(\mathbf{r}) e^{-i\omega t}, \tag{11.56}$$

where $\omega = \varepsilon/\hbar$. Generally one solves the eigenvalue equation for the Hamiltonian to obtain the energy levels accessible to the system, and the quantization of these energies arises due to the boundary conditions of the particular problem. Our particle isn't entirely free, but is contained within a box with volume V, so we need to introduce boundary conditions. However, let's argue that the box walls will not affect our particle noticeably, which will guide us to a particular set of boundary conditions. A rough measure of the "size" of the particle is its wavelength, $\lambda = 2\pi/|\mathbf{k}|$, and we assume the volume of the system is related to this by $\lambda \ll V^{1/3}$. Thus the probability that we will find our particle within a distance $\sim \lambda$ of one of the walls is $\sim \lambda/V^{1/3}$ and is negligible for many "ordinary" length scales.

Exercise 11.20 Consider an electron with a kinetic energy on the order of 1 keV. Show that if this were confined within a cube with volume 1 cm^3,

$$\frac{\lambda}{V^{1/3}} \sim 4 \times 10^{-9}.$$

Therefore the likelihood a particle will be near one of the walls is small enough that we can safely assume the particle is never near them (that is, we can ignore the walls).

Even though we can ignore the walls of the container our particle is in, we need to impose boundary conditions (the precise boundary conditions won't matter ultimately for the statistical problem) to obtain the expected quantized energy levels. There are many options, and we will use a simple set that also has applications beyond our purpose here.

Consider our system to in a rectangular solid of volume

$$V = L_x L_y L_z,$$

and we choose **periodic boundary conditions**. Any time a particle hits the boundary, it will move to the other side.[11] While these sorts of boundary conditions are not that realistic, they are useful because of how simple they are to implement. (Additionally, they are useful in numerical simulations of systems because they minimize numerical errors that arise when treating a system in a finite volume when it otherwise should be in an infinite volume.) The conditions can be stated as

$$\psi(x + L_x, y, z) = \psi(x, y, z), \tag{11.57}$$

$$\psi(x, y + L_y, z) = \psi(x, y, z), \text{ and} \tag{11.58}$$

$$\psi(x, y, z + L_z) = \psi(x, y, z). \tag{11.59}$$

So if we consider the x-direction,

$$e^{ik_x(x+L_x)+ik_y y+ik_z z} = e^{ik_x x+ik_y y+ik_z z}$$

$$e^{ik_x L_x} = 1$$

$$\Rightarrow k_x L_x = 2\pi n_x, \tag{11.60}$$

for n_x a non-zero integer (and it can be positive or negative). The same can be shown for the y- and z-directions, so that

$$k_x = \frac{2\pi n_x}{L_x}, \tag{11.61}$$

$$k_y = \frac{2\pi n_y}{L_y}, \text{ and} \tag{11.62}$$

11 For this reason, they are also sometimes called Pac Man boundary conditions.

$$k_z = \frac{2\pi n_z}{L_z}. \tag{11.63}$$

Thus the single-particle energy for our free particle is given by

$$\varepsilon_{n_x,n_y,n_z} = \frac{2\pi^2\hbar^2}{m}\left(\frac{n_x^2}{L_x^2} + \frac{n_y^2}{L_y^2} + \frac{n_z^2}{L_z^2}\right). \tag{11.64}$$

For a macroscopic volume, we assume that L_x, L_y, and L_z are each large enough so the energy levels are closely spaced and then we can treat the states as continuously spaced, as we have done plenty of times before. Recall that our sum over states is of the form[12]

$$\sum_R = \sum_{n_x,n_y,n_z} = \sum_{n_x=-\infty}^{\infty}\sum_{n_y=-\infty}^{\infty}\sum_{n_z=-\infty}^{\infty} \Delta n_x \Delta n_y \Delta n_z.$$

The explicit limits on the sum are shown for our particular example here, and we are using the fact that $\Delta n_x = \Delta n_y = \Delta n_z = 1$. Writing these in terms of the wave vector or momentum, we have

$$\Delta n_x = \frac{L_x}{2\pi}\Delta k_x = \frac{L_x}{2\pi\hbar}\Delta p_x. \tag{11.65}$$

If the length of each side is macroscopically large, then we can take the limit

$$\sum_{n_x=-\infty}^{\infty} \Delta n_x \to \int_{-\infty}^{\infty} \frac{L_x}{2\pi\hbar} dp_x,$$

and similarly for the y- and z-directions, so the sum over states becomes

$$\sum_R \to \frac{L_x L_y L_z}{(2\pi\hbar)^3}\int d^3\mathbf{p} = \frac{V}{h^3}\int d^3\mathbf{p}. \tag{11.66}$$

Classically we obtained the same result for the ideal gas. In that case, we started by integrating over *both* position and momentum, and the integral over positions gave the factor of the volume V. The difference is that to get to the point where we could integrate over the position and momentum (for example in Section 4.5.2), we had to introduce h_0, the volume of phase space for each direction. In that case, we obtained

$$\sum_R = \sum_R \frac{\Delta x \Delta p_x}{h_0}\frac{\Delta y \Delta p_y}{h_0}\frac{\Delta z \Delta p_z}{h_0} \to \frac{V}{h_0^3}\int d^3\mathbf{p}.$$

For our quantum system to agree with this in the classical limit, then we see that h_0 is no longer arbitrary! Instead, it must take on the value of Planck's constant,

$$h_0 = h. \tag{11.67}$$

We saw this was a problem when studying the law of mass action (specifically Problem 10.25): The equilibrium constant, for many systems, depended on h_0, but classically this was an arbitrary choice coming from how precisely we can measure positions and momenta, so physical results could be ambiguous. The problem is solved with quantum mechanics, as h_0 is not arbitrary at all—there is a fundamental choice for what this must be.

Exercise 11.21 It might worry you that our results relied on choosing periodic boundary conditions. One can work through our steps above with **Dirichlet boundary conditions**, where we

12 This is valid for any of our ensembles; the only difference is that we might need to include the appropriate exponential factor for the canonical or grand canonical ensembles, or a restriction on the sums given the restriction on the energy in the microcanonical ensemble.

require the wavefunction to vanish on the boundaries, so $\psi(0, y, z) = \psi(L_x, y, z) = 0$ (and similarly for the other directions). These are appropriate for the quantum particle in a box for example.

Show that the quantization conditions in Eqs. (11.61)–(11.63) become $k_x = \pi n_x / L_x$ with $n_x = 1, 2, \ldots$ (and similarly for the other directions). Work through what the allowed energies are (the updated version of Eq. (11.64)) and that Eq. (11.66) becomes

$$\sum_R \rightarrow \frac{L_x L_y L_z}{(\pi\hbar)^3} \int d^3\mathbf{p},$$

and argue this is the same as we had above. (Note the limits of integration range from $0 \leq p_{x,y,z} < \infty$ in this case, so be careful with factors of two.)

You might wonder why I decided on periodic boundary conditions instead of Dirichlet boundary conditions for this discussion. The reason is twofold. For one, this shows you that it doesn't matter what boundary conditions we choose, because for the statistical problem, the molecules in our system rarely interact with the boundary.

The other reason is that periodic boundary conditions are used often in many areas of theoretical physics. For example, when performing numerical simulations where we must approximate an infinite system by placing it in a finite box, we use periodic boundary conditions. These can be shown to have fewer *finite volume errors*, that is, they better simulate an infinite volume.

11.5 Quantum Partition Function in the Classical Limit

Now we are in a position to calculate the quantum partition function for a monatomic ideal gas in the classical limit. We start with the results of Section 11.4, specifically Eq. (11.53),

$$\ln Z = N\left(\ln \zeta - \ln N + 1\right).$$

The single-particle partition function involves a sum over integers n_x, n_y, and n_z, which we convert to sums over momenta using Eqs. (11.61)–(11.63) and Eq. (11.65)

$$\begin{aligned}\zeta &= \sum_{p_x, p_y, p_z} \exp\left[-\frac{\beta}{2m}\left(p_x^2 + p_y^2 + p_z^2\right)\right] \\ &= \left[\sum_{p_x} \exp\left(-\frac{\beta p_x^2}{2m}\right)\right]\left[\sum_{p_y} \exp\left(-\frac{\beta p_y^2}{2m}\right)\right]\left[\sum_{p_z} \exp\left(-\frac{\beta p_z^2}{2m}\right)\right].\end{aligned}$$

Converting these sums to integrals for macroscopically large systems, we obtain

$$\sum_{p_x} \exp\left(-\frac{\beta p_x^2}{2m}\right) = \sum_{p_x} \exp\left(-\frac{\beta p_x^2}{2m}\right)\frac{L_x}{2\pi\hbar} \rightarrow \frac{L_x}{h}\int_{-\infty}^{\infty} \exp\left(-\frac{\beta p_x^2}{2m}\right) dp_x. \tag{11.68}$$

This is a simple Gaussian integral, so

$$\sum_{p_x} \exp\left(-\frac{\beta p_x^2}{2m}\right) = \frac{L_x}{h}\sqrt{2\pi m k_B T},$$

and similarly for the y- and z-directions, so

$$\zeta = \frac{L_x L_y L_z}{h^3}\left(2\pi m k_B T\right)^{3/2} = \frac{V}{h^3}\left(2\pi m k_B T\right)^{3/2}. \tag{11.69}$$

Putting this into our partition function, we get

$$\ln Z = N\left[\ln\frac{V}{N} - \frac{3}{2}\ln\beta + \frac{3}{2}\ln\left(\frac{2\pi m}{h^3}\right) + 1\right], \tag{11.70}$$

and from this we get for the mean energy,

$$\bar{E} = -\frac{\partial \ln Z}{\partial \beta} = \frac{3}{2}\frac{N}{\beta} = \frac{3}{2}Nk_BT. \tag{11.71}$$

The entropy follows from Eq. (8.30),

$$S = Nk_B\left[\ln\frac{V}{N} - \frac{3}{2}\ln\beta + \frac{3}{2}\ln\left(\frac{2\pi m}{h^3}\right) + \frac{5}{2}\right]. \tag{11.72}$$

We can write this as

$$S = Nk_B\left(\ln\frac{V}{N} - \frac{3}{2}\ln T + \sigma_0\right), \tag{11.73}$$

with $\sigma_0 = \frac{3}{2}\ln\left(2\pi mk_B/h^2\right) + \frac{5}{2}$. We see this is exactly what we obtained in Eq. (8.47) using fudged classical statistics but with h the volume of phase space.

If we wanted to consider a particle which has spin s, then things would become more complicated. To a first approximation, the energy levels are independent of spin, so there would be $2s + 1$ possible spin components for a particle with spin-s, and the entropy would merely be multiplied by this factor. For more complicated problems, we would merely have to determine the energy levels from scratch and then recalculate these results.

A few things that we should point out here:

1. We obtain the correct dependence on the volume and number of particles to ensure that the entropy is an extensive function. Thus, the classical limit of the quantum problem automatically avoids the Gibbs paradox without having to *fudge* the calculation of Z.
2. We no longer have any arbitrary constants in the problem: h_0 is replaced by the well-defined physical constant h.

As a quick note, we could go back to Problem 10.25 and recalculate the equilibrium constants. We would get the same result as in that problem, but with $h_0 \to h$, and then we can say that it is a completely valid expression, because now we can unambiguously predict the relative concentrations of the different substances from first principles (hence this instruction for part (f) of that problem)! This only strengthens our argument that for any problem where we are studying the classical limit of a quantum system, we merely need to set $h_0 \to h$ and use results obtained with fudged classical statistics; no need to redo the calculations.

Exercise 11.22 Show that the result for the entropy would be the same if you use the Dirichlet boundary conditions of Exercise 11.21 to set up the problem.

11.6 Vapor Pressure of a Solid

In Section 10.3.2 we used the Clausius–Clapeyron equation to determine an approximate expression for the vapor pressure of a substance (solid or liquid) whose gaseous form can be approximated as ideal. This relied on treating the latent heat as constant over the temperatures of interest. For general systems this is not as accurate as we would like, because we used classical results (which we know are not valid at low temperatures), but now we can take a more realistic approach. For simplicity we will still use the classical limit of our system (so the density is low enough that classical physics is valid), but with the justification of Section 11.5, we will have no arbitrary parameters (such as h_0).

11.6.1 General Expression for the Vapor Pressure

We consider a solid of monatomic molecules, for example the noble gases listed in Table 11.1. The melting point for these is quite low at ~ 1 atm,[13] and as such we would expect a quantum description to be required. To determine the vapor pressure of the solid, we of course assume the solid and vapor forms are in equilibrium, so their chemical potentials are equal,

$$\mu_g = \mu_s. \tag{11.74}$$

Let's assume that the vapor is not too dense, so that we can treat it as ideal, and we'll use the result we obtained in Eq. (10.54), but with $h_0 \to h$,[14]

$$\mu_g = -k_B T \ln\left[\frac{V_g}{N_g}\left(\frac{2\pi m k_B T}{h^2}\right)^{3/2}\right]. \tag{11.75}$$

For the solid phase, we need the partition function, Z_s, to calculate the chemical potential,

$$\mu_s = \left(\frac{\partial F}{\partial N_s}\right)_{T,V_s} = -k_B T\left(\frac{\partial \ln Z_s}{\partial N_s}\right)_{T,V_s}. \tag{11.76}$$

If we have a fundamental way to determine Z_s, then we can easily obtain the chemical potential, which we will do using the Einstein model for a solid in Section 11.6.2. But first, let's assume that we do not have a solution to the microscopic theory; instead we have an expression (either from theory or experiment) for the specific heat per molecule of the solid $c_s(T)$[15] and show that we can obtain Z_s from this information.

If we know the specific heat, then the mean energy of the system can be written as

$$\bar{E}_s(T) = -N_s\varepsilon_0 + N_s\int_0^T c_s(T')dT' \tag{11.77}$$

because $N_s c_s(T) = (\partial \bar{E}_s/\partial T)$, and $-\varepsilon_0$ is the ground state energy of an atom in the solid phase (it is negative because it must be a bound state), as measured relative to the same zero-point energy of

Table 11.1 The melting points for the six naturally occurring noble (and thus monatomic) gases.

Gas	Melting point (°C)	Gas	Melting point (°C)
Helium	−272.2	Krypton	−157.3
Neon	−248.67	Xenon	−118.8
Argon	−189.3	Radon	−71

Note that helium is never solid at ordinary pressures, so the melting point here is listed for a high pressure of around 25 atm.

13 Although helium is not solid at 1 atm, one has to go to extremely high pressures, $\bar{p} \sim 25$ atm, to reach a solid state.
14 For now we will ignore any spin effects.
15 The subscript denotes that this is the specific heat of the solid form; I don't distinguish between the specific heat at constant volume or pressure because remember, they are approximately equal for solids.

the vapor. This means we can consider ε_0 to be the latent heat (per molecule) of the solid at zero temperature.

Exercise 11.23 Argue that ε_0 is the latent heat at zero temperature. To do this, consider the third law of thermodynamics and our understanding of entropy as counting the number of accessible states in the system (in either phase).

We also know that

$$\bar{E}(T) = -\left(\frac{\partial \ln Z_s}{\partial \beta}\right)_{N_s,V_s} = k_B T^2 \left(\frac{\partial \ln Z_s}{\partial T}\right)_{N_s,V_s}. \tag{11.78}$$

For a solid, the volume is very close to constant so long as the variations in other parameters are small, so we treat $\bar{E}$ to be independent of the volume. We can then integrate this expression to obtain a result for Z_s,

$$\ln Z_s(T) - \ln Z_s(0) = \frac{1}{k_B}\int_0^T \frac{\bar{E}_s(T')}{T'^2} dT', \tag{11.79}$$

choosing the lower limit of the integral to be zero for convenience. We discussed in Section 6.7, and we can see in Eq. (11.77) that we can determine $\bar{E}$ from the heat capacity (or specific heat).

As $T \to 0$, the partition function for the solid must go to

$$Z_s \to Z_{s0} = \Omega_0 e^{N_s \beta \varepsilon_0},$$

where Ω_0 is the degeneracy of the ground state. Generally Ω_0 is a small number, of order one, which then allows us to write,

$$\ln Z_{s0} = \ln \Omega_0 + N_s \beta \varepsilon_0 \approx N_s \beta \varepsilon_0.$$

This is infinite as $T \to 0$ ($\beta \to +\infty$), so in Eq. (11.79) we can replace $Z_s(0) = Z_{s0}$ with $Z_s(T_0)$, where

$$\ln Z_s(T_0) = \frac{N_s \varepsilon_0}{k_B T_0}, \text{ as } T_0 \to 0.$$

This means that when we calculate $Z_s(T)$, we have to take T to zero carefully to ensure it remains finite in this limit. Putting Eq. (11.78) into Eq. (11.79) and integrating,

$$\begin{aligned}
\ln Z_s(T) &= \lim_{T_0\to 0}\frac{N_s\varepsilon_0}{k_B T_0} - \frac{N_s\varepsilon_0}{k_B}\lim_{T_0\to 0}\int_{T_0}^{T}\frac{1}{T'^2}dT' + \frac{N_s}{k_B}\int_0^T dT'\frac{1}{T'^2}\int_0^{T'} c_s(T'')dT'' \\
&= \lim_{T_0\to 0}\frac{N_s\varepsilon_0}{k_B T_0} + \frac{N_s\varepsilon_0}{k_B}\frac{1}{T} - \frac{N_s\varepsilon_0}{k_B}\lim_{T_0\to 0}\frac{1}{T_0} + \frac{N_s}{k_B}\int_0^T dT'\frac{1}{T'^2}\int_0^{T'} c_s(T'')dT'' \\
&= \frac{N_s\varepsilon_0}{k_B T} + N_s\int_0^T dT'\frac{1}{k_B T'^2}\int_0^{T'} c_s(T'')dT''.
\end{aligned} \tag{11.80}$$

Notice in the second line, the first and third terms (which are divergent as $T_0 \to 0$) cancel.[16]

Using this expression, the chemical potential of a solid can be written as

$$\begin{aligned}
\mu_s &= -k_B T\left(\frac{\partial \ln Z_s}{\partial N_s}\right)_{T,V_s} \\
&= -\varepsilon_0 - T\int_0^T dT'\frac{1}{T'^2}\int_0^{T'} c_s(T'')dT''.
\end{aligned} \tag{11.81}$$

16 Phew!

We see if we used our classical (constant) approximation for the specific heat of a solid, via the equipartition theorem, then it would lead to an infinite chemical potential. This is not surprising—our classical results will generally always break down as we incorporate zero-temperature results.

If our vapor and solid are in equilibrium, we equate chemical potentials and solve for the pressure of the gas to obtain

$$\overline{p}_g = \left(\frac{2\pi m}{h^2}\right)^{3/2}(k_BT)^{5/2}\exp\left[-\frac{\epsilon_0}{k_BT} - \int_0^T dT'\frac{1}{k_BT'^2}\int_0^{T'} c_s(T'')dT''\right]. \tag{11.82}$$

Planck's constant is essential here, as without it we could not obtain an unambiguous expression for the vapor pressure. Since we know that the heat capacity is positive, increases with temperature for a proper theory, and that it should vanish rapidly enough as $T \to 0$, the double integral here will be a positive, increasing function of the temperature and it should remain finite. (Again it is clear this would not be valid if we used the classical result for the specific heat.) Finally, while part of this result can be obtained from the Clausius–Clapeyron equation, that approximation does not allow us to get the second term in the exponential nor the pre-factor exactly. This expression also allows us to understand that ϵ_0 is in fact the latent heat of the system (per atom) for the zero-temperature limit as argued.

Exercise 11.24 Fill in the steps using Eqs. (11.75) and (11.81) to obtain Eq. (11.82). To get $\overline{p}_g$ into the expression, remember that the vapor is treated as an ideal gas.

11.6.2 Vapor Pressure of a Solid in the Einstein Model

We'll now use the results from Section 11.6.1 to get an explicit expression for the vapor pressure in the context of the Einstein model for solids. Recall in this model, the system is treated as N_s atoms connected to each other (in three dimensions) via a Hookian interaction, specifically such that the spring constant is the same for all of the normal modes. We know this is not as accurate as say, the Debye model, but as it is simpler and allows for a closed-form solution for the vapor pressure, it's useful to understand the qualitative features that arise.

We can take two approaches, both using the results of Problem 8.3. The result for the partition function from that problem is

$$Z_s = N_s\left[\frac{3}{2}\beta\hbar\omega - 3\ln\left(e^{\beta\hbar\omega}-1\right) - \ln N_s + 1\right], \tag{11.83}$$

and for the specific heat per atom,

$$c_s(T) = 3k_B\left(\frac{\hbar\omega}{k_BT}\right)^2\frac{e^{\hbar\omega/k_BT}}{(e^{\hbar\omega/k_BT}-1)^2}. \tag{11.84}$$

We could either determine the vapor pressure either by differentiating Z_s to get μ_s, and then setting that equal to μ_g, or we could use the result in Eq. (11.82) of Section 11.6.1. We'll use the latter of these methods (and allow you to use the other method in Problem 11.9). Let's consider the necessary integrals we have to do one at a time. First we have

$$\int_0^{T'} c_s(T'')dT'' = 3k_B\int_0^{T'}\left(\frac{\hbar\omega}{k_BT''}\right)^2\frac{e^{\hbar\omega/k_BT''}}{(e^{\hbar\omega/k_BT''}-1)^2}dT''. \tag{11.85}$$

At first glance this seems like quite a painful integral to evaluate, but it's not as bad as it seems. We first write $y = e^{\hbar\omega/k_BT''}$, so

$$dT'' = -\frac{k_BT'^2}{\hbar\omega}\frac{dy}{y},$$

which makes this integral trivial to evaluate,

$$\int_0^{T'} c_s(T'')dT'' = -3\hbar\omega \int_\infty^{\exp[\hbar\omega/(k_BT')]} \frac{1}{(y-1)^2}dy = \frac{3\hbar\omega}{e^{\hbar\omega/(k_BT')}-1}. \tag{11.86}$$

Note that this is merely the mean energy per atom from Problem 8.3. The second integral, over T', is

$$\int_0^T dT' \frac{1}{k_BT'^2}\int_0^{T'} c_s(T'')dT'' = \int_0^T \frac{1}{k_BT'^2}\frac{3\hbar\omega}{e^{\hbar\omega/(k_BT')}-1}dT', \tag{11.87}$$

and in this case, we will now set $x = e^{\hbar\omega/(k_BT')}$, and then

$$\int_0^T dT' \frac{1}{k_BT'^2}\int_0^{T'} c_s(T'')dT'' = -3\int_\infty^{\exp[\hbar\omega/(k_BT)]} \frac{1}{x(x-1)}dx. \tag{11.88}$$

This is somewhat tricky, because this integral is not well-defined at the lower limit of integration, so we will reintroduce T_0 to write it as

$$\int_0^T dT' \frac{1}{k_BT'^2}\int_0^{T'} c_s(T'')dT'' = -3\lim_{T_0\to 0}\int_{\exp[\hbar\omega/(k_BT_0)]}^{\exp[\hbar\omega/(k_BT)]} \frac{1}{x(x-1)}dx, \tag{11.89}$$

so that

$$\int_0^T dT' \frac{1}{k_BT'^2}\int_0^{T'} c_s(T'')dT'' = -3\lim_{T_0\to 0}\ln\left[\frac{e^{\hbar\omega/(k_BT)}-1}{e^{\hbar\omega/(k_BT)}}\frac{e^{\hbar\omega/(k_BT_0)}}{e^{\hbar\omega/(k_BT_0)}-1}\right]. \tag{11.90}$$

Taking the limit $T_0 \to 0$, the second factor inside the logarithm goes to one, and we then put this into our expression for the vapor pressure,

$$\overline{p}_g = \left(\frac{2\pi m}{h^2}\right)^{3/2}(k_BT)^{5/2}e^{-\epsilon_0/(k_BT)}\left\{\left[e^{\hbar\omega/(k_BT)}-1\right]^3 - e^{3\hbar\omega/(k_BT)}\right\}. \tag{11.91}$$

We see here that while a realistic determination of the vapor pressure can be more complicated than with our earlier approximations, they are not impossible to obtain.

Exercise 11.25 Fill in the steps that lead to Eq. (11.91).

11.7 Partition Function of Ideal Polyatomic Molecules

Let us now expand our study of ideal gases to those which are not monatomic. To do so, we need to have more information about the internal degrees of freedom of our system, and we will again do this for the classical limit of the quantum problem. Calculating the partition function for a purely quantum problem is straightforward; however, it is much more complex and does not lend itself to a better understanding of the system. Thus we start with the partition function

$$Z = \frac{\zeta^N}{N!},$$

with the single-particle partition function given by

$$\zeta = \sum_s e^{-\beta\epsilon(s)}.$$

The energy for a given single-particle state s has four different sources:

Translational states, s_t: These come from the kinetic energy of the center of mass, $\varepsilon_t(s_t)$. Classically s_t would correspond to the momentum of the molecule, and quantum mechanically it would be the set of quantum numbers n_x, n_y, and n_z relevant for a particle in a box (or even the free particle with periodic boundary conditions).

Electronic states, s_e: These states arise when considering the quantum energy levels, $\varepsilon_e(s_e)$, of the outer electrons of the molecule. To determine these, we would need a viable model to describe these states, which is quite tricky to formulate, so we will of course use an approximation.

Rotational states, s_r: These do not occur for monatomic molecules given the spherical symmetry, but the atoms in non-monatomic molecules can rotate about the various axes that pass through the center of mass giving an energy $\varepsilon_r(s_r)$. This is not difficult to include because we know the rotational kinetic energy of the system quantum mechanically.

Vibrational states, s_ν: Last but not least, oscillations of the relative positions of the atoms in the molecule will give rise to these energies, $\varepsilon_\nu(s_\nu)$. If we consider our atoms to vibrate according to Hooke's law, the energy levels here will correspond to the simple harmonic oscillator energies we have already seen many times.

With each of these, the energy of a state $s = \{s_t, s_e, s_r, s_\nu\}$ can be written as a sum,

$$\varepsilon(s) = \varepsilon_t(s_t) + \varepsilon_e(s_e) + \varepsilon_r(s_r) + \varepsilon_\nu(s_\nu).$$

All of these are easily determined from results we have already derived, except for the electronic energies ε_e. Solving for these energies requires knowing the relevant Hamiltonian which describes the electronic states, but as already mentioned, we will use a simple approximation.

Since these energies are additive, we can simply break up the single-particle partition function to get

$$\begin{aligned}\zeta &= \sum_{s_t,s_e,s_r,s_\nu} e^{-\beta[\varepsilon_t(s_t)+\varepsilon_e(s_e)+\varepsilon_r(s_r)+\varepsilon_\nu(s_\nu)]} \\ &= \left(\sum_{s_t} e^{-\beta\varepsilon_t(s_t)}\right)\left(\sum_{s_e} e^{-\beta\varepsilon_e(s_e)}\right)\left(\sum_{s_r} e^{-\beta\varepsilon_r(s_r)}\right)\left(\sum_{s_\nu} e^{-\beta\varepsilon_\nu(s_\nu)}\right) \\ &= \zeta_t\zeta_e\zeta_r\zeta_\nu.\end{aligned} \tag{11.92}$$

and we can examine each of these in turn. For this, we will focus on the specific case of a diatomic molecule with atoms of masses m_1 and m_2.

11.7.1 Translational Motion of the Center of Mass

The energy of any problem with more than one particle usually can be written in terms of the kinetic energy of the center of mass plus the energies associated with the motion relative to the center of mass. The center-of-mass energy takes the (classical) form

$$\varepsilon_t = \frac{\mathbf{p}^2}{2(m_1 + m_2)}, \tag{11.93}$$

where $\mathbf{p}$ is the momentum of the center of mass of the system. There is no need to evaluate the single-particle partition function for this case as we have already done so many times, so

$$\zeta_t = \frac{V}{h^3}\left[2\pi\left(m_1 + m_2\right)k_B T\right]^{3/2}. \tag{11.94}$$

This particular result is easily generalized to any type of molecule, with $m_1 + m_2$ replaced with the total mass of the molecule.

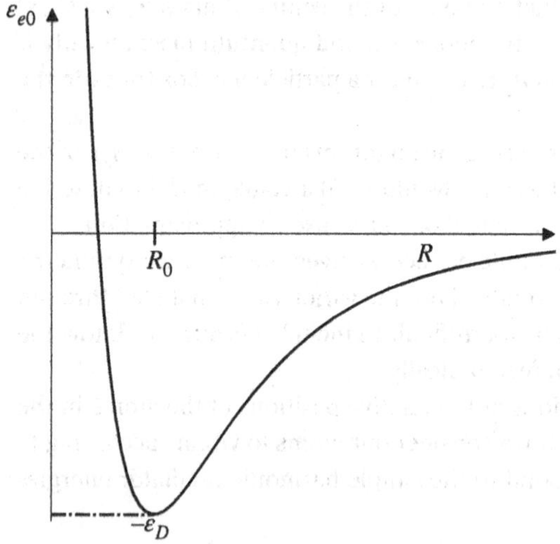

Figure 11.9 A typical dependence of the ground state electronic energy as a function of the internuclear distance R.

11.7.2 Electronic States

The first of the internal degrees of freedom that we'll consider comes from the possible states of the electrons in the atoms. To determine the ground state of the electrons as they orbit the nuclei in the molecule, one must solve a rather complicated Schrödinger equation, and for any realistic problem one must rely on approximation methods (such as perturbation theory or as is most common nowadays, numerical techniques). What results is often a determination of a given energy level as a function of R, the distance between the two nuclei in the molecule. The result of such calculations leads to an expression like that shown in Figure 11.9, which shows the ground-state energy as a function of R. The result here is the Lennard–Jones potential, which we've seen before in Eq. (8.56) when considering non-ideal gases.

When the two nuclei are a distance R_0 apart, they are in equilibrium (we will discuss oscillations about this distance shortly). We can make a very rough approximation that only the lowest electronic energy level $-\varepsilon_D$ contributes significantly at this temperature (the subscript D comes from this often being referred to as the **Dirac energy**). The electronic single-particle partition function would be given by

$$\zeta_e = \Omega_D e^{\beta\varepsilon_D} + \cdots, \tag{11.95}$$

where Ω_D is the degeneracy of this state and we neglect higher-order terms.

For most electronic systems, the first excited state is a few electron volts higher energy than the ground state, hence the omission of any higher energy level. If we get to the point where this is not the case, then the classical description breaks down, and we would have to consider a more complete description, which is beyond the scope of this text.

11.7.3 Rotation

Now considering the rotational degrees of freedom, we'll model the diatomic molecule to a first approximation as a rigid rotor (shown in Figure 11.10)—the two masses are connected by a rigid rod which rotates about an axis that passes through the center of mass. The moment of inertia is given by

$$I = m_{\text{red}} R_0^2,$$

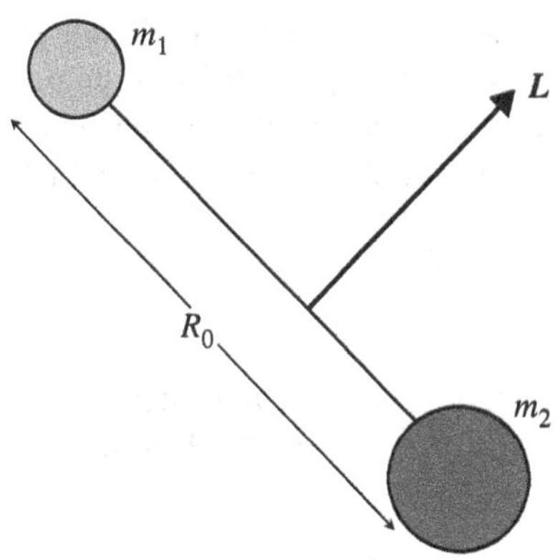

Figure 11.10 A diatomic molecule as a rigid rotor, which rotates with angular momentum **L** about an axis passing through the center of mass.

where the **reduced mass** of the system is given by

$$m_{\text{red}} = \frac{m_1 m_2}{m_1 + m_2},$$

and R_0 is the (equilibrium) atomic distance between the two atoms. The rotational energy of the molecule has orbital angular momentum **L** is given by

$$\varepsilon_r = \frac{\mathbf{L}^2}{2I}. \tag{11.96}$$

But as we've discussed before, orbital angular momentum can only take integer values, so

$$\varepsilon_r = \frac{\ell(\ell+1)\hbar^2}{2I} \quad \ell = 0, 1, \ldots .$$

For each ℓ there are $2\ell + 1$ distinct states with this same energy, given by the possible projections of **L** on the axis of rotation. Thus,

$$\zeta_r = \sum_{\ell=0}^{\infty} (2\ell + 1) e^{-\beta\hbar^2 \ell(\ell+1)/(2I)}.$$

As before, we will consider this in the classical limit, where the temperature is high enough such that compared with the energy scale, we have

$$\frac{\hbar^2 \ell(\ell+1)}{2Ik_B T} \ll 1, \tag{11.97}$$

so we can treat the sum for ζ_r as an integral. We define

$$u = \ell(\ell+1) \text{ and } \Delta u = (2\ell+1)\Delta\ell,$$

where $\Delta\ell = 1$, and $\sum \Delta u \to \int du$ if Eq. (11.97) holds, so

$$\begin{aligned} \zeta_r &= \int_0^{\infty} e^{-[\beta\hbar^2/(2I)]u} du \\ &= \frac{2Ik_B T}{\hbar^2}. \end{aligned} \tag{11.98}$$

Exercise 11.26 Verify our claim that Eq. (11.97) is valid for a nitrogen molecule at room temperature. You should find this ratio is around 0.01.

There is one issue that we have to contend with however. For a molecule with identical atoms (like H_2), we have to be concerned with overcounting, just as we did when considering the Gibbs paradox. For example, in Figure 11.10, if both atoms were the same, then interchanging the two atoms would give a different state, but we should count them as the same, quantum mechanically. However, this would not be true if they were different atoms. Thus we'll introduce a fudge factor of two for identical particles so that

$$\zeta_r = \frac{2Ik_BT}{\sigma\hbar^2}, \tag{11.99}$$

with

$$\sigma = \begin{cases} 1 & \text{distinct atoms} \\ 2 & \text{identical atoms} \end{cases}. \tag{11.100}$$

11.7.4 Vibration

Finally we consider the energy arising from the vibrational modes, where the atoms in the molecule are free to vibrate about their equilibrium separation R_0. For small oscillations, the potential energy of the atoms as a function of the internuclear distance R can be expanded as a simple harmonic potential about the ground state electronic energy,

$$U(R) = -\epsilon_D + \frac{1}{2}m_{\text{red}}\omega^2(R - R_0)^2 + \cdots. \tag{11.101}$$

Exercise 11.27 Show that the Lennard–Jones potential from Eq. (8.56) has the form in Eq. (11.101) when expanding about the minimum.

The kinetic energy of the vibrational modes is given by

$$K = \frac{1}{2}m_{\text{red}}\left(\frac{dR}{dt}\right)^2.$$

Solving this quantum mechanical simple harmonic oscillator we get the energies

$$\epsilon_v = \left(n + \frac{1}{2}\right)\hbar\omega$$

with $n = 0, 1, 2, \ldots$. Thus

$$\zeta_v = \sum_{n=0}^{\infty} e^{-\beta\hbar\omega(n+1/2)},$$

which we have already evaluated,

$$\zeta_v = \frac{e^{-\beta\hbar\omega/2}}{1 - e^{-\beta\hbar\omega}}.$$

For most diatomic molecules $\beta\hbar\omega \gg 1$, so for simplicity we can write (essentially keeping only the ground state energy in the sum for ζ_v)

$$\zeta_v = e^{-\beta\hbar\omega/2} \tag{11.102}$$

to a good approximation. This can certainly not be treated classically (recall the equipartition theorem).

Exercise 11.28 Consider the case where $\hbar\omega \sim 0.1$ eV for an oscillator at room temperature. Show that our approximate expression for ζ_v is only about 2% smaller than the full expression.

Putting all of these together, we have the single-particle partition function for a diatomic molecule

$$\zeta = \left\{ \frac{V}{h^3}[2\pi(m_1 + m_2)k_B T]^{3/2} \right\} \left(\Omega_D e^{\beta \epsilon_D}\right) \left(\frac{2Ik_B T}{\sigma \hbar^2} \right) \left(e^{-\beta\hbar\omega/2}\right),$$

or assuming no degeneracy of the electronic ground state,

$$\zeta = \left[\frac{2I(m_1 + m_2)}{(2\pi)^{3/2}\sigma\hbar^5} \right] \frac{V}{\beta^{5/2}} e^{\beta(\epsilon_D - \hbar\omega/2)}. \tag{11.103}$$

Some of these factors would be the same with molecules with more than two atoms given our assumptions. The real complications would arise in the rotational and vibrational contributions.

11.7.5 Molar Specific Heat of a Diatomic Molecule

Using the single-particle partition function in Eq. (11.103), we can calculate the molar specific heat at constant volume of a diatomic gas at temperature T in the limit used to obtain that expression (recall that implies that rotational levels are closely spaced and that the vibrational levels are largely spaced). We use $\bar{E} = -(\partial \ln Z/\partial\beta)_V$ to obtain

$$\bar{E} = N\left(\frac{5}{2}k_B T - \epsilon_D + \frac{\hbar\omega}{2} \right). \tag{11.104}$$

Note this is very similar to the monatomic case (with 5/2 instead of 3/2), but with additional terms from shifting the zero-point energy. With $C_V = (\partial\bar{E}/\partial T)_V$, we get

$$C_V = \frac{5}{2}Nk = \frac{5}{2}\nu R, \tag{11.105}$$

so the molar specific heat is

$$c_V = \frac{5}{2}R. \tag{11.106}$$

This was claimed in Exercise 6.29, so it's nice to finally see it to be valid. Additionally, we can now consider diatomic systems when determining the relative concentrations of molecules using the law of mass action instead of only monatomic systems.

Exercise 11.29 Fill in the steps to obtain Eqs. (11.104) and (11.106).

11.8 Summary

- A system with varying numbers of particles can be described by the grand canonical ensemble, which considers the system in contact with a heat and particle reservoir. For large systems, this is equivalent to treating the system in the canonical ensemble or the microcanonical ensemble.
- The requirement that we treat quantum particles as indistinguishable imposes symmetry restrictions which divide all particles into one of two classes: fermions and bosons. More than one fermion cannot be in the same state, while this is not true for bosons.
- The quantum statistical mechanics problem amounts to evaluating the occupation number, or quantum distribution function, from which all statistical quantities can be calculated.
- The different quantum partition functions and occupation numbers for both fermions and bosons reduce to the results obtained using fudged classical statistics, which was originally an ad hoc solution to the Gibbs paradox.

- With these tools, a proper calculation of the vapor pressure of a solid as well as the specific heat of polyatomic molecules can be performed. These will match classical results in the appropriate limits but also match thermodynamic expectations at zero temperature.

Problems

11.1 As you may have guessed, we can get the grand canonical ensemble starting from the initial assumption that we maximize the missing information in our theory, consistent with any constraints, just as in Problem 8.1 for the canonical ensemble. Follow the steps of that problem here, but in addition to maximizing the Shannon entropy,

$$S = -k_B \sum_r P_r \ln P_r,$$

with the constraints

$$\sum_r P_r = 1 \text{ and } \bar{E} = \sum_r P_r E_r,$$

also require the mean number of particles to be fixed,

$$\overline{N} = \sum_r P_r N_r.$$

11.2 Perform the Taylor expansion required to obtain Eq. (11.17).

11.3 A system consists of two particles, each of which can occupy any of three single-particle states with energies 0, ε, and 2ε, respectively. At a temperature such that $e^{-\beta\varepsilon} = 1/3$, calculate the mean number of particles in each orbital, $\overline{n}_0$, $\overline{n}_\varepsilon$, $\overline{n}_{2\varepsilon}$ for each of the following cases:

(a) Maxwell–Boltzmann particles,
(b) Bose–Einstein particles, and
(c) Fermi–Dirac particles.

11.4 Consider a system with two energy levels with energies 0 and ε in contact with a heat reservoir at temperature $T = 1/(k_B\beta)$. The temperature is chosen so that $e^{-\beta\varepsilon} = 1/2$. It is also in contact with a particle reservoir, whose chemical potential is chosen so that the *mean* number of particles in the system is exactly one.

(a) Assuming the particles obey Maxwell–Boltzmann statistics, find the mean numbers of particles $\overline{n}_0$ and $\overline{n}_\varepsilon$ in the states with energy 0 and ε, respectively.
(b) Repeat part (a) assuming the particles obey Bose–Einstein statistics. The quadratic equation you have to solve to get the answer of course has two roots. Explain clearly, with a *physical argument* why can you throw out one of them.
(c) Repeat part (a) assuming the particles obey Fermi–Dirac statistics. As in part (b), explain physically why you can throw out one of the solutions to the quadratic equation.

11.5 Return to the grand potential from Problem 10.18,

$$\Omega = \bar{E} - TS - \mu N. \tag{11.107}$$

(a) Show that you can determine Ω directly from the grand partition function,

$$\Omega = -k_B T \ln \mathcal{Z}. \tag{11.108}$$

(b) Calculate Ω for both Bose–Einstein and Fermi–Dirac statistics.

11.6 (a) Show that

$$\beta\varepsilon_r + \alpha = \ln(1 \mp \overline{n}_r) - \ln \overline{n}_r,$$

where ε_r is the energy of the single-particle state r, $\overline{n}_r$ is the mean occupation number, and the upper (lower) signs are for FD (BE) statistics, respectively.

(b) Show that you can write the grand partition function as

$$\ln \mathcal{Z} = \mp \sum_r \ln(1 \mp \overline{n}_r).$$

(c) Recall $S = k_B(\ln \mathcal{Z} - \beta\mu\overline{N} + \beta\overline{E})$ for the following.

i. Write down an expression for the entropy S of an ideal Fermi–Dirac gas. Express your answer solely in terms of $\overline{n}_r$, the mean number of particles in state r.

ii. Write a similar expression for the entropy S of a Bose–Einstein gas.

iii. What do these expressions for the entropy become in the classical limit when $\overline{n}_r \ll 1$?

(d) Show that the answer to (c)iii is what you would expect for an ideal gas with fudged classical statistics.

11.7 Recall that one can show that for an ideal monatomic (and non-relativistic) gas, $\overline{p} = 2\overline{E}/(3V)$. However, often one has to *assume* that, in equilibrium, the mean pressure on every wall of the rectangular container was the same. Now you can *prove* that fact under rather general conditions. Proceed as follows:

(a) Show that the mean pressure on the wall perpendicular to the x-direction is given by

$$\overline{p}_x = \sum_r \overline{n}_r(\epsilon_r, \alpha, \beta) \frac{\hbar^2\pi^2}{mV} \left(\frac{n_x}{L_x}\right)^2,$$

where the orbital r has energy ϵ_r and is specified by the quantum numbers n_x, n_y, and n_z. $\overline{n}_r(\epsilon_r, \alpha, \beta)$ is the mean occupation number (in either MB, FD, or BE statistics) of orbital r. Corresponding expressions hold for $\overline{p}_y$ and $\overline{p}_z$.

(b) Assume that the temperature is high enough that a large number of orbitals are occupied. In that case, n_x, n_y, and n_z may be treated as continuous variables. Use this fact to rewrite $\overline{p}_x$ above as an integral over n_x, n_y, and n_z.

(c) Change variables in your integral to eliminate any explicit reference to L_x, L_y, and L_z from the integrand. Now argue that this shows that $\overline{p}_x = \overline{p}_y = \overline{p}_z$.

(d) The integral expression for p_x in (c) looks, at first glance, like it says that $\overline{p}_x$ depends only on the temperature, not on the volume or the total number of particles. Explain where the dependence on N/V (which we know must be there from the classical ideal gas equation $\overline{p} = Nk_BT/V$) is hiding.

(e) In part (b) we had to assume that the temperature is high enough. Is this just a technical assumption (to make it easy to do the problem), or is it really true that at low enough temperature the pressures in the three directions may not necessarily be equal? Explain.

Hint: Consider as a test case a BE gas at $T = 0$ in an asymmetrical box ($L_x > L_y > L_z$). There is only one microstate in this case, so it's easy to explicitly calculate the pressures in the three directions.

11.8 (a) Considering the classical approximation for the partition function, write down the chemical potential μ for an ideal gas of N atoms of mass m in a volume V at absolute temperature T.

(b) Now consider a two-dimensional system where a gas of N' weakly interacting particles freely moves around a surface of area A on which they are free to move. Such a gas can be treated as ideal with energy $\mathbf{p}^2/2m - \varepsilon_0$, where $\mathbf{p}$ denotes its two-component momentum vector and ε_0 is the binding energy which holds a molecule on the surface. Calculate the chemical potential μ' of this ideal gas. The partition function can again be evaluated in the classical approximation.

(c) These two problems can be combined by considering a gas in a cubic container such that one side of the cube (of area $A = V^{2/3}$) attracts molecules of the gas. As such, the gas is free to move around the container while a fraction of the gas adsorbs on the surface. The equilibrium condition between the three- and two-dimensional gases in parts (a) and (b), respectively, can be expressed by, as usual, equating the chemical potentials. What is the mean number of molecules per unit area ($n' = N'/A$) on the surface when the surrounding gas is at pressure $\overline{p}$?

11.9 Return to the vapor pressure of a solid in the context of the Einstein model. Differentiate the expression for the partition function of the solid, Eq. (11.83), to obtain μ_s, and then setting that equal to μ_g. Solving for $\overline{p}_g$, show that your result is the same as Eq. (11.91).

11.10 Calculate the equilibrium constant from Problem 10.24 using our expressions for the partition function for a diatomic molecule, Eq. (11.103). Compare to your result when we used the monatomic expressions for ζ.

Reference

1 F. Reif. *Fundamentals of Statistical and Thermal Physics*. McGraw Hill, Tokyo, 1965.

12

Applications of Quantum Statistics

In Chapter 11 we laid out the formalism for the statistical mechanics of quantum systems. In this chapter, we will apply these techniques to three different problems: First we will use the Planck distribution to study blackbody radiation, then we will use the Bose–Einstein distribution to study a Bose–Einstein condensate, and finally we will use the Fermi–Dirac distribution to study a gas of fermions. In this way, we will explore each of the three quantum distribution functions we derived as well as study some new techniques when solving thermodynamic problems.

After finishing this chapter, you should be able to

- understand an ideal photon gas and its application to blackbody radiation,
- understand what a Bose–Einstein condensate is and how it arises from the study of the Bose–Einstein distribution, and
- calculate the contribution of electrons (due to their motion) to the specific heat of a conductor, when treated as a free Fermi gas.

12.1 Blackbody Radiation

Around the turn of the twentieth century, one of the problems that helped lead to the development of quantum mechanics was known as the **ultraviolet catastrophe**.[1] This is studied in detail in most modern physics courses as a motivation for a new paradigm beyond classical physics, which predicts a divergent power spectrum from an object emitting electromagnetic radiation in the high-energy region. The term *ultraviolet* is used because high-energy waves have short wavelengths, and the ultraviolet region of the electromagnetic spectrum has lower wavelength than visible light. Planck solved this "catastrophe" in a way that led to a quantum description of electromagnetic radiation, and in this section we will work through a detailed derivation of the power spectrum of a **blackbody**, which is a perfect emitter of electromagnetic radiation.

12.1.1 From E&M to Photons

We can understand **photons**, the massless quanta of the electromagnetic field, with little quantum mechanics, as many of their basic properties come directly from Maxwell's equations. Maxwell's equations in free space (with no charges or currents) are given by

1 As with the word paradox, we also like to use the word catastrophe sometimes when theory and experiment do not agree. One wouldn't think of physicists as melodramatic, but there you have it.

Statistical Thermodynamics: An Information Theory Approach, First Edition. Christopher Aubin.

Companion website: www.wiley.com/go/Aubin/StatisticalThermodynamics

$$\nabla \cdot \mathbf{E} = 0, \tag{12.1}$$

$$\nabla \cdot \mathbf{B} = 0, \tag{12.2}$$

$$\nabla \times \mathbf{E} = -\frac{\partial \mathbf{B}}{\partial t}, \text{ and} \tag{12.3}$$

$$\nabla \times \mathbf{B} = \frac{1}{c^2}\frac{\partial \mathbf{E}}{\partial t}. \tag{12.4}$$

The last two of these are Faraday's law[2] and the Ampère–Maxwell law,[3] which can be combined to obtain the wave equations for the electric and magnetic fields,

$$\nabla^2 \mathbf{E} = \frac{1}{c^2}\frac{\partial^2 \mathbf{E}}{\partial t^2} \text{ and } \nabla^2 \mathbf{B} = \frac{1}{c^2}\frac{\partial^2 \mathbf{B}}{\partial t^2}. \tag{12.5}$$

The solutions to these equations can be written as (the real parts of) complex plane waves,

$$\mathbf{E}(\mathbf{r}, t) = \mathbf{E}_0 e^{i(\mathbf{k}\cdot\mathbf{r}-\omega t)} \text{ and } \mathbf{B}(\mathbf{r}, t) = \mathbf{B}_0 e^{i(\mathbf{k}\cdot\mathbf{r}-\omega t)}. \tag{12.6}$$

These two laws also lead to the requirement that **E** and **B** must be perpendicular to each other, and the magnitudes of the two fields are related by the speed of light,

$$|\mathbf{E}_0| = c|\mathbf{B}_0|. \tag{12.7}$$

In Eq. (12.6), **k** is the **wave vector** that points in the direction of the traveling wave and ω is the angular frequency. They satisfy the condition

$$k = |\mathbf{k}| = \frac{2\pi}{\lambda} = \frac{\omega}{c}, \tag{12.8}$$

where λ is the wavelength and c (the speed of light) is the speed of the wave. Equation (12.6) corresponds to a monochromatic wave (one with a single wavelength), and while most waves are not monochromatic, any wave can be written as a linear combination of such plane waves thanks to the principle of superposition. The momentum and energy of the wave are related to the wave vector and angular frequency via

$$\mathbf{p} = \hbar\mathbf{k},$$

and

$$\varepsilon = \hbar\omega.$$

The other two of Maxwell's equations, Eqs. (12.1) and (12.2) (the first of which is Gauss's law and the second is often unnamed, or referred to as Gauss's law for magnetism), provide additional constraints on our electromagnetic wave. Specifically, we see by taking the gradient of Eq. (12.6), they require our wave to be **transverse,**

$$\mathbf{k} \cdot \mathbf{E}_0 = \mathbf{k} \cdot \mathbf{B}_0 = 0,$$

so the electric and magnetic fields are perpendicular to the direction of motion. This, combined with the constraints above, reduces our six components of the electric and magnetic fields down to just two components. These two components are the **polarizations** of the electromagnetic wave, which are familiar from the field of optics.

2 Michael Faraday (1791–1867).
3 André-Marie Ampère (1775–1836).

Exercise 12.1 Show that Maxwell's equations in free space do in fact require the transversality and mutual orthogonality of the electric and magnetic fields and that they are related by Eq. (12.7). Also from this, argue that there are only two independent components (or polarizations). If you need a refresher on electromagnetic waves, check Chapter 9 of Ref. [1].

The transition to the quantum description of the electromagnetic wave is not obvious, but can be understood from our discussion above. If we consider a slightly more general solution beyond a plane wave, we can write (we'll focus on the electric field for now without losing generality as we know it is directly related the magnetic field)

$$\mathbf{E}(\mathbf{r}, t) = \mathbf{E}_0 e^{i\mathbf{k}\cdot\mathbf{r}} f(t), \tag{12.9}$$

so that it has a definite wavelength, which is a **normal mode** (which we discussed in Chapter 9 when discussing the Einstein model for specific heats). Putting this into the wave equation we find

$$\frac{d^2 f}{dt^2} = -\omega^2 f(t),$$

which is just the equation for a simple harmonic oscillator. We see that each normal mode acts like a simple harmonic oscillator, so we can use these modes as the basis states that we use to describe quantized electromagnetic fields, or photons.

The polarizations of the electric field correspond to the spin components of the photon: the photon is spin-1, and can have a spin projection that has one of two values, $m = \pm 1$.[4] Whether or not we treat our fields classically or quantum mechanically, we will have to account for both polarizations as will be clear in Section 12.1.2.

12.1.2 Photon Gas

Now let's apply our ideas from Section 12.1 to a **photon gas**, or a container that is filled with electromagnetic radiation.[5] These photons can be absorbed by the walls of the container and emitted continuously, so the number of photons is not fixed. We considered this in Section 11.3.2, where we found the mean number of photons with energy ε_r in a container with temperature $T = 1/(k_B\beta)$ to be

$$\overline{n}_r = \frac{1}{e^{\beta\varepsilon_r} - 1}. \tag{12.10}$$

The energy levels ε_r come from the solutions to the wave equations shown in Eq. (12.6). To determine a discrete set of quantized energies as in Chapter 11, we assume the container is large enough such that we can ignore the walls, and thus will impose periodic boundary conditions. The wave vector is restricted as in Eqs. (11.61)–(11.63),

$$k_x = \frac{2\pi n_x}{L_x},\ k_y = \frac{2\pi n_y}{L_y},\ \text{and}\ k_z = \frac{2\pi n_z}{L_z},$$

where the integers n_x, n_y, and n_z can take on positive or negative values. As with other applications, we will not use the quantized energy levels, but instead assume the temperature is high enough that we can consider them continuous variables. However, the boundary conditions allow us to obtain

4 Because the photon is massless, the $m = 0$ component is not allowed.

5 Again, this might seem odd to think of a "box of photons," but we live with this everyday. The room you are sitting in hopefully is not pitch black, so there are photons all around you. You are sitting in such a box of photons.

the proper factor involving the volume of the photons in phase space that we should get when converting the sums over n_x, n_y, and n_z to an integral over $\mathbf{k}$,

$$\sum_{n_x,n_y,n_z} \rightarrow V\int \frac{d^3\mathbf{k}}{(2\pi)^3}. \tag{12.11}$$

We will follow a similar discussion as with the Maxwell velocity distribution in Section 9.4.1. In that vein, we will first consider the number of photons,

$$dN = Vf(\mathbf{k})d^3\mathbf{k}.$$

The volume appears immediately here, unlike with the Maxwell velocity distribution, where there was a factor of $d^3\mathbf{r}$ due to the classical nature of that system. Specifically this is the number of photons with a wave vector in a range from $\mathbf{k}$ to $\mathbf{k} + d^3\mathbf{k}$. $\overline{n}_r$ in Eq. (12.10) is the mean number of photons with a *definite* value of $\mathbf{k}$, or given Eq. (12.8), a definite value of $\omega = \varepsilon/\hbar$. This allows us to write

$$dN = Vf(\mathbf{k})d^3\mathbf{k} = \frac{1}{e^{\beta\hbar\omega} - 1}\frac{V\,d^3\mathbf{k}}{(2\pi)^3}. \tag{12.12}$$

While we've written f as a function of $\mathbf{k}$, we can see that it is only a function of the magnitude k (again, just like the Maxwell velocity distribution, which was only dependent on the speed but not the velocity). In fact, this expression for f can be shown to be independent not only of the direction of the photons and their positions (note $\mathbf{r}$ does not appear in our expression), but also the polarization, as well as the size and shape of the container (for a nice discussion of this, see section 9–14 of Ref. [2]).

Each photon, regardless of its wave vector, has two allowed polarizations, which the energy does not depend upon. To account for this, we multiply Eq. (12.12) by two, for the number of polarizations. Additionally, it is conventional to consider the **photon density**, the number of photons that exist per unit volume, dN/V, so we will do that from now on. Additionally, we'll write the integration measure in spherical coordinates, $d^3\mathbf{k} = 4\pi k^2 dk$, and with $dk = d\omega/c$, we can write

$$2f(\mathbf{k})(4\pi k^2 dk) = \frac{1}{e^{\beta\hbar\omega} - 1}\frac{8\pi\omega^2 d\omega}{(2\pi)^3 c^3}. \tag{12.13}$$

If we denote the mean **energy density** (energy per unit volume) by $\bar{u}(\omega, T)d\omega$ with a frequency in the range $\omega \rightarrow \omega + d\omega$, we can calculate this by multiplying Eq. (12.13) by $\hbar\omega$ to get

$$\bar{u}(\omega, T)d\omega = \hbar\omega\left[2f(\mathbf{k})(4\pi k^2 dk)\right] = \frac{\hbar}{\pi^2 c^3}\frac{\omega^3 d\omega}{e^{\beta\hbar\omega} - 1}. \tag{12.14}$$

To make this easier to analyze, let's define the dimensionless parameter $\eta \equiv \hbar\omega/(k_B T)$, so this becomes

$$\bar{u}(\omega, T)d\omega = \left(\frac{k_B^4}{\pi^2 c^3 \hbar^3}\right)\left(\frac{\eta^3 d\eta}{e^\eta - 1}\right)T^4. \tag{12.15}$$

The integrand in the second factor of this expression has a maximum at $\eta = \eta_{\max} \approx 2.82$, and we show this in Figure 12.1.

The maximum of Eq. (12.15) always occurs at $\eta_{\max}$ regardless of the temperature (that only affects that maximum value of the function itself). This implies that as we change the temperature of the system, the most probable frequency of the photons in the container also changes, and these are related by

$$2.82 \approx \frac{\hbar\omega_1}{k_B T_1} = \frac{\hbar\omega_2}{k_B T_2} \tag{12.16}$$

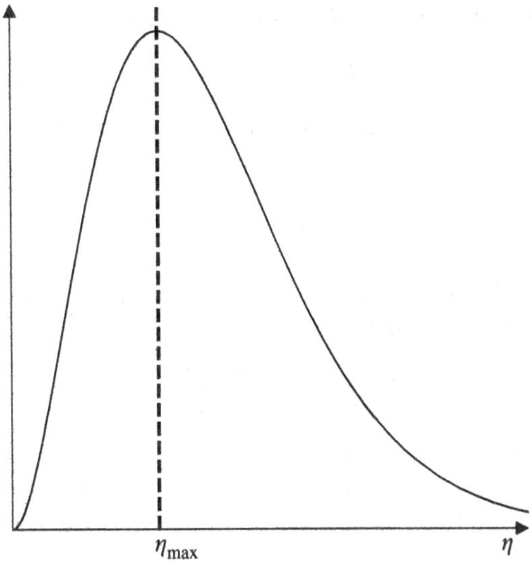

Figure 12.1 The integrand in Eq. (12.15), showing the peak at $\eta = \eta_{max}$.

or in terms of the frequency ν, with $\omega = 2\pi\nu$,

$$\frac{\nu_1}{T_1} = \frac{\nu_2}{T_2}. \tag{12.17}$$

This is known as **Wien's displacement law**,[6] which is also a result of classical thermodynamics. Once we apply these results to blackbody radiation, this is how we relate the temperature of a star to its color, for example.

Exercise 12.2 Show that the maximum of

$$\frac{\eta^3}{e^\eta - 1}$$

occurs at $\eta_{max} \approx 2.82$. After setting the derivative to zero, you will obtain a transcendental equation to solve for η_{max}, so to determine the solution, you can use a graphical or numerical approach.

For a given temperature, we can integrate over the frequencies ω to determine the mean energy density,

$$\bar{u}_0(T) = \int_0^\infty \bar{u}(\omega, T)d\omega = \left[\left(\frac{k_B^4}{\pi^2 c^3 \hbar^3}\right)\int_0^\infty \frac{\eta^3 d\eta}{e^\eta - 1}\right] T^4. \tag{12.18}$$

The integral is just a number, and while we can (and will) evaluate it, remember a large part of our work entails making predictions without knowing everything about the system. We can see even without knowing this particular number that the energy density scales quartically with the temperature,

$$\bar{u}_0(T) \propto T^4\,.$$

With this, we could measure the energy density of a system as a function of the temperature to determine empirically what the proportionality constant is.

6 Wilhelm Wien (1864–1928).

Exercise 12.3 By writing the integral

$$I = \int_0^\infty \frac{\eta^3 d\eta}{e^\eta - 1} = \int_0^\infty \frac{e^{-\eta}\eta^3 d\eta}{1 - e^{-\eta}},$$

expand the denominator using $1/(1-x) = 1 + x + x^2 + \cdots$ and show that you can write

$$I = \sum_{n=1}^{\infty} \frac{6}{n^4}. \tag{12.19}$$

The result of Exercise 12.3 allows us to determine the constant exactly in front of the T^4 factor for $\bar{u}_0$. We obtain

$$\int_0^\infty \frac{\eta^3 d\eta}{e^\eta - 1} = \sum_{n=1}^{\infty} \frac{6}{n^4} = \frac{\pi^4}{15},$$

so

$$\bar{u}_0(T) = \frac{\pi^2 k_B^4}{15(\hbar c)^3} T^4. \tag{12.20}$$

Exercise 12.4 Evaluate the coefficient in front of T^4 in Eq. (12.20) and show that it has the value

$$\frac{\pi^2 k_B^4}{15(\hbar c)^3} = 4752 \frac{\text{eV}}{\text{m}^3 \cdot \text{K}^4}.$$

There are several ways to evaluate the infinite sum $\sum_{n=1}^{\infty} 1/n^4$, and I'll leave it as an exercise for you to evaluate this sum using Fourier series in Problem 12.1.

12.1.3 Radiation Pressure

The photon gas, as any other gas, provides a pressure inside the container.[7] The mean pressure is given by Eq. (11.27),

$$\bar{p} = \sum_r \bar{n}_r \left(-\frac{\partial \varepsilon_r}{\partial V}\right). \tag{12.21}$$

For a given state, assuming the box is a cube with sides of length $L = V^{1/3}$, we have (with periodic boundary conditions),

$$\varepsilon_r = \frac{2\pi\hbar c}{V^{1/3}} \sqrt{n_x^2 + n_y^2 + n_z^2}, \tag{12.22}$$

so that with Eq. (12.21), we find

$$\bar{p} = \frac{\bar{E}}{3V}. \tag{12.23}$$

The right-hand side is just one-third of the energy density,

$$\bar{p} = \frac{\bar{u}_0}{3}. \tag{12.24}$$

This is true when L_x, L_y, and L_z are different, using the fact that the mean pressure in each direction is the same, see Problem 12.5.

Exercise 12.5 Fill in the steps to obtain Eq. (12.23) from Eqs. (12.21) and (12.22).

7 Because photons have momentum, they can impart that momentum on the walls of the container and thus provide a force.

12.1.4 Radiation from a Hot Object

To be able to apply our results above to blackbodies, we now wish to imagine a hot object emitting electromagnetic radiation. This does not seem like an equilibrium situation (in our standard sense), but we can actually still apply our equilibrium results of Section 12.1.2 by understanding what we mean by this term here.

We assume the system inside the container to be in equilibrium at temperature T, and it is emitting energy in the form of power radiating in some range of frequency ω to $\omega + d\omega$. The power emitted per unit area in this energy (or frequency) range, or **intensity**, is denoted as[8]

$$\mathcal{P}_e(\omega)d\omega. \tag{12.25}$$

Additionally, we denote the power per unit area that is incident on the object as

$$\mathcal{P}_i(\omega)d\omega, \tag{12.26}$$

along with the power per unit area that is absorbed by it,

$$\mathcal{P}_a(\omega)d\omega. \tag{12.27}$$

In general, not all of the energy striking the object will be absorbed, as some may get reflected; these are related by

$$\mathcal{P}_a(\omega) = \xi\mathcal{P}_i(\omega), \tag{12.28}$$

with $\xi \leq 1$ is the **absorption coefficient**: the fraction of radiation incident on an object that is absorbed. In general this may be dependent on the incident wave vector (including the direction) as well as the polarization (different polarizations may reflect better than others). A **blackbody** is specifically defined as an object which has $\xi = 1$; it is a perfect absorber. We will make this assumption for the remainder of our discussion for simplicity. It is straightforward to keep this parameter throughout the calculation, and thus arise at a more general result, but it does not give any additional insight into the problem.

Blackbodies (or near blackbodies) are part of everyday life. Some examples include old-school incandescent light bulbs (which you may still have in your home), many electric space heaters, or the filaments in a toaster oven. In the galaxy, our sun and other stars are near perfect blackbodies, and planets actually act fairly close to being blackbodies. Notice that the term itself is misleading—it needn't actually be the color black!

To ensure an equilibrium situation, we impose a condition known as **principle of detailed balance**. This states that the power emitted from our hot object must be the same as the power absorbed by it. This is the same "what comes in must go out" rule that we discussed when discussing the equilibrium condition for effusion in Section 9.4.2. This allows for an equilibrium situation even though there are regular changes (the absorption and emission of photons) occurring; this is still a **steady state**. This can be obtained if we consider our object at temperature T placed inside of a container that is maintained at this same temperature. In terms of the incident and emitted power, the principle of detailed balance takes the form

$$\mathcal{P}_e(\omega) = \mathcal{P}_a(\omega),$$

and for a blackbody we have the additional requirement

$$\mathcal{P}_a(\omega) = \mathcal{P}_i(\omega),$$

8 Don't confuse $\mathcal{P}$ in this section with a probability density.

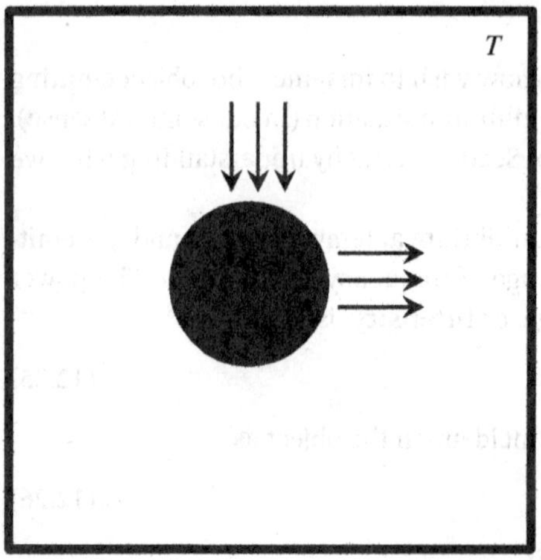

Figure 12.2 A blackbody in a container filled with a photon gas at temperature T. The amount of radiation absorbed is the same as that it emits.

so we can say

$$\mathcal{P}_e(\omega) = \mathcal{P}_i(\omega).$$

This connection between $\mathcal{P}_e$ and $\mathcal{P}_a$ is an example of **Kirchhoff's law**,[9] which states that a good emitter of radiation is a good absorber and vice versa. This is something we could apply outside of equilibrium and is generally valid (because it depends on the properties of *the object itself*, not necessarily the situation of its state). I show a picture of a blackbody in a container filled with a photon gas at temperature T in Figure 12.2. All of the radiation incident on it is absorbed and then (eventually) re-emitted.

We'll calculate the incident power in precisely the same way we calculated the number density of classical particles in the Maxwell velocity distribution. We consider a small area A near the surface and the number of photons that are striking that surface in time dt is given by[10]

$$2(Acdt\cos\theta)f(k)d^3\mathbf{k}. \tag{12.29}$$

These photons are coming in at some angle θ and must be a distance cdt away (just as in Figure 9.11 with $v \to c$), hence the volume factor $V = Acdt\cos\theta$. Multiplying this by the photon energy $\hbar\omega$ gives the energy that reaches the surface, and dividing by Adt results in the power per unit area, or intensity,

$$\mathcal{P}_i(\omega)d\omega d\Omega = 2\hbar\omega(c\cos\theta)f(k)d^3\mathbf{k},$$

where we include $d\Omega$ on the left-hand side, as this only includes those photons arriving within this solid angle. Writing $d^3\mathbf{k}$ in spherical coordinates and then the frequency ω again, we have

$$d^3\mathbf{k} = k^2dkd\Omega = \frac{\omega^2}{c^3}d\omega d\Omega,$$

so

$$\mathcal{P}_i(\omega) = 2\frac{\hbar\omega^3}{c^2}f(k)\cos\theta.$$

The principle of detailed balance for a blackbody allows us to say this is the emitted intensity,

$$\mathcal{P}_e(\omega)d\omega d\Omega = \frac{\hbar\omega^3}{c^2}f(k)\cos\theta d\omega d\Omega.$$

9 This is the same Gustav Kirchhoff (1824–1887) from the "Kirchhoff's rules" used to study electric circuits.
10 I am including a factor of two here to account for summing over the two possible polarizations.

Now we integrate over the solid angle to get all of the forward pointing photons that are being absorbed,

$$\mathcal{P}_e(\omega)d\omega = 2\frac{\hbar\omega^3}{c^2}f(k)d\omega \int_{\phi=0}^{2\pi}\int_{\theta=0}^{\pi/2} \cos\theta d\Omega = 2\pi\frac{\hbar\omega^3}{c^2}f(k)d\omega. \tag{12.30}$$

The right side is just proportional to the mean energy density $\bar{u}(\omega)d\omega$, where

$$\bar{u}(\omega)d\omega = \left[2f(k)\left(4\pi k^2 dk\right)\right]\hbar\omega = \left[2f(k)\frac{4\pi\omega^2 d\omega}{c^3}\right]\hbar\omega,$$

so

$$\mathcal{P}_e(\omega)d\omega = \frac{1}{4}c\bar{u}(\omega)d\omega, \tag{12.31}$$

or putting in our expression for $\bar{u}$,

$$\mathcal{P}_e(\omega)d\omega = \frac{\hbar}{4\pi^2c^2}\frac{\omega^3}{e^{\beta\hbar\omega}-1}d\omega. \tag{12.32}$$

This is the expression Planck empirically came up with to fix the ultraviolet catastrophe. He showed that this ensured the intensity did not blow up at small wavelengths (large ω) with the $1/(e^{\beta\hbar\omega}-1)$ factor.

Exercise 12.6 Perform the integral in the first equality of Eq. (12.30) to obtain the second equality.

We can integrate over all frequencies to get the total emitted intensity

$$\mathcal{P}_e^{(0)} = \frac{\hbar}{4\pi^2c^2}\int \frac{\omega^3}{e^{\beta\hbar\omega}-1}d\omega.$$

This is merely the integral we had in Eq. (12.15) with the right change of variables, and we thus obtain the **Stefan–Boltzmann law**,[11]

$$\mathcal{P}_e^{(0)} = \frac{c\bar{u}_0}{4} = \sigma T^4, \tag{12.33}$$

with the Stefan–Boltzmann constant[12]

$$\sigma = \frac{\pi^2}{60}\frac{k_B^4}{c^2\hbar^3} \approx 5.67\times10^{-8}\,\frac{\text{J}}{\text{s}\cdot\text{m}^2\cdot\text{K}^4}.$$

This, along with the other results of this section, is applicable to many situations, from astrophysics to climate models, and some examples are included in the problems. We have only (barely) scratched the surface of what is understood about blackbody radiation, but this gives you the basic tools from which to build more complex models of realistic systems. For more information on the thermodynamics of blackbody radiation, see Ref. [3].

12.2 Bose–Einstein Condensation

Now let's consider bosons in general beyond the special case of photons. Because any number of bosons can exist in the same state, a special state of matter arises for a system of bosons known as a

---

11 Josef Stefan (1835–1893).

12 Because all of the constants in σ are now defined exactly, we can actually write this exactly as well as

$$\sigma = \frac{5\,454\,781\,984\,210\,512\,994\,952\,000\,000\pi^5}{29\,438\,455\,734\,650\,141\,042\,413\,712\,126\,365\,436\,049}\,\frac{\text{J}}{\text{s}\cdot\text{m}^2\cdot\text{K}^4}.$$

This is not very useful in practice of course.

Bose–Einstein condensate, where the ground state of a bosonic system becomes overwhelmingly preferred under certain conditions. This was predicted by Bose and Einstein in the 1920s, and first seen experimentally with rubidium atoms in 1995 [4].

We'll consider an ideal system of bosons in a volume V at temperature T that is large enough that we can treat the mean number of particles as just the number of particles, $\overline{N} \approx N$. The mean number of bosons in a state with energy ε_r is given by the Bose–Einstein distribution from Chapter 11,

$$\overline{n}_r = \frac{1}{e^{\beta(\varepsilon_r-\mu)} - 1}.$$

The number of bosons with a wave vector in a range from $\mathbf{k}$ to $\mathbf{k} + d^3\mathbf{k}$ can be described similarly to how we treated the photon gas, as

$$\left[f(\mathbf{k})V\right] d^3\mathbf{k} = \frac{1}{e^{\beta(\varepsilon_r-\mu)} - 1} \frac{Vd^3\mathbf{k}}{(2\pi)^3}.$$

For a free particle, the single-particle energy is given by (dropping the subscript r)

$$\varepsilon = \frac{\mathbf{p}^2}{2m} = \frac{\hbar^2\mathbf{k}^2}{2m},$$

and while we know that the energy is quantized as discussed in Chapter 11, as we have done several times already, we are going to treat the energies as continuous. For now we will also assume that we are dealing with spin-0 bosons so that we won't need to include factors corresponding to different spin states.[13] We then have, writing $d^3\mathbf{k} = k^2dkd\Omega$,

$$\begin{aligned}\left[f(\mathbf{k})V\right] &= \frac{1}{e^{\beta(\varepsilon-\mu)} - 1} \frac{Vk^2dkd\Omega}{(2\pi)^3} \\ &= \frac{\sqrt{2m^3}V}{h^3} \frac{\varepsilon^{1/2}}{e^{\beta(\varepsilon-\mu)} - 1} d\varepsilon d\Omega.\end{aligned} \tag{12.34}$$

Exercise 12.7 Change variables from $k \to \varepsilon$ to obtain the result in Eq. (12.34).

As was the case for blackbody radiation, the energy is independent of the direction of the particles' motion, so let's integrate over the solid angle to get the number of particles with energy from ε to $\varepsilon + d\varepsilon$, with $\mathcal{N}$ the number of particles per unit energy,

$$\mathcal{N}(\varepsilon)d\varepsilon = 2\pi V\left(\frac{2m}{h^2}\right)^{3/2} \frac{\varepsilon^{1/2}}{e^{\beta(\varepsilon-\mu)} - 1} d\varepsilon. \tag{12.35}$$

If we integrate this over the energy we get the total number of bosons,

$$N = 2\pi V\left(\frac{2m}{h^2}\right)^{3/2} \int_0^\infty \frac{\varepsilon^{1/2}}{e^{\beta(\varepsilon-\mu)} - 1} d\varepsilon. \tag{12.36}$$

At this point we now have to determine the chemical potential μ which, for a rough estimate, we could use a classical approximation (see Problem 12.6). Instead though, let's consider the zero temperature limit, as $\beta \to +\infty$. In this case, only the ground state will be occupied, so we set $\varepsilon \to 0$ and obtain

$$N = \overline{n}_0 = \frac{1}{e^{-\beta\mu} - 1},$$

13 For an ideal gas, these would just lead to an overall factor of the allowed spin states, as in this case there is no spin-dependent term in ε.

which gives us

$$\mu = -k_B T \ln\left(1 + \frac{1}{N}\right). \tag{12.37}$$

We see this is in fact negative as mentioned in Chapter 11. For large N, we can approximate this as

$$\mu \approx -\frac{k_B T}{N},$$

and we'll assume we're at low enough temperatures so that with large enough N,

$$e^{-\beta\mu} \approx 1.$$

This essentially is the same as assuming that $\mu \ll k_B T$ so that it can be neglected in Eq. (12.36). The number of bosons is then given by

$$N = 2\pi V\left(\frac{2m}{h^2}\right)^{3/2} \int_0^\infty \frac{\varepsilon^{1/2}}{e^{\beta\varepsilon} - 1} d\varepsilon,$$

or changing integration variables to $\eta = \beta\varepsilon$,

$$N = 2\pi V\left(\frac{2mk_B T}{h^2}\right)^{3/2} \int_0^\infty \frac{\eta^{1/2}}{e^{\eta} - 1} d\eta. \tag{12.38}$$

This expression, while not obvious at first glance, omits any contribution from the ground state energy $\varepsilon = 0$, because the density vanishes like $\varepsilon^{1/2}$. If we want to see the effects of the ground state, we have to break N up into two terms,

$$N = N_0 + N_{\text{ex}},$$

where

$$N_{\text{ex}} = 2\pi V\left(\frac{2mk_B T}{h^2}\right)^{3/2} \int_0^\infty \frac{\eta^{1/2}}{e^{\eta} - 1} d\eta$$

is the expression we derived above and is the number of bosons in excited states and N_0 is the number of bosons in the ground state. The integral in N_{ex} is nothing other than the Riemann zeta function, where

$$\int_0^\infty \frac{\eta^{1/2}}{e^{\eta} - 1} d\eta = \frac{1}{2}\sqrt{\pi}\zeta\left(\frac{3}{2}\right) \approx \frac{2.61}{2}\sqrt{\pi} \approx 2.32.$$

The number of particles in excited states then is

$$N_{\text{ex}} = \zeta(3/2) V\left(\frac{2\pi mk_B T}{h^2}\right)^{3/2}. \tag{12.39}$$

At this point we define the **Bose temperature,** T_B, as the temperature which is high enough such that there are no particles in the ground state, $N_0 = 0$, and thus

$$N_{\text{ex}} = N = \zeta(3/2) V\left(\frac{2\pi mk_B T_B}{h^2}\right)^{3/2}.$$

Solving for the Bose temperature, we get

$$T_B = \frac{h^2}{2\pi mk_B}\left[\frac{1}{\zeta(3/2)}\frac{N}{V}\right]^{2/3}. \tag{12.40}$$

For temperatures less than the Bose temperature, then some fraction of our bosons are in the ground state, and this fraction is given by

$$\frac{N_0}{N} = 1 - \frac{N_{\text{ex}}}{N} = 1 - \left(\frac{T}{T_B}\right)^{3/2} \quad \text{if } T \leq T_B. \tag{12.41}$$

This shows that if we are at a temperature lower than the Bose temperature, we have some number of particles in the ground state, and as $T \to 0$, $N_0 \to N$, so that *all* of the bosons are in the ground state. If $T > T_B$, then

$$\frac{N_0}{N} = 0,$$

by definition.

Exercise 12.8 Use the expressions for N_{ex} and N to show that

$$\frac{N_{\text{ex}}}{N} = \left(\frac{T}{T_B}\right)^{3/2}$$

in Eq. (12.41).

Exercise 12.9 Consider a mole of ^{4}He atoms in a volume of $(2\ \text{cm})^3$. Show that the Bose temperature is $T_B = 3.17$ K (just below the point at which helium liquefies, 4.2 K).

Many images of Bose–Einstein condensates, including those from the first experimental observation of them, are quite colorful, for example the image produced by NASA and the Jet Propulsion Laboratory, which you can see online [5]. Given that these are purely quantum systems, you can't "see" the condensate with your own eyes, so such imaging is required. A nice visualization can be seen with a numerical simulation of the Bose–Einstein condensate, for example that discussed in Refs. [6, 7]. One result from these simulations (with 10 000 particles) is shown in Figure 12.3. This compares the occupation number of the ground state as a function of temperature for bosons and classical particles. In the classical case, we see as the temperature decreases, only when T precisely reaches zero does the ground state become completely occupied. However for the quantum case, we see the ground state become noticeably occupied for non-zero (but small) temperature, similar to as predicted by Eq. (12.41). This is merely one useful result from this reference, which allows us to visualize something that is not possible to see classically.

Bose–Einstein condensates may not seem to have many applications in our everyday lives, and many of the applications are in ultracold experiments, often to understand fundamental

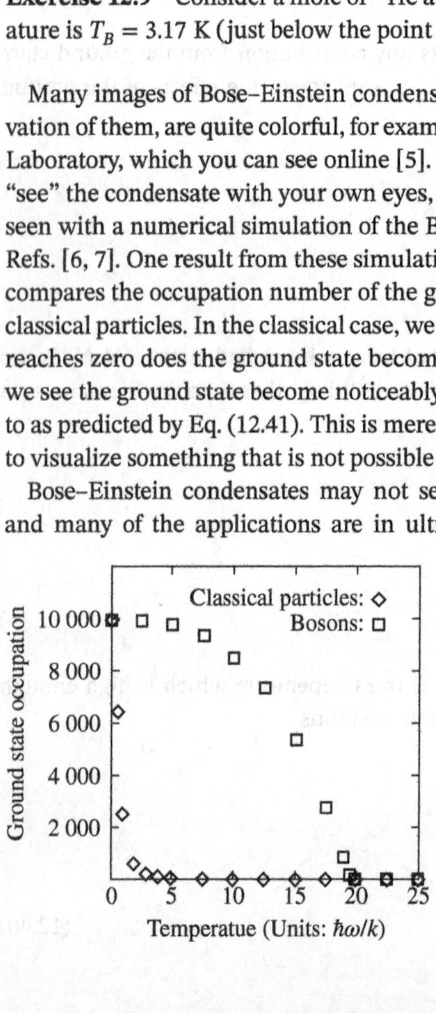

Figure 12.3 A comparison of the ground state occupation number for classical particles in a harmonic trap to that of a Bose–Einstein condensate as a function of temperature. Reproduced from [6]/with permission of AIP Publishing.

interactions of quantum many-body systems. However, they have also been used to advance atomic clock measurements, which are crucial for properly calibrated global positioning systems (GPS)—very important for using your smartphone to find your way around! Additionally, they can be used to be create atomic lasers and improve the sensitivity of various sensors (e.g., gravitational sensors or magnetic sensors).

12.3 Fermi Gas

As a final application of quantum statistics, we will consider a gas of fermions to see how this can affect the specific heat of a metal. The classical determination, from the equipartition theorem, showed that the molar specific heat of a solid was constant,

$$c_V^{(\nu)} = 3R.$$

This came from the vibrational modes of the molecules, so in this expression I have added the superscript (ν) to remind us of this origin. For a metal, there will be **conduction electrons**, which behave as free particles as they move through the metal, and therefore, these free electrons would, at least *classically*, contribute to the molar specific heat another constant amount,

$$c_V^{(e)} = \frac{3}{2}R,$$

using the equipartition theorem. We know that neither of these contributions are valid quantum mechanically. A correct quantum mechanical treatment, however, should result in these in the classical limit. The vibrational contribution has already been treated quantum mechanically via the Einstein and Debye models in Chapter 9, and now we will consider the electronic contribution to the specific heat.

We start with the Fermi–Dirac distribution,

$$\overline{n}_r = F(\varepsilon_r) = \frac{1}{e^{\beta(\varepsilon_r - \mu_F)} + 1}$$

with the usual requirement that the total number of electrons is fixed,

$$\sum_r \overline{n}_r = N, \tag{12.42}$$

in a volume V. As is conventional, I have defined the **Fermi function,**

$$F(\varepsilon) \equiv \frac{1}{e^{\beta(\varepsilon - \mu_F)} - 1}, \tag{12.43}$$

where, for non-zero temperature, the function drops to zero in an energy interval on the order $k_B T$, as in Figure 11.7.

At zero temperature, how do we determine the Fermi energy, μ_F? We use the energy of an electron with wave number k,

$$\varepsilon(k) = \frac{\hbar^2 k^2}{2m}. \tag{12.44}$$

We put electrons into the lowest energy states (one per state, as required by the exclusion principle), until we run out of electrons. The highest energy with an electron corresponds to the Fermi energy, or $\overline{n}_r = 0$ if $\varepsilon > \mu_F$. So

$$\mu_F = \frac{\hbar^2 k_F^2}{2m}, \tag{12.45}$$

with k_F the maximum allowed wave number. This corresponds to a sphere of radius $p_F = \hbar k_F$ in momentum space, with p_F the **Fermi momentum**. As we saw before, when we considered the classical approximation of the quantum problem, the sum over the discrete energy levels becomes an integral over momentum and we gain a factor of the volume when we convert the sum to an integral, so

$$\sum_r \to V \int \frac{d^3\mathbf{p}}{h^3} = V \int \frac{d^3\mathbf{k}}{(2\pi)^3}.$$

Electrons are spin-1/2 particles, and given that to a first approximation, the energy is independent of the spin, there are two spin states per energy. And as before, to account for this we will multiply any sums/integrals we perform by a factor of 2. The sum in Eq. (12.42) becomes[14]

$$\sum_r \bar{n}_r \to 2V \int \frac{d^3\mathbf{k}}{(2\pi)^3} = N,$$

and this integral is merely the volume of a sphere of radius k_F in k-space, so it is simple to evaluate as

$$\frac{2V}{(2\pi)^3}\frac{4}{3}\pi k_F^3 = N,$$

or

$$k_F = \left(3\pi^2 \frac{N}{V}\right)^{1/3}. \tag{12.46}$$

Thus, at zero temperature, the Fermi energy is

$$\mu_{F0} = \frac{\hbar^2}{2m}\left(3\pi^2 \frac{N}{V}\right)^{2/3} = \frac{h^2}{8m}\left(\frac{3}{\pi}\right)^{2/3}\left(\frac{N}{V}\right)^{2/3}, \tag{12.47}$$

with the subscript 0 added to denote this is only at zero temperature. Finally, given that $k_B T$ has units of energy, we can define the **Fermi temperature** which corresponds to this Fermi energy,

$$\mu_{F0} \equiv k_B T_F. \tag{12.48}$$

For a real Fermi gas, only those electrons with energy around the Fermi energy (in that range $\sim \mu_F \pm k_B T$) contribute significantly to the specific heat, because those are the electrons that are freely able to move around the metal (they aren't "trapped" in those lowest energy states). A quick qualitative argument allows us to see these electrons give a dependence to $c_V^{(e)}$ that is linear in the temperature. That region of width $\sim k_B T$ contains roughly $N(k_B T)/(k_B T_F) = N(T/T_F)$ electrons. These electrons each increase the energy of the system by $\sim k_B T$, so the total energy of these electrons is

$$\bar{E} \sim N\left(\frac{T}{T_F}\right)(k_B T) = N k_B \frac{T^2}{T_F}. \tag{12.49}$$

This leads to a molar specific heat of

$$c_V^{(e)} \sim \left(\frac{2R}{T_F}\right) T, \tag{12.50}$$

which, as I mentioned above, is linear in T. Note that this vanishes quickly enough as $T \to 0$ as is required from the third law.

14 For a general spin-s system, we would have a factor of $(2s + 1)$ in front.

Exercise 12.10 Flesh out the argument that leads to the statement that the region of width $\sim k_B T$ about the fermi energy contains roughly $N(T/T_F)$ electrons. Additionally, fill in the steps that lead to Eq. (12.50).

Exercise 12.11 Estimate the Fermi energy and Fermi temperature for a mole of electrons in a cubic centimeter volume. You should find $\mu_F \sim 26$ eV $\sim 10^{-18}$ J and $T_F \sim 3 \times 10^5$ K. You can see in table 1 of chapter 6 of Ref. [8] a list of Fermi energies and temperatures for common metals. The results for these tend to be on the order of $1 - 10$ eV and $10^4 - 10^5$ K, respectively.

For a more exact calculation, we consider the average energy of the system, using the same steps we have used before,

$$\begin{aligned}\bar{E} &= 2V \int \frac{d^3\mathbf{k}}{(2\pi)^3} \frac{\varepsilon}{e^{\beta(\varepsilon-\mu_F)}+1} \\ &= \frac{4\pi V}{h^3}(2m)^{3/2} \int_0^\infty d\varepsilon \frac{\varepsilon^{3/2}}{e^{\beta(\varepsilon-\mu_F)}+1} \\ &= \frac{4\pi(2m)^{3/2}V}{h^3} \frac{1}{\beta^{5/2}} \int_0^\infty dx \frac{x^{3/2}}{e^{x-y}+1}.\end{aligned}$$

In the last step we set $x = \beta\varepsilon$ and $y = \beta\mu_F$. We can integrate this by parts and then change variables to $x' = x - y$ to obtain

$$\begin{aligned}\bar{E} &= \frac{8\pi(2m)^{3/2}V}{5h^3} \frac{1}{\beta^{5/2}} \int_0^\infty dx \left(\frac{d}{dx}x^{5/2}\right) \frac{1}{e^{x-y}+1} \\ &= \frac{8\pi(2m)^{3/2}V}{5h^3} \frac{1}{\beta^{5/2}} \int_0^\infty dx x^{5/2} \frac{e^{x-y}}{(e^{x-y}+1)^2} \\ &= \frac{8\pi(2m)^{3/2}V}{5h^3} \frac{1}{\beta^{5/2}} \int_{-y}^\infty dx' (x'+y)^{5/2} \frac{e^{x'}}{(e^{x'}+1)^2}.\end{aligned}$$

We can assume safely that the lower limit can be replaced with $-y \to -\infty$, because of the $e^{x'}$ factor in the integrand; this will be negligible for x less than $-y$. Thus the mean energy can be written

$$\bar{E} = \frac{8\pi(2m)^{3/2}V}{5h^3} \frac{1}{\beta^{5/2}} \int_{-\infty}^\infty dx' (x'+y)^{5/2} \frac{e^{x'}}{(e^{x'}+1)^2}. \tag{12.51}$$

and we would like to Taylor expand $(x'+y)^{5/2}$ about $x' = 0$,[15]

$$(x'+y)^{5/2} = y^{5/2} + \frac{5}{2}x' y^{3/2} + \frac{15}{8}x'^2 y^{1/2} + \cdots,$$

and then

$$\bar{E} = \frac{8\pi(2m)^{3/2}V}{5h^3} \frac{1}{\beta^{5/2}} \int_{-\infty}^\infty dx' \left(y^{5/2} + \frac{5}{2}y^{3/2}x' + \frac{15}{8}y^{1/2}x'^2\right) \frac{e^{x'}}{(e^{x'}+1)^2}$$

These integrals can be evaluated straightforwardly (see Problem 12.7) to obtain

$$\int_{-\infty}^\infty dx' \frac{e^{x'}}{(e^{x'}+1)^2} = 1, \tag{12.52}$$

$$\int_{-\infty}^\infty dx' \frac{x' e^{x'}}{(e^{x'}+1)^2} = 0, \tag{12.53}$$

15 This corresponds to an expansion about $x = y$, or $\varepsilon = \mu_F$, the Fermi energy. As we expect only those fermions near this energy to contribute to the specific heat this is a valid assumption.

$$\int_{-\infty}^{\infty} dx' \frac{x'^2 e^{x'}}{(e^{x'}+1)^2} = \frac{\pi^2}{3}. \tag{12.54}$$

After some algebra, we get for the mean energy,

$$\bar{E} = \frac{3N}{5}\mu_F\left(1 + \frac{5\pi^2}{8}\frac{k_B^2T^2}{\mu_F^2}\right). \tag{12.55}$$

However, recall that μ_F is a function of the temperature, so we have to go through a similar calculation using the requirement that summing over $\overline{n}_r$ gives us the number of fermions in the system,

$$2V\int \frac{d^3\mathbf{k}}{(2\pi)^3}\frac{1}{e^{\beta(\varepsilon-\mu_F)}-1} = N. \tag{12.56}$$

This is somewhat simpler than the calculation for $\bar{E}$, and you will show in the problems (see Problem 12.8) that evaluating the integral on the left-hand side (to quadratic order in the temperature) gives us

$$N = N\left(\frac{\mu_F}{\mu_{F0}}\right)^{3/2}\left(1 + \frac{\pi^2k_B^2T^2}{8\mu_F^2}\right), \tag{12.57}$$

which we can invert, for low temperature, to obtain (see Problem 12.9)

$$\mu_F = \mu_{F0}\left(1 - \frac{\pi^2k_B^2T^2}{12\mu_{F0}^2}\right). \tag{12.58}$$

Substituting this into Eq. (12.55), we can write the mean energy to quadratic order in the temperature in terms of the zero-temperature Fermi energy,

$$\bar{E} = \frac{3N}{5}\mu_{F0}\left(1 + \frac{5\pi^2}{12}\frac{k_B^2T^2}{\mu_{F0}^2}\right), \tag{12.59}$$

and from this we calculate the contribution to molar specific heat from conduction electrons,

$$c_V^{(e)} = \frac{\pi^2}{2}\frac{R}{T_F}T. \tag{12.60}$$

Exercise 12.12 While the full calculation of Eqs. (12.58) and (12.59) are relegated to the problems, calculate the molar specific heat Eq. (12.60) from the latter of these equations.

We can see from Eq. (12.60) that the specific heat is linear in the temperature when we focus on the electrons near the Fermi energy. We can write the electronic contribution to the specific heat as

$$c_V^{(e)} = \gamma T,$$

and with the Debye model, we found the contribution from the vibrational modes of the lattice was cubic in the temperature, or

$$c_V^{(L)} = AT^3.$$

The complete expression for the specific heat of a metal can be written as

$$\frac{c_V}{T} = \gamma + AT^2. \tag{12.61}$$

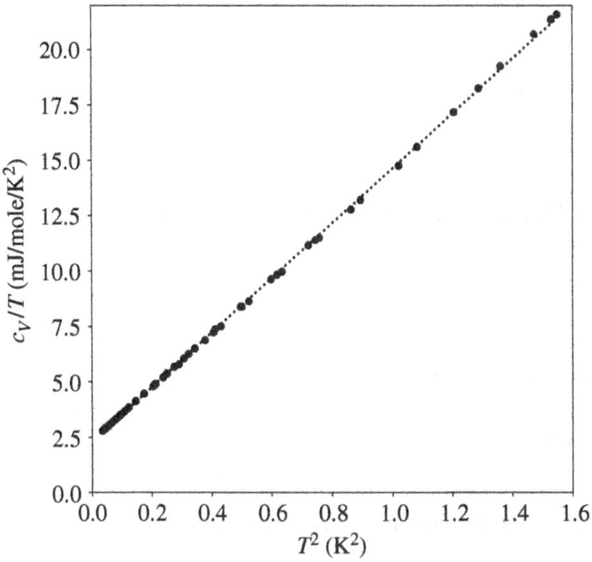

Figure 12.4 c_V/T as a function of T^2 for rubidium, data from Ref. [9]. The dotted line is a fit to Eq. (12.61).

In Figure 12.4, we show the molar specific heat of rubidium using data from Ref. [9], compared with this result. Specifically, this figure shows c_V/T vs. T^2, and the dotted line is a fit to Eq. (12.61). The result of this fit gives

$$\gamma = 2.29 \text{ mJ/mol/K}^2, \qquad A = 12.35 \text{ mJ/mol/K}^4.$$

The result for γ from Eq. (12.60), with $T_F = 2.15 \times 10^4$ K for rubidium, is 1.9 mJ/mol/K^2, not too far from the fit. In Ref. [9] they perform a fit to the expression

$$\frac{c_V}{T} = \gamma + AT^2 + BT^4,$$

and obtain

$$\gamma = 2.41 \text{ mJ/mol/K}^2, \qquad A = 11.40 \text{ mJ/mol/K}^4, \qquad B = 0.636 \text{ mJ/mol/K}^6.$$

This additional term can be seen as the next term omitted from our expansions of Eq. (12.51) above. What is nice to see is that our model works very well experimentally, and it is straightforward to improve upon this result.

12.4 Summary

- By applying the Planck distribution to a photon gas, a study of blackbody radiation is straightforward. Applications to astrophysics and climate models are simple to explore.
- Bosons have a unique state of matter known as a Bose–Einstein condensate, where at low enough temperatures, all bosons occupy the ground state of the theory. This is a simple application of the Bose–Einstein distribution of Chapter 11.
- We can calculate the contribution that free electrons in a metal make to the specific heat of a solid. This, which uses the Fermi–Dirac distribution, can be combined with the Debye model (which considers the vibrational contributions) to obtain a useful and accurate model for the total specific heat.

Problems

12.1 Let's evaluate the sum

$$\sum_{n=1}^{\infty} \frac{1}{n^4} = \frac{\pi^4}{90},$$

using a simple version of **Parseval's theorem**,[16] which states that if we have a Fourier series,

$$a(x) = \sum_{n=-\infty}^{\infty} a_n e^{inx},$$

then the Fourier coefficients satisfy the expression

$$\sum_{n=-\infty}^{\infty} a_n a_n^* = \frac{1}{2\pi} \int_{-\pi}^{\pi} a(x) a(x)^* dx.$$

(a) Consider the function $a(x) = x^2$, and calculate the Fourier coefficients a_n.
(b) Show that you can write

$$\sum_{n=1}^{\infty} \frac{1}{n^4} = \frac{1}{8} \left(\sum_{n=-\infty}^{\infty} a_n a_n^* - \frac{\pi^4}{9} \right).$$

(c) Evaluate the integral in Parseval's theorem, and use this to evaluate the sum we wanted. [Note, you can also use contour integration to evaluate this, if you are familiar with this technique, and of course sums are readily calculated now using Mathematica or other software.]

12.2 Let's study the first law of thermodynamics as applied to a photon gas. Here we can write the mean energy as $\bar{E} = V\bar{u}(T)$ where $\bar{u}(T)$ is independent of the volume V, and the radiation pressure is $\bar{p} = \frac{1}{3}\bar{u}$. Treat $\bar{u}$ as an *unknown* function of T for this problem.
(a) Write the entropy as a function of temperature and volume, and express dS in terms of dT and dV. Explicitly evaluate the partial derivatives in terms of T, V, $\bar{u}(T)$, and $d\bar{u}/dT$.
(b) Show that the equality of the mixed second derivatives of S gives immediately a differential equation for $\bar{u}$ which can be integrated to yield the Stefan–Boltzmann law. (In this case, the constant σ is unknown and would have to be determined experimentally.)

12.3 Suppose electromagnetic radiation initially at temperature T_1 fills a cavity, and undergoes a quasistatic expansion to a volume eight times larger than its initial volume, resulting in a final temperature T_2. Ignore the heat capacity of the cavity walls.
(a) Calculate T_2 if the expansion is adiabatic.
(b) Now calculate T_2 if the radiation undergoes a free expansion (with the same change in volume).
(c) Why is the final temperature in (b) greater than the final temperature in (a)?

12.4 Let's apply our results for blackbody radiation to our sun. The surface temperature of the sun is $T_0 = 5800\,\text{K}$ and its radius is $R = 6.96 \times 10^5$ km. The electromagnetic radiation travels

16 Marc-Antoine Parseval (1755–1836).

from the sun a distance of $L = 1.5 \times 10^8$ km and is incident on the Earth, which has a radius of $r = 6\,371$ km. To a first approximation, let's consider both the sun and the Earth as blackbodies.

The Earth has reached a steady state so that its mean temperature T does not change in time despite the fact that the earth constantly absorbs and emits radiation.

(a) Assuming the Earth has reached an equilibrium state, so that the temperature is constant, calculate an (approximate) expression for the temperature T of the Earth in terms of the parameters above.

(b) What is this temperature T numerically?

12.5 The energy of the photon orbital r in a box of dimensions L_x, L_y, and L_z and periodic boundary conditions is given by

$$\varepsilon_r = \hbar\omega = 2\hbar c\pi \left(\frac{n_x^2}{L_x^2} + \frac{n_y^2}{L_y^2} + \frac{n_z^2}{L_z^2} \right)^{1/2},$$

where quantum numbers n_x, n_y, and n_z specify the state.

(a) By following the method of Problem 4.8, show that, for an ideal gas of photons,

$$\overline{p} = \frac{1}{3}\frac{\bar{E}}{V} = \frac{1}{3}\bar{u}_0(T),$$

where $\bar{u}_0 = \bar{E}/V$, the energy per unit volume of the photon gas. Include an explanation of why the results of parts (a)–(c) of the above-referenced problem also apply in this case, so that $\overline{p}_x = \overline{p}_y = \overline{p}_z$.

(b) For photons we see that $\overline{p} = \bar{u}_0(T)/3$ shows that $\overline{p}$ does just depend on the temperature, not the volume or number of photons. Why does part (d) of the Problem 11.7 not apply here?

12.6 Evaluate the integral in Eq. (12.36) to obtain the chemical potential in the classical limit of a Bose–Einstein condensate. That is, use Eq. (11.48) to simplify the integrand to show that

$$\beta\mu = \ln\left[h^3 \left(\frac{\beta}{2m\pi}\right)^{3/2} \frac{N}{V}\right].$$

12.7 Evaluate the integrals in Eqs. (12.52)–(12.54).

Hints:

(a) For Eq. (12.52), a simple substitution $z = e^{x'}$ works to put the integral in a doable form.

(b) For Eq. (12.53), show that the integrand is an odd function.

(c) For Eq. (12.54), show that the integrand is an even function, so that you can write it as twice the integral from $0 \to \infty$. Then, integrate by parts to simplify the integral, realizing

$$\frac{d}{dx}\left(\frac{1}{e^x + 1}\right) = \frac{e^x}{(e^x + 1)^2}.$$

12.8 Using Eq. (12.56) and the methods to obtain $\bar{E}$ for the Fermi gas, derive Eq. (12.57).

12.9 From your results from Problem 12.8, use a Taylor expansion to derive Eq. (12.58).

References

1 D. J. Griffiths. *Introduction to Electrodynamics*. Cambridge University Press, 4th edition, 2017.

2 F. Reif. *Fundamentals of Statistical and Thermal Physics*. McGraw Hill, Tokyo, 1965.

3 R. E. Kelly. Thermodynamics of blackbody radiation. *American Journal of Physics*, 49(8):714–719, 1981.

4 M. H. Anderson, J. R. Ensher, M. R. Matthews, C. E. Wieman, and E. A. Cornell. Observation of Bose-Einstein condensation in a dilute atomic vapor. *Science*, 269(5221):198–201, 1995.

5 NASA/JPL-Caltech. URL https://www.jpl.nasa.gov/images/pia22561-bose-einstein-condensate-graph. Accessed: 20 March 2023.

6 M. Ligare. Numerical analysis of Bose–instein condensation in a three-dimensional harmonic oscillator potential. *American Journal of Physics*, 66(3):185–190, 1998.

7 M. Ligare. Comment on "Numerical analysis of Bose–Einstein condensation in a three-dimensional harmonic oscillator potential," by Martin Ligare [Am. J. Phys. 66 (3), 185–190 (1998)]-an extension to anisotropic traps and lower dimensions. *American Journal of Physics*, 70(1):76–78, 2002.

8 C. Kittel. *Introduction to Solid State Physics*. Wiley, 8th edition, 2005.

9 W. H. Lien and N. E. Phillips. Low-temperature heat capacities of potassium, rubidium, and cesium. *Physical Review*, 133:A1370–A1377, Mar 1964.

13

Black Hole Thermodynamics

After developing a significant amount of formalism for both classical and quantum statistical mechanics, which can readily be applied to a host of applications, let us now take a slight detour, to black holes. These have long since been a fascinating topic for both scientists and science fiction writers for their mystery and odd nature, and this fascination only grew after the first image was captured in 2019.

I want to spend this chapter briefly discussing the thermodynamics of black holes to show not only how powerful this subject is when being applied to any area of physics, but to see how some rather bizarre thermodynamic features can arise. This will require a brief introduction to black holes, as they are not often discussed in undergraduate courses in detail, except in the rare case of an undergraduate general relativity course.

After finishing this chapter, you should be able to

- understand the basics of black holes,
- apply these concepts to thermodynamics and understand how to modify what we have done thus far to the topic of black holes, and
- calculate the heat capacity of a black hole, showing that in many cases, they are inherently unstable phenomena.

13.1 Brief Introduction to General Relativity

In order to discuss the thermodynamics of black holes, we need to first cover some of the basics of these phenomena, which means that we have to discuss a little bit about general relativity. We start with the set of units used in most discussions, and then move to a cursory discussion of the features of black holes, as best we can understand (without needing to study differential geometry).

13.1.1 Geometrized Units

In the field of general relativity, as is often the case with advanced theoretical fields, we use a set of units that often feels unnatural, because we set certain fundamental constants to unity. This allows equations to become less cluttered, making it easier to understand the salient features of what we

Statistical Thermodynamics: An Information Theory Approach, First Edition. Christopher Aubin.

Companion website: www.wiley.com/go/Aubin/StatisticalThermodynamics

derive. As long as we do not do this to too many constants, there is always a unique way to get back any expression in SI units. The four constants that are standard candidates for this approach are

- k_B, the Boltzmann constant,
- $\hbar$, the reduced Planck's constant,
- G, Newton's gravitational constant, and
- c, the speed of light.

We can set any *two* of the three latter constants to one unambiguously, and in addition to those three, often when thermodynamics is involved, we can also set $k_B = 1$.

Exercise 13.1 Note that everywhere the temperature appears in thermodynamic problems, the Boltzmann constant tends to also be there, as it "converts" temperature units to energy units. Setting $k_B = 1$ allows one to measure temperature in joules (or also eV). What is 1 J as a temperature? What about 1 eV?

The units used in general relativity are called **geometrized units**, where we set

$$G = c = 1.$$

Setting the speed of light to unity is more natural for us than you might think at first. This is precisely what is done when we talk about a **light-year**, or the distance light travels in one year (in a vacuum). In this case, a *year* is not a unit of time but a unit of distance.

Exercise 13.2 The average distance between the sun and the Earth is $1\ \text{AU} = 1.5 \times 10^{11}$ m. What is the distance to the sun in minutes?

Exercise 13.3 Setting other speeds to one is quite common in everyday life. Consider the following:

1. The distance between two classrooms on campus—do you actually know the *distance*, or roughly how long it takes you to travel between the two? What speed are you setting to one?
2. The distance between two cities, say New York and Boston. How far apart are they in hours if you set some reference speed to one?
3. The distance between two cities that are very far, say Beijing and New York. How far are they in hours? What reference speed are you using?

Setting $c = 1$ thus allows us to interchange meters and seconds for units of distance or time. Since G has SI units of

$$[G] = \frac{\text{m}^3}{\text{kg} \cdot \text{s}^2},$$

after setting $c = 1$, then meters and seconds have the same dimension. Thus we can say dimensionally,

$$[G] = \frac{\text{m}}{\text{kg}}.$$

Once we set $G = 1$, then mass, energy, and momentum all have the same dimensions as distance and time. Thus, all of these quantities can be measured in meters (or kilometers), and other quantities will have dimensions of distance to some power.

Example 13.1 It is straightforward to convert between these geometrized units and SI units, so long as we know what the SI units are supposed to be. Consider an energy given in meters, say $E = 1$ km. We convert this to SI units by multiplying the energy (in meters) by the correct powers of G and c. What is this energy in joules? We write

$$E(\text{in J}) = E(\text{in m})\ G^{\alpha} c^{\beta},$$

where α and β are going to be chosen to make the units work out.[1] Forgetting the numbers for a minute we can write this equation in terms of just units,

$$\frac{\text{kg}\cdot\text{m}^2}{\text{s}^2} = \text{m}\left(\frac{\text{m}^3}{\text{kg}\cdot\text{s}^2}\right)^{\alpha}\left(\frac{\text{m}}{\text{s}}\right)^{\beta}.$$

Setting the exponents of each unit on both sides equal, we obtain three equations,

$$\begin{aligned} -\alpha &= 1\,[\text{kg}], \\ 1 + 3\alpha + \beta &= 2\,[\text{m}], \text{ and} \\ -2\alpha - \beta &= -2\,[\text{s}]. \end{aligned}$$

These three equations are not independent, but we do have a unique solution,

$$\alpha = -1, \ \beta = 4,$$

so that

$$E(\text{in J}) = E(\text{in m})\ \frac{c^4}{G}.$$

Putting in $c = 3 \times 10^8$ m/s and $G = 6.67 \times 10^{-11}$ m^3/(kg · s^2), we get

$$E = 1.2 \times 10^{47} \text{ J}.$$

A "kilometer of energy" is quite large, but actually is roughly the scale of the energy of astrophysical systems, so quoting energy in kilometers will make our lives simpler.

Exercise 13.4 Look up the mass of the sun and use $E = mc^2$ to determine this mass in energy units. You should find it is close to 1 km.

Again, I want to stress that there will always be a unique way to convert these natural units to something that is more useful in physical applications. Several useful conversions are relegated to the problems.

13.1.2 Black Holes

The formation of a black hole generally occurs when matter collapses into such a small volume that gravity becomes so strong that it traps all matter and energy inside a certain region. The most common example of this is the collapse of a star. In this scenario, a star reaches the end of its life as it runs out of fuel to continue the nuclear fusion reactions. These reactions create a repulsion in the core of the star that balances the gravitational force that would (and thus ultimately does) cause the star to collapse. The result is a supernova explosion after which what remains is a nebula, neutron star, or if $M > 8M_{\odot}$, where $M_{\odot}$ is the mass of our sun, then a black hole remains.[2] The

1 If we were to be more rigorous about this, we would write everything in terms of *dimensions* and not units. We would write [L] and [T] for the dimensions of length and time, and not write m and s for meters and seconds. But I feel it's easier to understand this way, so I would rather just do it as such.

2 There are hypothetical black holes, called **primordial black holes**, that could be leftover from the Big Bang, which do not have this restriction.

gravitational effects are so strong due to this mass being concentrated at a **singularity**, or a single point in space.

Even though the mass of a black hole is located at a single point, there is a relevant length scale, known as the **event horizon**. This is a distance R such that if something moves to a point $\mathbf{r}$ such that $r \leq R$, it cannot escape. The term black hole comes from the fact that even light cannot escape the clutches of the gravitational attraction. For a spherically symmetric system, the event horizon is easily described by this radius R, which can be found when solving Einstein's equations for general relativity.[3] For more complicated black holes, it is not as simple as this, but we will still use the term "radius" for the event horizon.

The simplest black hole, known as a **Schwarzschild black hole**, has no net charge or angular momentum, and the radius is related to its mass by the equation[4]

$$R = 2M. \tag{13.1}$$

Exercise 13.5 How would you reintroduce the constants G and c in Eq. (13.1) to write it in SI units?

We are not just interested in the radius itself: Given that we can approach the black hole on all sides, we also would like to determine the **surface area** of the black hole, which is given (for a spherical system) by

$$A = 4\pi R^2 = 16\pi M^2. \tag{13.2}$$

Black holes come with many other properties, and for our purposes (which you may have guessed given our definition of a Schwarzschild black hole), the most important properties to consider are the charge Q and angular momentum $\mathbf{L}$. We can classify black holes by whether or not either or both of these are non-zero:

Schwarzschild black hole: $Q = L = 0$,
Kerr black hole: $Q = 0$ and $L \neq 0$,[5]
Reissner–Nordström black hole: $L = 0$ and $Q \neq 0$,[6] and
Kerr–Newman black hole: $Q \neq 0$ and $L \neq 0$.[7]

For a general black hole, the radius of the event horizon is given by [2]

$$r_+ = M + \sqrt{M^2 - Q^2 - \mathbf{a}^2}, \tag{13.3}$$

where Q is the total charge of the black hole and

$$\mathbf{a} \equiv \frac{\mathbf{L}}{M}, \tag{13.4}$$

is the angular momentum in units of the black hole mass. The surface area is not simply 4π times this radius squared, but rather [2]

$$\begin{aligned} A &= 4\pi\left(r_+^2 + \mathbf{a}^2\right) \\ &= 4\pi\left[\left(M + \sqrt{M^2 - Q^2 - \mathbf{a}^2}\right)^2 + \mathbf{a}^2\right]. \end{aligned} \tag{13.5}$$

3 Which, as you can guess, we will not be doing here. Ref. [1] is a nice introductory textbook to the subject.
4 Determined by Karl Schwarzschild (1873–1916), who was the first to exactly solve Einstein's equations for a spherically symmetric mass.
5 Named after Roy Kerr (born 1934).
6 Named after Hans Reissner (1874–1976) and Gunnar Nordström (1881–1923).
7 Named after Roy Kerr and Ezra Newman (1929–2021).

The Schwarzschild radius clearly results when we take the limit $Q = 0$ and $\mathbf{L} = 0$. Sometimes we will use **rationalized surface area**,[8]

$$\alpha = \frac{A}{4\pi}, \tag{13.6}$$

just to avoid extra factors of 4π and again to reduce clutter in equations.

It's important to allow ourselves to consider these different classes of black holes in thermodynamics. Even if initially we had a Schwarzschild black hole, any absorption of particles would necessarily change either its charge or angular momentum (or both). In fact, in some sense, you can never have a purely Schwarzschild black hole. Additionally, while we often state that nothing can escape a black hole, this isn't entirely true, as there is a constant stream of radiation that comes from black holes, as we'll discuss in Section 13.1.3.

13.1.3 Hawking Radiation

Hawking radiation is a purely quantum phenomenon of electromagnetic radiation near black holes, first predicted by Stephen Hawking (1942–2018) [3]. We won't go into rigorous detail (given the lack of general relativistic formalism), but we'll discuss the important features.

In quantum mechanics, the "vacuum" isn't truly empty—particle–antiparticle pairs can pop up, seemingly violating energy conservation. This violation generally can only occur on a short timescale, via a version of Heisenberg's uncertainty principle,[9]

$$\Delta t \sim \frac{\hbar}{\Delta E}.$$

ΔE is the amount by which we are violating energy conservation, and Δt is the timescale on which this would be allowed. However, suppose this occurs near the event horizon of a black hole: Specifically let's assume two photons are produced from the vacuum. One of these photons may get absorbed by the black hole as it crosses the event horizon while the other is moving away (and as it's not within the event horizon, it can escape), thus producing the appearance of radiation from the black hole![10]

This radiation gives the appearance of the black hole as a blackbody system with a temperature given by the results of Section 12.1. But also as many different particles, charged or not, can be absorbed or "emitted" by this effect, the charge and angular momentum can also change.

13.2 Black Hole Thermodynamics

The thermodynamics of black holes follows in many ways by considering analogies to classical thermodynamics instead of a straightforward application. Many of the results below have been discussed in great deal in the literature (for example, but not limited to Refs. [2, 4–6]), and if you're interested, I would encourage you to delve deeper into the subject. I will work through some of the basics in this section to give you a flavor for what is out there.

For example, instead of the energy of a system, we consider the mass of the black hole. That is, for an isolated black hole, the mass is constant,

$$M = \text{const},$$

8 This is a common notation here but do not confuse this with the coefficient of thermal expansion.

9 Werner Heisenberg (1901–1976).

10 This is a severe oversimplification, and in some ways is an incorrect picture of what is actually occurring. However, we would need to delve into quantum field theory as well as general relativity to describe precisely what is going on here. However, for our purposes let us just state that there are methods by which this can occur.

and we need to determine how the mass changes dM when there is any change in other parameters of the system dM (in other words, we will need to determine an expression for the first law). Any energy that comes into (or out of) the black hole will contribute to a change in the mass.[11]

13.2.1 Black Hole Heat Engine

To begin, in order to identify the temperature of a black hole, we will consider a *gedankenexperiment* (thought experiment) that requires some imagination, largely taken from Ref. [7]. We will construct a four-stage black hole heat engine using a Schwarzschild black hole as the sink (the cold reservoir) as follows.

1. We start with a box of photons (a blackbody) at temperature T and lower it some distance until it is close to the event horizon of a black hole, thereby doing work on the box.
2. Once right near the horizon, we open the box, allowing it to emit some radiation into the black hole (heat flows from the box to the black hole).
3. We then close the box and raise it back to its original position. Now that the box has less energy, the box is "lighter," so less work is needed to raise it than was needed to lower it originally. Thus the net work is positive because we lost some energy in the form of heat to the sink.
4. To make it a proper, cyclical engine, we imagine at this point the box is put into contact with a source of photons to bring it back to its original state by allowing heat to flow into the system. I won't worry about this step for our purposes.

For now, we will ignore gravitational radiation due to the quasistatic process of lowering and raising the box. Additionally we will consider the box to be lowered only to a point where the center is a distance ℓ from the black hole radius, with ℓ the same scale of the size of the box.[12] Suppose the initial energy of the box of blackbody radiation is m, and as we lower the box a distance ℓ, then the work done is given by

$$W_{\text{down}} = m(1 - g\ell), \tag{13.7}$$

with g the **surface gravity** of the black hole at the horizon (otherwise known as the acceleration due to gravity in a classical system).[13] This work is positive, as we assume $g\ell \ll c^2 = 1$, where for a brief moment we return to ordinary units to see clearly what this is.

Exercise 13.6 A Newtonian estimate for a non-rotating black hole for g is

$$g = \frac{1}{4M}. \tag{13.8}$$

Reintroduce factors of G and c to this equations to write it in SI units. Consider a black hole with a mass 10 times that of our sun and calculate g in SI units.

When we open the box, some energy dm is radiated, and then we lift it back up. The work to lift the box is now

$$W_{\text{up}} = -(m - dm)(1 - g\ell), \tag{13.9}$$

11 Using mass instead of energy may see odd, but relativistically speaking, mass and energy are interchangeable.
12 Incidentally, calculations have been made on how large ℓ must be if the rope attached to the box is not to break. One can show that $\ell > (e^2 - 1)^{-1}R_s \approx 0.16R_s$, with R_s the Schwarzschild radius [8].
13 But for a black hole, g is most definitely not 9.8 m/s ^{2}!

so the net work is just the sum of these two,

$$W = W_{\text{down}} + W_{\text{up}} = dm(1 - g\ell). \tag{13.10}$$

The heat is given by the energy which enters the black hole, $Q = dm$, so the efficiency of this heat engine can be written as

$$\eta = \frac{W}{Q} = 1 - g\ell. \tag{13.11}$$

For the highest efficiency, we need ℓ to be as small as possible, which is limited by quantum mechanics, and we need it to be large enough to be able to apply thermodynamics. Thus it must accommodate the wavelengths where the bulk of the energy lies in the radiation, so that

$$\ell \geq \lambda_{\text{max}} = \frac{2\pi c}{\omega} \propto \frac{2\pi c\hbar}{k_B T} = \frac{\hbar}{2\pi k_B T}$$

where we have used Wien's displacement law, Eq. (12.16). The last equality arises from a quantum mechanical argument that you can read about in Ref. [7].

We see the efficiency satisfies the inequality

$$\eta \leq 1 - \frac{\hbar g}{2\pi k_B T},$$

and if we assume this is an ideal (Carnot) engine,

$$\eta \leq 1 - \frac{T_c}{T_h},$$

and we can identify the temperature of the black hole (the cold reservoir),

$$T_{\text{BH}} = \frac{\hbar g}{2\pi k_B} = \frac{\hbar}{8\pi k_B M_{\text{BH}}}. \tag{13.12}$$

The factor of $\hbar$ shows that this is purely quantum mechanical phenomenon, and we see the interesting effect that an *increase* in the mass of the black hole *decreases* its temperature! Note also that a form of the second law of thermodynamics appears here (that we cannot have a heat engine with $\eta = 1$), but in terms of limitations of other parameters.

Exercise 13.7 Reintroduce G and c into Eq. (13.12) and calculate the temperature in kelvin of a black hole of mass $M = \xi M_{\odot}$, where ξ is some number greater than eight. You should find

$$T_{\text{BH}} \sim \left(\frac{10^{-7}}{\xi}\right) \text{ K}.$$

The entropy of the box of electromagnetic radiation decreases because the mass of the box decreases from $m \rightarrow m - dm$, and this change is given by

$$dS_{\text{box}} = -\frac{dm}{T}. \tag{13.13}$$

Additionally, there is an increase in the mass of the black hole, dM_{BH}, with a corresponding change in the black hole entropy,

$$dS_{\text{BH}} = \frac{dM_{\text{BH}}}{T_{\text{BH}}}.$$

The mass ejected from the box leads to a change in the mass of the black hole in the form of gravitational potential energy, or $dM_{\text{BH}} = dmg\ell > 0$. This allows us to write the change in the black hole entropy as

$$dS_{\text{BH}} = \left(\frac{g\ell}{T_{\text{BH}}}\right) dm.$$

Putting in the temperature of the black hole from above, and imposing the inequality of the efficiency η above, we get

$$dS_{\text{BH}} = \frac{2\pi k_B \ell}{\hbar} dm \geq \frac{dm}{T} = \left|dS_{\text{box}}\right|.$$

We see that the black hole indeed gains more entropy than the box loses, satisfying the second law of thermodynamics. If we write the entropy change using the assumption that $dM_{\text{BH}} \ll M_{\text{BH}}$, and using our result for T_{BH} above,

$$\begin{aligned} dS_{\text{BH}} &= \frac{8\pi k_B M_{\text{BH}} dM_{\text{BH}}}{\hbar} \\ \Rightarrow S_{\text{BH}} &= \frac{4\pi k_B M_{\text{BH}}^2}{\hbar} = \frac{k_B}{4\hbar} A_{\text{BH}}, \end{aligned} \tag{13.14}$$

where we used Eq. (13.2) in the last equality.

Exercise 13.8 Fill in the steps that lead to Eq. (13.14).

This sets us up to see the connections between the parameters relevant to a black hole and those in thermodynamics. In Section 13.2.2 we will treat this more systematically.

13.2.2 The Math of Black Hole Thermodynamics

To more formally develop the thermodynamics of black holes, we need a version of the first law of thermodynamics. Following Ref. [2], I will treat $M = M(\alpha, L, Q)$,[14] where $\alpha = A/(4\pi)$ is the rationalized area of the black hole (which is related to the entropy via Eq. (13.14), up to a factor). With this, only considering rotation around one axis for simplicity, we can write,

$$dM = \theta d\alpha + \omega dL + \Phi dQ. \tag{13.15}$$

In this equation, θ plays a role similar to the temperature (since α is the analog to entropy) so the first term is the heat flowing into the system. ω is the angular velocity, and Φ is the electrostatic potential, so the second and third terms are different forms of work done by the black hole. As the area, angular momentum, and charge are the "easy" things to change, these are the naturally independent variables. L and Q are the external parameters that can be changed (doing work), so ω and Φ are the generalized forces conjugate to these external parameters (as pressure is the generalized force conjugate to volume).

We can find these terms by considering the work required to change the angular momentum of an object with moment of inertia I. The kinetic energy is given by

$$K = \frac{L^2}{2I} \Rightarrow dK = \frac{LdL}{I},$$

14 As we will not be discussing the heat explicitly here, there shouldn't be confusion when using Q for the charge of the black hole.

and for a rigid object, $L = I\omega$ and thus

$$dK = \omega dL = đW, \tag{13.16}$$

where the second equality comes from the work-kinetic energy theorem.

Exercise 13.9 Argue that ΦdQ is the electrostatic work done by the system when the charge changes by an amount dQ.

From the first law, we can read off the derivatives

$$\left(\frac{\partial M}{\partial \alpha}\right)_{L,Q} = \theta, \tag{13.17}$$

$$\left(\frac{\partial M}{\partial L}\right)_{\alpha,Q} = \omega, \text{ and} \tag{13.18}$$

$$\left(\frac{\partial M}{\partial Q}\right)_{L,\alpha} = \Phi. \tag{13.19}$$

Exercise 13.10 Derive the three Maxwell relations that result from Eqs. (13.17)–(13.19).

If we solve the rationalized area from Eqs. (13.5) and (13.6) for the mass, keeping in mind $\mathbf{a} = \mathbf{L}/M$, we get the fundamental relation,

$$M = \frac{\sqrt{4L^2 + (\alpha + Q^2)^2}}{2\sqrt{\alpha}}. \tag{13.20}$$

From this, using Eq. (13.17) we can show

$$\theta = \frac{1}{2M}\left(\frac{\alpha^2 - 4L^2 - Q^4}{4\alpha^2}\right). \tag{13.21}$$

In general we can use Eqs. (13.17)–(13.19) to determine (see Problem 13.6)

$$\theta = \frac{1}{4\alpha}\left(r_+ - r_-\right), \tag{13.22}$$

$$\omega = \frac{\mathbf{a}}{\alpha}, \text{ and} \tag{13.23}$$

$$\Phi = \frac{Qr_+}{\alpha}. \tag{13.24}$$

r_- is defined similarly to r_+, which we previously defined, where

$$r_- = M - \sqrt{M^2 - Q^2 - \mathbf{a}^2}. \tag{13.25}$$

Just as ordinary thermodynamic variables ($\overline{p}$, V, and T) are not actually completely independent but are treated as such for calculations, the same is true for these variables. Eqs. (13.22)–(13.24) are the equations of state for our black hole.

As mentioned before, θ corresponds to the temperature of the black hole, which we can show by using our expression for the entropy,

$$S_{\text{BH}} = \frac{\pi k_B}{\hbar}\alpha,$$

so that

$$d\alpha = \frac{\hbar}{\pi k} dS_{\mathrm{BH}},$$

and thus

$$T_{\mathrm{BH}} = \frac{\hbar\theta}{\pi k_B}. \tag{13.26}$$

That is, T_{BH} and θ are directly related to each other.

Just as we did in Chapter 6, we can define the Legendre transforms

$$F_{\mathrm{BH}} = M - \theta\alpha, \tag{13.27}$$

$$G_{\mathrm{BH}} = M - \theta\alpha - L\omega - Q\Phi, \text{ and} \tag{13.28}$$

$$H_{\mathrm{BH}} = M - L\omega - Q\Phi, \tag{13.29}$$

which allow us to obtain a host of Maxwell relations. I will leave this for an exercise for you in the problems.

We can extend the terms involving the angular momentum to three dimensions, so that

$$dM = \theta d\alpha + \boldsymbol{\omega} \cdot d\mathbf{L} + \Phi dQ.$$

The relations involving vector quantities involve gradient operators, such as

$$\nabla_\omega = \hat{\imath}\frac{\partial}{\partial\omega_x} + \hat{\jmath}\frac{\partial}{\partial\omega_y} + \hat{k}\frac{\partial}{\partial\omega_z},$$

and so we have, for example

$$(\nabla_{\mathbf{L}} M)_{\alpha,Q} = \boldsymbol{\omega}.$$

These make the calculations trickier but are not necessarily that difficult. This extra generality is useful to keep in mind for more complicated problems but will not be needed here.

With these basics, we can now ask any questions about the thermodynamic properties of the system, like we did before, but with these different variables. For example, we could ask: How does the mass of a black hole change when we change the temperature while keeping the angular momentum and charge fixed? You can show in Problem 13.5 that

$$\left(\frac{\partial M}{\partial\theta}\right)_{L,Q} = \theta\left(\frac{\partial\alpha}{\partial\theta}\right)_{L,Q}. \tag{13.30}$$

Given that $\theta d\alpha$ is akin to TdS in our thermodynamic analogy, the right-hand side is essentially a heat capacity at constant angular momentum and charge. This is the quantity we will focus on in Section 13.3, although you can read up on many other interesting topics in the aforementioned references.

13.3 Heat Capacity of a Black Hole

The heat capacity at constant y for a quasistatic process is

$$C_y = T\left(\frac{\partial S}{\partial T}\right)_y.$$

As mentioned at the end of Section 13.2, with the connection between the entropy (temperature) and area (θ) of a black hole,

$$S_{\rm BH} = \frac{\pi k_B}{\hbar}\alpha, \qquad T_{\rm BH} = \frac{\hbar}{\pi k_B}\theta$$

we will consider

$$C'_y = \theta\left(\frac{\partial \alpha}{\partial \theta}\right)_y \tag{13.31}$$

where $C'_y = \frac{\hbar}{\pi k_B}C_y$. It'll be simpler to use C'_y, α, and θ, rather than C_y, $S_{\rm BH}$, and $T_{\rm BH}$ for our calculations, but we can easily convert back later.

Writing θ from above in terms of α, L, and Q, we have

$$\theta = \frac{\alpha^2 - 4L^2 - Q^4}{4\alpha^{3/2}\sqrt{4L^2 + (Q^2+\alpha)^2}}. \tag{13.32}$$

After a simple (albeit algebraically messy) calculation, we find

$$\left(\frac{\partial \theta}{\partial \alpha}\right)_{L,Q} = \theta\left[\frac{2\alpha}{\alpha^2 - 4L^2 - Q^4} - \frac{3}{2}\frac{1}{\alpha} - \frac{Q^2+\alpha}{4L^2 + (Q^2+\alpha)^2}\right], \tag{13.33}$$

from which we can show that

$$\begin{aligned} C'_{L,Q} &= \theta\left(\frac{\partial \alpha}{\partial \theta}\right)_{L,Q} \\ &= \left[\frac{2\alpha}{\alpha^2 - 4L^2 - Q^4} - \frac{3}{2}\frac{1}{\alpha} - \frac{Q^2+\alpha}{4L^2 + (Q^2+\alpha)^2}\right]^{-1}. \end{aligned} \tag{13.34}$$

This is relatively complicated, so let's look at the case where $L = 0$ and $Q = 0$, where heat capacity becomes a much simpler expression,

$$C'_{L,Q} = -2\alpha, \tag{13.35}$$

or in terms of the temperature,

$$C_{L,Q} = -2\left(\frac{\pi k_B}{\hbar}\frac{A_{\rm BH}}{4\pi}\right) = -\frac{\hbar}{4\pi k_B T_{\rm BH}^2}. \tag{13.36}$$

This heat capacity is negative: If we add energy to a black hole, it gets colder! This is a well-known result of black hole thermodynamics, and from our discussion for stable equilibrium in Chapter 10, this implies that a black hole of this sort would be thermodynamically unstable. Of course, the full expression in Eq. (13.34) can be positive or negative, so this would imply that the angular momentum and charge of a black hole cannot be zero if it's in stable equilibrium. More importantly, this expression also violates the third law of thermodynamics when both the charge and angular momentum are zero. (Not only does it not vanish at zero temperature, but it blows up!)

One can easily continue an analysis by looking at other heat capacities (holding ω or Φ constant, for example), or even studying our expression for $C_{L,Q}$ for the more general cases of charged and/or rotating black holes. Given the formalism we have developed throughout this text, this is relatively simple to do with what we have laid out in this chapter. Many interesting results have been derived in this formalism, and it just goes to show that statistical mechanics and thermodynamics can truly be applied to any physics topic.

13.4 Summary

- The thermodynamics of black holes is a rich field that we have only barely scratched the surface of. With the formalism we have developed, studying this different system is no serious challenge, especially with the analogies that connect black hole thermodynamic quantities with standard thermodynamic quantities.
- The specific heat of a Schwarzschild black hole is negative, so such a black hole is unstable. This is also true for more general black holes with certain ranges of charge and angular momentum.

Problems

13.1 Perform the following conversions in geometrized units:
(a) kg to m,
(b) kg to s,
(c) momentum (kg·m/s) to m.

13.2 Consider force in geometrized units.
(a) Show that force is dimensionless in these units.
(b) What is a force of 2 in SI units?
(c) What is the force of the sun on the Earth in geometrized units?

13.3 Derive all of the Maxwell relations that result from Eqs. (13.27)–(13.29). For example,

$$\left(\frac{\partial \theta}{\partial L}\right)_{\alpha,Q} = \left(\frac{\partial \omega}{\partial \alpha}\right)_{L,Q}$$

13.4 Examine the Legendre transforms where we only transform one of the work terms at a time:

$$I_1 = M - L\omega,$$

and

$$I_2 = M - Q\Phi.$$

Determine the differentials of these new potentials as well as the new Maxwell relations that arise.

13.5 Derive Eq. (13.30).
Hint: Solve dM for $d\alpha$, then calculate dM using $M = M(\theta, L, Q)$.

13.6 Use Eqs. (13.17)–(13.19) and Eq. (13.20) to derive Eqs. (13.22), (13.23), and (13.24).

13.7 Derive Eq. (13.33) from Eq. (13.32).

13.8 Consider the heat capacity in Eq. (13.34) in the case where $L = 0$ but $Q \neq 0$. Simplify this expression and define a new "charge,"

$$Q' = \frac{Q}{\sqrt{\alpha}}.$$

In this form, for what values of Q' (and thus Q) is the heat capacity always positive, thereby giving a stable black hole?

References

1 B. Schutz. *A First Course in General Relativity*. Cambridge University Press, 2009.

2 B. R. Parker and R. J. McLeod. Black hole thermodynamics in an undergraduate thermodynamics course. *American Journal of Physics*, 48(12):1066–1070, 1980.

3 S. W. Hawking. Particle creation by black holes. *Communications in Mathematical Physics*, 43:199–220, 1975. [Erratum: *Communications in Mathematical Physics* 46:206, 1976].

4 J. D. Bekenstein. Black holes and entropy. *Physical Review D*, 7:2333–2346, Apr 1973.

5 P. S. Custódio and J. E. Horvath. Thermodynamics of black holes in a finite box. *American Journal of Physics*, 71(12):1237–1241, 2003.

6 S. F. Ross. Black hole thermodynamics. Feb 2005.

7 D. W. Sciama. Black holes and their thermodynamics. *Vistas in Astronomy*, 19:385–401, 1976.

8 G. W. Gibbons. On lowering a rope into a black hole. *Nature Physical Science*, 240(100):77–77, Nov 1972.

Appendix A

Important Constants and Units

In statistical mechanics and thermodynamics, as in every area of physics, there are several important constants that will arise. As of May 2019, many fundamental constants were given exact values so as to define the base units in the new International System of Units (SI). In this appendix I discuss several of them while giving some useful conversions between common units used in the subject. The base units that are specifically relevant here are defined as follows.

Mole: This denotes the amount of substance, where one mole is defined as have N_A elementary entities, with N_A **Avogadro's number**. The new SI defines this to be an exact value, given by

$$N_A \equiv 6.022\,140\,76 \times 10^{23}\,\text{mole}^{-1}. \tag{A.1}$$

The previous definition of the mole was given by the number of atoms in 12 g of carbon-12. This previous definition, while not as precise as the new definition, is still approximately correct (definitely for everyday use).

Kelvin: As discussed in detail in Section 5.3.2, this is the SI unit of temperature. In the new SI, this is now defined so as to make the **Boltzmann constant**, k_B exact,

$$k_B \equiv 1.380\,649 \times 10^{-23}\,\text{J/K}. \tag{A.2}$$

Table A.1 Base and some derived SI units.

Dimension	Unit	Symbol
Length	meter	m
Mass	kilogram	kg
Time	second	s
Amount of substance	mole	mole
Temperature	kelvin	K
Electric current	ampere	A
Luminous intensity	candela	cd
Force	newton	$\text{N} = \text{kg}\cdot\text{m/s}^2$
Pressure	pascal	$\text{Pa} = \text{N/m}^2$
Energy	joule	$\text{J} = \text{N}\cdot\text{m}$
Electric charge	coulomb	$\text{C} = \text{A}\cdot\text{s}$

Statistical Thermodynamics: An Information Theory Approach, First Edition. Christopher Aubin.

Companion website: www.wiley.com/go/Aubin/StatisticalThermodynamics

The other base units, shown in Table A.1, are also defined to give fundamental constants exact values, and while they are important as well, these are the two units that are specifically new to thermodynamics. You can visit the National Institute of Standards and Technology website [1] along with the nice summaries in Refs. [2, 3] to read more about the new definitions of the base units. An extensive discussion on SI units can be found in Ref. [4].

Table A.2 Important fundamental constants that we will need. Those which are exact are marked with an asterisk (*) and those that are not have their uncertainties included.

Constant	Symbol	Value
Boltzmann constant*	k_B	$1.380\,649 \times 10^{-23}$ J/K
Avogadro's number*	N_A	$6.022\,140\,76 \times 10^{23}$ mole^{-1}
Gas constant*	R	8.314 462 618 153 24 J/K/mole
Planck's constant*	h	$6.626\,070\,15 \times 10^{-34}$ J·s
Speed of light*	c	299 792 458 m/s
Elementary charge*	e	$1.602\,176\,634 \times 10^{-19}$ C
Newton's gravitational constant	G	$6.674\,3(15) \times 10^{-11}$ m^3/kg/s^2
Electron mass	m_e	$9.109\,383\,701\,5(28) \times 10^{-31}$ kg
Proton mass	m_p	$1.672\,621\,923\,69(51) \times 10^{-27}$ kg

Table A.3 Some important unit conversions, where I include not just metric units but also imperial units (important for US engineering applications).

Mass	
1 u	1.661×10^{-27} kg
1 lb	0.4536 kg
Volume	
1 m^3	1000 L
1 cm^3	1 mL
1 m^3	35.3147 ft^3
1 gallon	3.78541 L
Pressure	
1 atm	$1.01\,325 \times 10^5$ Pa
1 mmHg	133.322 Pa
1 lb/in^2	6894.76 Pa
1 bar	10^5 Pa
1 torr	1 mmHg

Temperature	
T (in K)	T(in °C) + 273.15
T (in °C)	$\frac{5}{9}T$(in °F) − 32
T (in R)	T(in °F) + 459.67
T (in R)	$\frac{5}{6}T$(in K)
Energy	
1 eV	1.602×10^{-19} J
1 erg	10^{-7} J
1 cal	4.184 J
1 kW-hr	3.6×10^6 J
1 BTU	1054 J
1 ft-lb	1.355818 J

One other constant that arises in thermodynamics, which often is more familiar to students is the **gas constant**, R, defined as the product of Avogadro's number and the Boltzmann constant,

$$R \equiv N_A k_B = 8.314\,462\,618\,153\,24\,\mathrm{J/(mole\,K)}. \tag{A.3}$$

Because it's a product of two exact constants, R is also exact, although we will rarely use it (or the other constants) with this level of precision.

The constants defined in Eqs. (A.1)–(A.3), along with other important fundamental constants, are shown in Table A.2. Additionally, while SI units are standard for most physics applications, there are many other units used in this text and elsewhere for thermodynamic applications, so I give some relevant conversion factors in Table A.3.

References

1 National Institute of Standards and Technology (NIST). https://www.nist.gov/si-redefinition. Accessed: 23 January 2023.

2 D. B. Newell. A more fundamental International System of Units. *Physics Today*, 67(7):35–41, 2014.

3 M. A. Martin-Delgado. The new SI and the fundamental constants of nature. *European Journal of Physics*, 41(6):063003, 2020.

4 E. O. Göbel and U. Siegner. *The New International System of Units (SI): Quantum Metrology and Quantum Standards*. John Wiley & Sons, Inc., 2019.

Appendix B

Periodic Table of Elements

The periodic table of elements, from R.L. Workman et al. (Particle Data Group), Prog. Theor. Exp. Phys. **2022**, 083C01 (2022) and 2023 update. The upper left number in each entry is the atomic number, corresponding to the number of protons in the nucleus. The number below the element name is the atomic mass in atomic mass units (u), which corresponds to the molar mass in grams/mole.

PERIODIC TABLE OF THE ELEMENTS

1 IA	2 IIA	3 IIIB	4 IVB	5 VB	6 VIB	7 VIIB	8	9 VIII	10	11 IB	12 IIB	13 IIIA	14 IVA	15 VA	16 VIA	17 VIIA	18 VIIIA
1 H hydrogen 1.008																	2 He helium 4.002602
3 Li lithium 6.94	4 Be beryllium 9.012182											5 B boron 10.81	6 C carbon 12.0107	7 N nitrogen 14.007	8 O oxygen 15.999	9 F fluorine 18.998403163	10 Ne neon 20.1797
11 Na sodium 22.98976928	12 Mg magnesium 24.305											13 Al aluminum 26.9815385	14 Si silicon 28.085	15 P phosphorus 30.973761998	16 S sulfur 32.06	17 Cl chlorine 35.45	18 Ar argon 39.948
19 K potassium 39.0983	20 Ca calcium 40.078	21 Sc scandium 44.955908	22 Ti titanium 47.867	23 V vanadium 50.9415	24 Cr chromium 51.9961	25 Mn manganese 54.938044	26 Fe iron 55.845	27 Co cobalt 58.933195	28 Ni nickel 58.6934	29 Cu copper 63.546	30 Zn zinc 65.38	31 Ga gallium 69.723	32 Ge germanium 72.630	33 As arsenic 74.921595	34 Se selenium 78.971	35 Br bromine 79.904	36 Kr krypton 83.798
37 Rb rubidium 85.4678	38 Sr strontium 87.62	39 Y yttrium 88.90584	40 Zr zirconium 91.224	41 Nb niobium 92.90637	42 Mo molybdenum 95.95	43 Tc technetium (97.907212)	44 Ru ruthenium 101.07	45 Rh rhodium 102.90550	46 Pd palladium 106.42	47 Ag silver 107.8682	48 Cd cadmium 112.414	49 In indium 114.818	50 Sn tin 118.710	51 Sb antimony 121.760	52 Te tellurium 127.60	53 I iodine 126.90447	54 Xe xenon 131.293
55 Cs caesium 132.90545196	56 Ba barium 137.327	57–71 LANTHA- NIDES	72 Hf hafnium 178.49	73 Ta tantalum 180.94788	74 W tungsten 183.84	75 Re rhenium 186.207	76 Os osmium 190.23	77 Ir iridium 192.217	78 Pt platinum 195.084	79 Au gold 196.966569	80 Hg mercury 200.592	81 Tl thallium 204.38	82 Pb lead 207.2	83 Bi bismuth 208.98040	84 Po polonium (208.98243)	85 At astatine (209.98715)	86 Rn radon (222.01758)
87 Fr francium (223.01974)	88 Ra radium (226.02541)	89–103 ACTINIDES	104 Rf rutherford. (267.12169)	105 Db dubnium (268.12567)	106 Sg seaborgium (269.12863)	107 Bh bohrium (270.13336)	108 Hs hassium (269.13375)	109 Mt meitnerium (278.15631)	110 Ds darmstadt. (281.16451)	111 Rg roentgen. (282.16912)	112 Cn copernicium (285.17712)	113 Nh nihonium (286.18221)	114 Fl flerovium (289.19042)	115 Mc moscovium (290.19598)	116 Lv livermorium (293.20449	117 Ts tennessine (294.21046)	118 Og oganesson (294.21392)

Lanthanide series	57 La lanthanum 138.90547	58 Ce cerium 140.116	59 Pr praseodym. 140.90766	60 Nd neodymium 144.242	61 Pm promethium (144.91276)	62 Sm samarium 150.36	63 Eu europium 151.964	64 Gd gadolinum 157.25	65 Tb terbium 158.92535	66 Dy dysprosium 162.500	67 Ho holmium 164.93033	68 Er erbium 167.259	69 Tm thulium 168.93422	70 Yb ytterbium 173.054	71 Lu lutetium 174.9668
Actinide series	89 Ac actinium (227.02775)	90 Th thorium 232.0377	91 Pa protactinium 231.03588	92 U uranium 238.02891	93 Np neptunium (237.04817)	94 Pu plutonium (244.06420)	95 Am americium (243.06138)	96 Cm curium (247.07035)	97 Bk berkelium (247.07031)	98 Cf californium (251.07959)	99 Es einsteinium (252.08298)	100 Fm fermium (257.09511)	101 Md mendelevium (258.09844)	102 No nobelium (259.10103)	103 Lr lawrencium (262.10961)

Statistical Thermodynamics: An Information Theory Approach, First Edition. Christopher Aubin.

Companion website: www.wiley.com/go/Aubin/StatisticalThermodynamics

Appendix C

Gaussian Integrals

The Gaussian distribution is given by Eq. (2.28), and involves the function

$$\mathcal{P}(x)dx = C\exp\left[-\frac{(x-\mu)^2}{2\sigma^2}\right]dx. \tag{C.1}$$

The factor C is determined by requiring this to be normalized to one so that

$$\int_{-\infty}^{\infty} \mathcal{P}(x)dx = 1.$$

In this appendix I will derive this constant C and discuss several other useful relationships and tricks that we'll need for such integrals.

To evaluate this integral we use a common trick, where we define

$$I = \int_{-\infty}^{\infty} \exp\left[-\frac{(x-\mu)^2}{2\sigma^2}\right]dx$$

and then evaluate I^2. Because x is a dummy variable over which we are integrating, we can write this as

$$I^2 = \int_{-\infty}^{\infty}\int_{-\infty}^{\infty} \exp\left[-\frac{(x-\mu)^2}{2\sigma^2}\right] \exp\left[-\frac{(y-\mu)^2}{2\sigma^2}\right] dydx.$$

We change to polar coordinates, with

$$x = \mu + r\cos\theta \quad \text{and} \quad y = \mu + r\sin\theta,$$

and thus

$$I^2 = \int_0^{2\pi} d\theta \int_0^{\infty} \exp\left(-\frac{r^2}{2\sigma^2}\right) rdr.$$

The angular integral gives a factor of 2π, and the radial integrand is a total derivative. Thus,

$$I^2 = 2\pi\left[-\sigma^2 \exp\left(-\frac{r^2}{2\sigma^2}\right)\Big|_0^{\infty}\right] = 2\pi\sigma^2,$$

and we obtain

$$I = \int_{-\infty}^{\infty} \exp\left[-\frac{(x-\mu)^2}{2\sigma^2}\right]dx = \sqrt{2\pi\sigma^2}, \tag{C.2}$$

which gives $C = 1/\sqrt{2\pi\sigma^2}$ in Eq. (2.28).

Statistical Thermodynamics: An Information Theory Approach, First Edition. Christopher Aubin.

Companion website: www.wiley.com/go/Aubin/StatisticalThermodynamics

Because of the symmetry of the integration limits, we can consider a variable $z = x - \mu$ for our expressions which will make our calculations below easier later. In order to obtain mean values, one has to integrate functions of the form

$$I_n = \int_{-\infty}^{\infty} z^n \exp\left(-\frac{z^2}{2\sigma^2}\right) dz \text{ for } n \geq 0, \tag{C.3}$$

where $I_0 = I$ from above. In this form we can easily see that $I_n = 0$ for any odd value of n. Only even values of n will give a non-zero value for I_n, so we write $n = 2k$ for $k \geq 0$. Additionally, I will define $\alpha = 1/(2\sigma^2)$, so our integral can be written as

$$I_{2k} = \int_{-\infty}^{\infty} z^{2k} e^{-\alpha z^2} dz. \tag{C.4}$$

We could use integration by parts to evaluate this, but there is a simpler[1] trick. Consider the case $k = 1$, where we can write

$$I_2 = \int_{-\infty}^{\infty} z^2 e^{-\alpha z^2} dz = -\frac{\partial}{\partial \alpha} \int_{-\infty}^{\infty} e^{-\alpha z^2} dz. \tag{C.5}$$

The integral now, looking at Eq. (C.2), is simple, so

$$I_2 = -\frac{\partial}{\partial \alpha}\sqrt{\frac{\pi}{\alpha}} = \frac{\sqrt{\pi}}{2\alpha^{3/2}}, \tag{C.6}$$

or in terms of σ,

$$I_2 = \sqrt{2\pi}\sigma^3. \tag{C.7}$$

In general, we can just apply minus the partial derivative k times to determine I_{2k}, that is

$$I_{2k} = \underbrace{\left(-\frac{\partial}{\partial \alpha}\right)\cdots\left(-\frac{\partial}{\partial \alpha}\right)}_{k \text{ times}}\sqrt{\frac{\pi}{\alpha}}.$$

Exercise C.1 Another useful result that can be easily obtained from above is that

$$\int e^{-\alpha r^2} d^n\mathbf{r} = \left(\frac{\pi}{\alpha}\right)^{n/2}. \tag{C.8}$$

Show that this is true by considering the integral I from Eq. (C.2) and evaluate I^n.

Exercise C.2 Evaluate $\overline{x}$ and the standard deviation for x, σ_x, for the Gaussian distribution and show that they are equal to μ and σ. respectively.

Exercise C.3 Evaluate

$$\overline{(x-\mu)^{2k}}$$

explicitly for $k = 2$ and 3 for the Gaussian distribution.

1 And more fun!

Appendix D

Volumes in n-Dimensions

In statistical mechanics, we often have to do large dimensional integrals ($n = 3N$ where N is possibly on the order of 10^{24}), but luckily many of them are over n-dimensional spheres or ellipsoids. Thus it's useful to review how to perform integrals of volumes of these shapes in general.[1] We consider these shapes centered on the origin, with the n axes aligned with the axes of symmetry.

For spheres, the origin is a distance r from any point on the shape. For ellipsoids, we will define the distances R_i, for $i = 1, \dots, n$, which are the distances from the origin to the surface of the ellipsoid along the ith axis (in either direction as they line up with the symmetry axes).[2]

First, we wish to evaluate the volume of an n-dimensional sphere of radius R,

$$V_n = \int d^n\mathbf{r}, \tag{D.1}$$

where the integral is over the entire volume of the sphere. We can write this as the integral over the radial component and the n-dimensional solid angle,

$$V = \int_0^R r^{n-1} dr \int d\Omega_n. \tag{D.2}$$

The first integral can be evaluated easily,

$$\int_0^R r^{n-1} dr = \frac{1}{n} R^n. \tag{D.3}$$

The second integral is a little trickier, so we rely on using the result of Eq. (C.8),

$$\int e^{-\mathbf{r}^2} d^n\mathbf{r} = \pi^{n/2}. \tag{D.4}$$

Instead of evaluating these as before, we write $d^n\mathbf{r}$ as we did in Eq. (D.2),

$$\int e^{-\mathbf{r}^2} d^n\mathbf{r} = \int_0^\infty e^{-r^2} r^{n-1} dr \int d\Omega_n, \tag{D.5}$$

and the first integral here can be evaluated with a simple change of variables,

$$\int_0^\infty e^{-r^2} r^{n-1} dr = \frac{1}{2} \int_0^\infty e^{-y} y^{n/2-1} dy. \tag{D.6}$$

1 Usually when we have more than three dimensions, they are referred to as **hypervolumes** of **hyperspheres** or **hyperellipsoids**. However, I find it simpler to just use the terms volumes, spheres, and ellipsoids for any number of dimensions. The confusions that might arise are usually overcome after the initial strangeness of this terminology.
2 For two-dimensional ellipses, R_1 and R_2 would be called the semi-major and semi-minor axes if $R_1 > R_2$.

Statistical Thermodynamics: An Information Theory Approach, First Edition. Christopher Aubin.

Companion website: www.wiley.com/go/Aubin/StatisticalThermodynamics

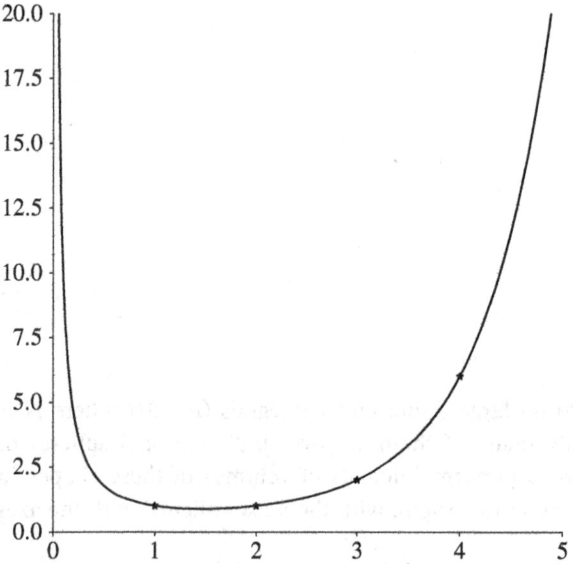

Figure D.1 The Γ function (curve) shown along with the values of the factorial (stars).

The right-hand side here is 1/2 times the Γ function,

$$\int_0^\infty e^{-r^2} r^{n-1} dr = \frac{1}{2}\Gamma(n/2). \tag{D.7}$$

$\Gamma(x)$ is shown in Figure D.1 and is defined so that for integer values of x, $\Gamma(n) = (n-1)!$ (marked by the stars in the figure), but this integral representation allows us to generalize this to non-integer values.

Exercise D.1 Show that the Γ function satisfies the same recurrence relation as the factorial function does, that is,

$$x\Gamma(x) = \Gamma(x+1).$$

While this is valid for any value of x, we only care about integers and half integers (for us, n is always an integer, and so is the argument of Γ). With this we can write the solid angle as

$$\int d\Omega_n = \frac{2\pi^{n/2}}{\Gamma(n/2)}. \tag{D.8}$$

And we have a general expression for the volume of an n-dimensional sphere of radius R,

$$V_n = \frac{2\pi^{n/2}}{n\Gamma(n/2)} R^n. \tag{D.9}$$

For odd numbers of dimensions, we use

$$\Gamma(1/2) = \sqrt{\pi},$$

and for even numbers of dimensions, we use

$$\Gamma(n) = (n-1)!.$$

These results (combined with the recurrence relation for odd dimensions) allow us to obtain $\Gamma(n/2)$ for any positive value of n.

Exercise D.2 Show that Eq. (D.9) gives the appropriate length, area, and volume of the appropriate system in one, two, and three dimensions, respectively.

Usually for our purposes, the prefactor is not going to be important for most physical quantities, and we will often just write Eq. (D.9) as

$$V_n = bR^n, \tag{D.10}$$

and not specify b.[3]

The volume of an n-dimensional ellipsoid is a simple generalization of the result in Eq. (D.9), by allowing the distances R_i from the origin to the surface along the ith axis to be different. We evaluate Eq. (D.1) but now we require that the values of r_i are restricted such that

$$\sum_{i=1}^{n} \frac{r_i^2}{R_i^2} \leq 1, \tag{D.11}$$

the equation of an n-dimensional ellipsoid. If we change our integration variables to

$$r_i' = \frac{r_i}{R_i},$$

then Eq. (D.1) becomes

$$V_n = \left(\prod_{i=1}^{n} R_i\right)\left(\int_{\text{unit}} d^n\mathbf{r}\right), \tag{D.12}$$

and the subscript "unit" on the integral denotes that this is the volume integral of a sphere of radius one. Thus we use the expression in Eq. (D.9) with $R = 1$ for the second factor and immediately obtain

$$V_n = \frac{2\pi^{n/2}}{n\Gamma(n/2)} \prod_{i=1}^{n} R_i. \tag{D.13}$$

Exercise D.3 Use the definition of the Γ function to evaluate the average values $\overline{(x-\mu)^{2k}}$ in Exercise C.3 exactly.

3 Recall that b usually drops out of physical calculations in thermodynamics.

Appendix E

Partial Derivatives in Thermodynamics

The application of calculus to thermodynamics is often confusing, because it seems very different from the vector calculus from other subjects [1]. In classical mechanics, we usually apply vector calculus to "ordinary" three-dimensional space, so the variables involved are the spatial coordinates x, y, and z, which are easily visualized. Thus the complexity, compared with introductory physics, is just in learning the new math (working with the gradient, divergence, etc.). Additionally, when this is later studied in E&M, even though abstract concepts such as the electric field are introduced, these concepts are still functions of x, y, and z; derivatives taken with respect to these variables have clear interpretations. This is not the case in thermodynamics, as the concepts *and* the variables are abstract quantities such as temperature or entropy.

In this appendix, I give a concrete example that gives a better understanding of how thermodynamic approaches can be used in other subjects. The point is not to say that we should use the thermodynamic approach in other subjects, but rather to show how the same results can be obtained with either of the two methods. I just want to make it clear that the unfamiliar approach is not actually a different method to calculate important quantities.

Let's start with a simple example that often comes up early in vector calculus and is simple to visualize. Suppose there is a hilly area, shown in Figure E.1,[1] the altitude of which can be described by the function

$$h(x,y) = \frac{9}{5}\left[\sin\left(\frac{x}{2}\right)\cos\left(\frac{y}{2}\right)+5\right], \tag{E.1}$$

where h is the height in meters above sea level, x and y are the distances (in kilometers) east and north of the origin, respectively (thus negative values of x and y correspond to west and south).

Common questions for such a function in introductory vector calculus include: Where are the top and bottom of the hill, what is the height of the hill, and how steep is the slope at some point? These questions are easy ways to understand and practice using the gradient. They are also concepts that are familiar to most students, and thus the hard part is just getting used to the mathematics.

In thermodynamics, though, there are many other types of questions that we tend to ask, which make sense here, but might be a little stranger when asking analogous questions about Eq. (E.1). For example, suppose we are asked, "How does the altitude change as you move due eastward?" This is of course straightforward: We take the partial derivative of h with respect to x (the eastward direction) while holding y constant, or

$$dh = \left(\frac{\partial h}{\partial x}\right)_y dx = \frac{9}{10}\cos\left(\frac{x}{2}\right)\cos\left(\frac{y}{2}\right)dx. \tag{E.2}$$

1 Generated using *Mathematica*, Version 12.2, Champaign, IL (2020).

Statistical Thermodynamics: An Information Theory Approach, First Edition. Christopher Aubin.

Companion website: www.wiley.com/go/Aubin/StatisticalThermodynamics

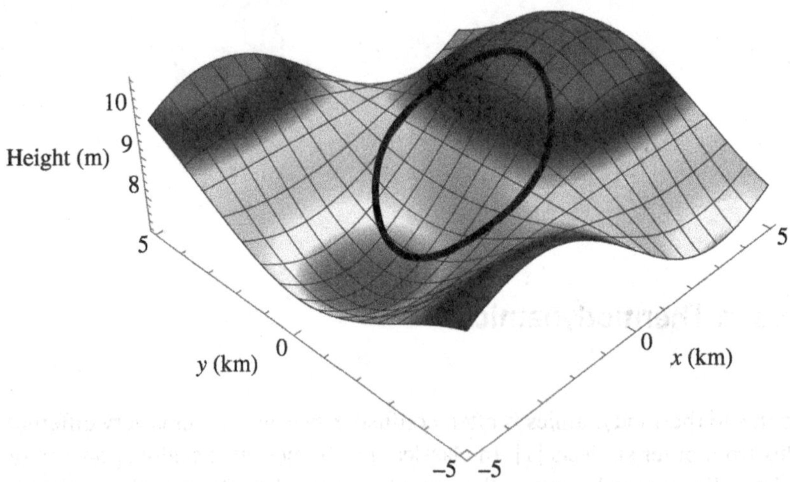

Figure E.1 The hilly region described by Eq. (E.1). The thin lines are lines of constant x or y, while the thick line shows a curve about the origin of constant radius of 2 km. A color version can be seen at the companion site.

As I have mentioned before, the variable held constant is not usually explicitly shown in vector calculus (or other physics classes) because it is clear: It is always the *other* variable that h is a function of. However, this is not always the case in thermodynamics, so it is best to specify the constant variable. This is relatively simple; one can easily see in Figure E.1 the thin lines point in the x- and y-directions, so this derivative tells us how we walk along one of (or parallel to) these lines in the x-direction.

Exercise E.1 How does the altitude change if we walked due north for a fixed value of x?

But what about the following question: "How does the altitude change as we try to go eastward but must remain a fixed distance from the origin?" You could imagine having a rope of fixed length attached to a pole at the origin, as you try to walk along one of the latitude (eastward) lines. The relevant derivative to consider would be

$$dh = \left(\frac{\partial h}{\partial x}\right)_r dx, \tag{E.3}$$

where I have made it clear that r, the distance from the origin, is what is being held constant. This is shown in Figure E.1 as the solid black curve (for $r = 2$ km)—note you cannot move strictly eastward if you maintain a constant distance from the origin. This is actually equivalent to moving in the θ-direction if we were to convert from cartesian to polar coordinates.

And in fact, the standard approach in a vector calculus class is to convert to polar coordinates, where r is given by

$$r = \sqrt{x^2 + y^2} \Rightarrow y = \pm\sqrt{r^2 - x^2}. \tag{E.4}$$

We then rewrite our altitude function as a function of x and r, to get

$$h(x, r) = \frac{9}{5}\left[\sin\left(\frac{x}{2}\right)\cos\left(\frac{\sqrt{r^2 - x^2}}{2}\right) + 5\right], \tag{E.5}$$

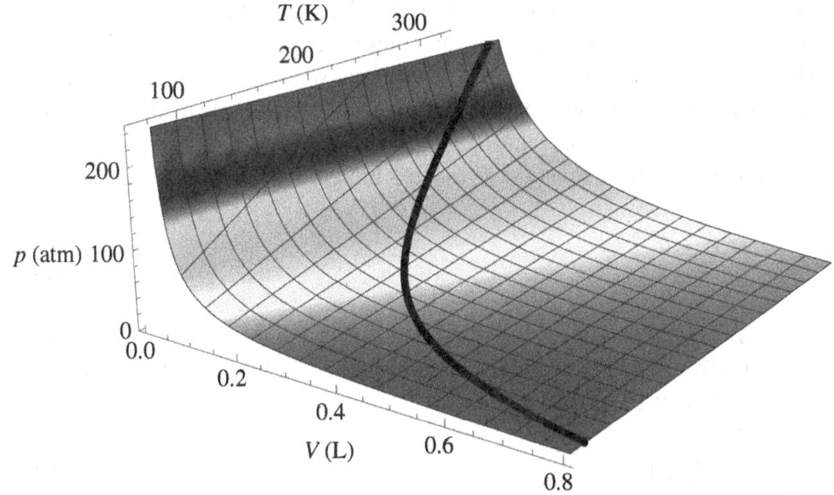

Figure E.2 Pressure as a function of temperature and volume for one mole of an ideal gas. The thin lines are lines of constant T or V, while the thick curve shows a line of constant entropy.

and then our result follows from straightforward (if a little tedious) differentiation:

$$\left(\frac{\partial h}{\partial x}\right)_r = \frac{9}{10}\left[\cos\left(\frac{x}{2}\right)\cos\left(\frac{y}{2}\right) + \frac{x}{y}\sin\left(\frac{x}{2}\right)\sin\left(\frac{y}{2}\right)\right]. \tag{E.6}$$

After differentiating, I put back in y to make it a little less cumbersome. $r(x, y)$ in Eq. (E.4) is another relationship among the variables x and y, such that if r is constant, then when y changes, so must x (and vice versa).

Comparing these equations to thermodynamic expressions, we could consider the altitude as our "equation of state," such as $\overline{p}(V, T)$, and then $r(x, y)$ could be analogous to the "entropy," $S(V, T)$.[2] The simplest examples of these are for the classical monatomic ideal gas, where from Eq. (6.27), if we had one mole of such a gas,

$$\overline{p}V = RT. \tag{E.7}$$

The pressure is shown in Figure E.2 as a function of volume and temperature.[3] The entropy in this case is given from Eq. (8.47),

$$S = R\left[\ln\left(\frac{V}{N}\right) + \frac{3}{2}\ln T + \sigma\right], \tag{E.8}$$

with

$$\sigma = \frac{3}{2}\ln\left(\frac{2\pi m k_B}{h^2}\right) + \frac{5}{2}.$$

Looking at Figure E.2, this can be thought of as a less interesting altitude function,[4] and any questions we can ask for $h(x, y)$ we can ask for $\overline{p}(V, T)$. For example, the derivative calculated in Eq. (E.2) in this case (where $x \to V, y \to T$, and $h \to \overline{p}$) is

$$d\overline{p} = \left(\frac{\partial \overline{p}}{\partial V}\right)_T dV = -\frac{RT}{V^2}dV = -\frac{\overline{p}}{V}dV, \tag{E.9}$$

2 The analogy is far from perfect; r is not a fundamental relation from which all quantities can be derived, but I won't worry about this—we aren't trying to compare these ideas physically, just mathematically.
3 Generated using *Mathematica*, Version 12.2, Champaign, IL (2020).
4 Or a fun rock climbing wall?

where in the last equality we used the ideal gas equation. The difference in this case is just physics: Here we are asking how the pressure changes with the volume for an isothermal process (discussed in Section 7.1), not how the altitude changes. The fact that we are specifying T being held constant seems as unnecessary as specifying constant y in $h(x, y)$, because as the ideal gas equation is presented, pressure is "only" a function of V and T.

But now consider the question: "How does the pressure change with the volume at constant entropy," or what is

$$d\overline{p} = \left(\frac{\partial \overline{p}}{\partial V}\right)_S dV? \tag{E.10}$$

This is the same as our second question above about how the altitude changes while walking eastward with a constant distance from the origin, and using the approach we took to evaluate Eq. (E.3), we would need to solve Eq. (E.8) for the temperature,

$$T = \exp\left(\frac{2S}{3Nk_B} - \frac{2}{3}\sigma\right)\left(\frac{V}{N}\right)^{-2/3}, \tag{E.11}$$

so the ideal gas equation is

$$\overline{p}V^{5/3} = N^{1/3}k_B \exp\left(\frac{2S}{3Nk_B} - \frac{2}{3}\sigma\right). \tag{E.12}$$

The derivative is straightforward now since we have the pressure explicitly as a function of the volume and entropy. In fact, for this case, it is quite simple because the entire right-hand side is constant, so we can immediately say $\overline{p}V^{5/3} =$ constant, which is how we wrote the pressure–volume relationship for an adiabatic expansion of the monatomic ideal gas in Section 7.1.

But this is not how we normally solve such a thermodynamic problem, since we do not always know $S(V, T)$ for the general case (and even if we did, solving explicitly for T may not be possible). We instead would *crush* this derivative and write it in terms of easily measurable (or calculable) derivatives: the heat capacities, the coefficient of thermal expansion α, and the isothermal compressibility κ (all defined in Section 6.2). Following the derivative crushing procedure of Section 6.4, we can write Eq. (E.10) as

$$d\overline{p} = -\frac{C_p}{C_V}\frac{dV}{V\kappa}, \tag{E.13}$$

which we saw in Eq. (7.5). Using the results for these derivatives for one mole of a monatomic ideal gas ($\alpha = 1/T$, $\kappa = 1/\overline{p}$, $C_V = \frac{3}{2}R$, and $C_p = \frac{5}{2}R$, we can write this as

$$d\overline{p} = -\frac{5}{3}\frac{\overline{p}}{V}dV, \tag{E.14}$$

which can be integrated to obtain the same result above.

We can crush the derivative in Eq. (E.9) to obtain

$$\left(\frac{\partial p}{\partial V}\right)_T = -\frac{1}{V\kappa}, \tag{E.15}$$

and with $\kappa = 1/\overline{p}$, this agrees with the above result. We have already obtained these results in the main text, and we know that what is useful about these expressions in terms of the derivatives $C_{\overline{p},V}$, α, and κ, is that they are general for *any* system. As has been discussed extensively, the thermodynamic approach, while more abstract, allows us to write down very general expressions for such derivatives.

Returning to the altitude example, let's approach that problem in a similar way as we do in thermodynamics. That is, we will manipulate the derivative we wish to evaluate to write it in terms of "standard derivatives" of $r(x, y)$ and $h(x, y)$ that are either known or easy to determine (either by measurement or calculation). We saw above with the thermodynamic approach that the actual functions are not needed (that is, we didn't need to know S or even $\overline{p}$ if we knew the derivatives $C_{V,p}$, α, and κ). I'll give these "standard" derivatives similar names as those in thermodynamics, just for the analogy.

First there are the "heat capacities," from Eq. (E.4),

$$C_x \equiv \left(\frac{\partial r}{\partial y}\right)_x = 2y, \tag{E.16}$$

$$C_y \equiv \left(\frac{\partial r}{\partial x}\right)_y = 2x. \tag{E.17}$$

Then I will define the additional derivatives,

$$\alpha_y \equiv \left(\frac{\partial h}{\partial x}\right)_y = \frac{9}{10}\cos(x/2)\cos(y/2), \tag{E.18}$$

$$\kappa_x \equiv \left(\frac{\partial h}{\partial y}\right)_x = -\frac{9}{10}\sin(x/2)\sin(y/2). \tag{E.19}$$

In all of these cases the second equalities are specific to our example above, with the "equations of state" $r(x, y)$ and $h(x, y)$ in our example. The derivative we considered in Eq. (E.2) is nothing more than α_y, so I won't consider that example explicitly as "crushing" it would be trivial.

But as for the second question, considering how the altitude changes with x and fixed r, let's manipulate these derivatives in a similar way as in thermodynamics:

$$\begin{aligned}\left(\frac{\partial h}{\partial x}\right)_r &= \left(\frac{\partial h}{\partial x}\right)_y\left(\frac{\partial x}{\partial x}\right)_r + \left(\frac{\partial h}{\partial y}\right)_x\left(\frac{\partial y}{\partial x}\right)_r \\ &= \alpha_y + \kappa_x\left(\frac{\partial y}{\partial x}\right)_r \\ &= \alpha_y - \kappa_x\frac{C_y}{C_x}.\end{aligned} \tag{E.20}$$

This result is true for *any* altitude function we were given, which would be useful for more complicated applications. However, with the results from Eqs. (E.16)–(E.19), it is simple to see that this result is exactly what we obtained in Eq. (E.6).

This method would generally not be used in such a problem, as it's not the most intuitive method and often not necessarily the quickest approach. As mentioned above, I have taken a relatively easy problem and unnecessarily overcomplicated it. The utility of this method is of course only apparent with problems that may not be solvable in the standard approach, or, as is often the case in thermodynamics, when we want to write a given derivative in terms of easily measurable quantities.

Reference

1 J. W. Cannon. Connecting thermodynamics to students' calculus. *American Journal of Physics*, 72(6):753–757, 2004.

Index

Statistical Thermodynamics: An Information Theory Approach, First Edition. Christopher Aubin.

Companion website: www.wiley.com/go/Aubin/StatisticalThermodynamics

t

u

v

w

z